SYSTEMS MANAGEMENT

People • Computers
Machines • Materials

Joseph C. Hassab

Director, Advanced Systems and Technologies
Lockheed Martin Corporation

CRC Press
Boca Raton   New York

Acquiring Editor: Ron Powers
Project Editor: Jennifer Richardson
Cover design: Denise Craig
PrePress: Greg Cuciak

Library of Congress Cataloging-in-Publication Data

Hassab, Joseph C., 1941–
     Systems management : people, computers, machines, materials /
Joseph C. Hassab.
          p.   cm.
     Includes bibliographical references and index.
     ISBN 0-8493-7971-7 (alk. paper)
     1. Production management. I. Title.
TS155.H3478 1997
658.5—dc21

96-47400
CIP

This book contains information obtained from authentic and highly regarded sources. Reprinted material is quoted with permission, and sources are indicated. A wide variety of references are listed. Reasonable efforts have been made to publish reliable data and information, but the author and the publisher cannot assume responsibility for the validity of all materials or for the consequences of their use.

Neither this book nor any part may be reproduced or transmitted in any form or by any means, electronic or mechanical, including photocopying, microfilming, and recording, or by any information storage or retrieval system, without prior permission in writing from the publisher.

The consent of CRC Press LLC does not extend to copying for general distribution, for promotion, for creating new works, or for resale. Specific permission must be obtained in writing from CRC Press LLC for such copying.

Direct all inquiries to CRC Press LLC, 2000 Corporate Blvd., N.W., Boca Raton, Florida 33431.

© 1997 by CRC Press LLC

No claim to original U.S. Government works
International Standard Book Number 0-8493-7971-7
Library of Congress Card Number 96-47400
Printed in the United States of America 1 2 3 4 5 6 7 8 9 0
Printed on acid-free paper

DEDICATION

TO NANCY AND JOEY

MY MOTIVATION

AUTHOR

Joseph C. Hassab, Ph.D., was born in Jounieh, Lebanon, on January 1, 1941. He received a B.S degree in Electrical Engineering, a B.S. degree in Civil Engineering, a M.S. degree in Electrical Engineering, and a Ph.D. degree in Electrical Engineering from Drexel University, Philadelphia, PA, in 1966, 1967, 1968, and 1970, respectively.

From 1970 to 1971, he was Assistant Professor of Physics at LaSalle University, Philadelphia, PA, where he taught electromagnetics, physics, and electronics.

From 1971 to 1985, he joined the Naval Undersea Warfare Center (NUWC), Newport, RI. While at NUWC, he was Manager of the Systems Analysis and Assessment Division, Development Division, Combat Control Systems Department (Acting), Chief Engineer/Scientist, and Systems Architecture and Targeting Division. He also conducted research and consulting, which resulted in approximately 100 journal publications in the varied aspects of sonar/radar signal and data processing, systems analysis and synthesis, wave propagation, electromagnetic scattering, ocean channel modeling, contact localization and motion analysis, weapon targeting, numerical analysis and expert systems.

From 1973 to 1985, he was also Adjunct Professor at Roger Williams University, Bristol, RI, and from 1979 to 1985 at University of Massachusetts, North Dartmouth, MA, where he taught courses on radar, sonar, speech and seismic signal processing, electromagnetics, engineering instrumentation, digital signal processing and control systems. At present, Dr. Hassab has been offering seminars in the U.S. and Europe on sonar systems, combat systems, and systems management.

In 1985, he joined RCA which merged with GE Aerospace, then Martin Marietta and now Lockheed Martin, where he has been Manager of ASW Shipboard Programs, Manager of Systems Engineering, Deputy Program Manager and Technical Director of a billion dollar system development contract, and Manager of Combat Systems Development and Advanced Programs. He has directed multiple business entries, horizontal expansions and developments of advanced systems e.g., combat systems, sonar, combat controls, electronic warfare and optronics.

In 1989, Dr. Hassab authored a book on Underwater Signal and Data Processing, published by CRC Press.

Dr. Hassab is a member of several honorary societies including Eta Kappa Nu, Phi Kappa Phi, Tau Beta Pi, Pi Mu Epsilon, Sigma Pi Sigma, and Chi Epsilon. He has been listed in Who's Who in the U.S., Marquis Who's Who, and Who's Who Registry Worldwide. He has been a referee to several technical journals, including: IEEE Journal of Oceanic Engineering, Journal of Acoustical Society of America, British Acoustical Journal of Sound and Vibration, IEEE Transactions on Acoustics, Speech and Signal Processing, IEEE Transactions on Automatic Controls, IEEE Transactions on Aerospace Engineering, IEEE Transactions on Computers, Radio Science Journal, and IEEE Transactions on Antennas and Propagation. In 1979, he organized and chaired the Office of Naval Research Conference on Time Delay Estimation and Applications. He has co-chaired the IEEE sponsored Workshop on Applications of Artificial Intelligence and Signal Processing to Underwater Acoustics and Geophysics problems. He has been Session Chairman at multiple IEEE and NATO conferences. He has conducted many seminars in the U.S.A., Canada, and Europe on Signal and Data Processing.

PREFACE

This book blends concepts and principles from varied fields in a unified approach to problems in systems management. Systems management enables effective transformation of input resources into desired output resources. The system processes that served us well during the agricultural revolution did not survive the industrial revolution; in due time, the system processes of the industrial revolution will not fare well in the ongoing information revolution. As the buggy whip that moved horse and carriage in the agricultural age would not start a car in the industrial age, similar industrial precepts do not carry into the information age. Effective processes need to be matched to the technology at hand where now the old requirement for a physical presence to make a product or provide a service is increasingly supplanted by an electronic presence; the latter triggers greater speed of action, increased flexibility and impressive productivity.

This book presents system processes mated to the computer technology for innovative management of resources during system conception, development and/or operation. Resources may be people, computers, machines and/or materials. Systems processes and computer implementation easily reduce complex works and integrate them into a coherent whole for productive transformation of input resources into desired output resources. Thus, the traditional foundations of a system's productive power require updated management processes of its four elements a) *natural resources,* b) *technology or human-made resources,* c) *work decomposition or specialization,* d) *work integration or exchange.* This book weaves through these four elements within the context of the ongoing information revolution.

A system view of management simplifies both the decomposition process and integration process of the parts and their sub-goals into a unified whole, thus leading to effective actualization of total goals. Each system or its derivative is consistently set in terms of an input, a process and an output, and this approach blends the contributions of the parts' productive powers into an effective and orderly whole. Thinking in terms of systems' patterns eases the application of the management processes of:

- Planning: What is to be done? Where will it be done? When? How?

- Organizing: Who is to do what and under what conditions?
- Activating: coordinating and implementing the plan and/or revising it to reach objectives.
- Controlling: seeing to it that planned work is yielding desired objectives.
- Communicating: linking of processes to effect system cooperative operations.

The foundations of a system's productive power rest on the effective management of its four cornerstones (Figure P-1): natural resources, technology or human-made resources, work decomposition or specialization, and work integration or exchange.

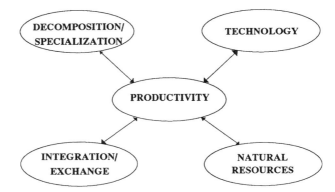

Figure P-1 Four inclusive cornerstones of productivity.

- Natural resources include land fit for productive crop use, timber, petroleum, metals, water; mineral deposits are vital to industrial production; scarcity of natural resources is an impediment to human productive capacity.
- Technology or human-made resources get the most out of the available natural resources. It provides the pathway up from poverty to riches; it improves the productivity of goods and services through production resources such as computers, machinery, equipment, buildings, etc. At present, computer technology is putting at human disposal powerful productive resources to outperform automatically any manual option; without technology, population rises faster than the manual productive capacity of more people, thus yielding a lowered standard of living. Improved technology leads to more productive workers who use processes and capital goods to offset scarcity in natural resources. Technology separates the haves from the have nots.

- Work decomposition is the systematic breakdown of work and its allocation to human/computer/machine for execution with the objective of increased productivity. Decomposition leads to specialization with resultant in-depth competence in a specialty, creation of new technology, and increased effectiveness in its practice, e.g., legal with specialty in criminal, tax, civil laws... ; engineering with specialty in structures, computers, communication... ; medical with specialty in neurology, cardiology, pediatrics... .
- Work integration is the systematic pulling together of specialties to satisfy consumer products/services; in fact, any specialist would have a hard time making it on his/her own, to satisfy his/her wants unless he/she can exchange his/her product/services.

While these four cornerstones are essential, it is their managed concurrency which allows increased productivity and a continual increase in the standard of living. Only higher productivity can provide for higher wages to workers and for profits to owners. While machinery brought about the higher technological productivity during the industrial revolution, it is the utilization of computer technology that is bringing the higher productivity during this evolving information revolution.

As one views human activities, progress has required decomposition of the total work and specialization; specialization lowers self-sufficiency and calls for integration, i.e., exchange of unitary products/services; exchange has required local proximity, otherwise the extended time expanded on the exchange can dissipate the accrued benefits from specialization. Technology has enabled the shortening of the exchange time, widening the physical locale, and extending its linkages to others. Through the mechanical engine, the industrial revolution shortened the time exchange process of materials over land, highways, waterways, and air ways. Through the electronic engine (computer), the digital revolution is circumventing the need for physical transport through symbolic transport, and enabling the real time exchange of data representation of materials over information highways; thus, the time and resource consuming requirement for physical presence is relegated to a second stage. Influence and action are thus executed at a greater spatial span and in a shorter time through transport or wave propagation of digits rather than materiel transport. The full impact of the digital revolution is yet to be realized; it promises to be the propeller of small, agile, and specialized businesses availing themselves of the facilitated exchange mechanism through secure electronic networks and/or processes now set in software programs; it could be the harbinger of the virtual corporation constituted of specialized dynamic units and the demise of the large, sluggish, and burdened by overhead businesses that have overgrown in an attempt to provide a self-contained exchange; already, small enterprises account for more than half the value added to all products made.

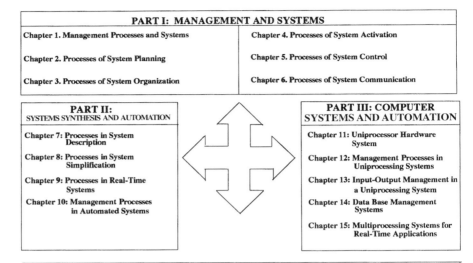

Figure P-2 Text Structure.

Structurally, the text is divided into four parts (Figure P-2) where particular attention is called to the following main features:

- Coverage of a wide variety of topics in Management, Systems, and Computers, specifically reflecting their synergy.
- Self-contained parts and self-contained chapters within parts, thus enabling maximum teaching and reading flexibility.
- Development and integration of principles and ideas from various fields with emphasis on resource management and decision-making.
- Stress of the common thread in work processes (decomposition and integration) and resources (natural and human-made) underlying all the apparent differences.
- Dissection and simplification of system elements and their interactions to get jobs done on-time, just-in-time, or real-time; this deciphers the principal competitive edge today, speed of action in the right direction velocity.

Part I, *Management,* describes the system processes for bringing means to reach desired ends; it delineates the processes that affect judicious application of an enterprise's productive powers in the speedy accomplishment of desired goals; it capitalizes on the similarities and highlights the differences in the management of human and nonhuman resources. A unified management of human and nonhuman resources is presented as system processes for bringing means to reach desired ends. The decomposition and integration processes for managing resources include those of planning, organizing, activating, controlling, communicating; they are expanded upon respectively in Chapters 1 through 6. The processes are linked in a system structure for effective accomplishment of the goals. The presentation stresses the view of management as processes fundamental to system activities, be it human, computer and/or machine systems. The unified view of management principles is prompted by the increased introduction of computer systems to better human productivity through automation. While automation does not compensate for a "broken" human organization, it is definitely the productivity multiplier to an effective human organization.

Part II, *Systems,* applies system processes to simplify work and work flow and develops: a) the analysis processes to define the goals or purposes and to discover operations for their most effective realizations; b) the synthesis processes to compose the combined elements or parts thereof, however diverse, into a coherent whole. A systems approach is presented to simplify decomposition and integration of resource management and to frame the interrelatedness and interdependency of the management processes and their contribution to the whole. Processes for system representation, analysis, design, and implementation of the management processes are described in Chapter 7. Processes for system simplification are explained in Chapter 8 to take the complexity out of the management processes and in turn their non-productive activities. Enabling processes for speedy systems operation are explained in Chapter 9 to shorten time use of the managed resources and in turn their costs. Finally, the processes are integrated in Chapter 10 to delineate basic automation of a system.

Part III, *Computers,* expounds on the facilitated implementation of systems management processes through automated means to realize a profound effect on productivity. Human management activities, as represented by natural language instructions, are executed by the computer through a series of decomposition–integration processes; human instructions are transformed into macroinstructions then microinstructions, and in turn into a combination of on–off switching activities, representative of the microinstructions, which the computer hardware then executes at tremendous speed. The computer as a uniprocessor is explained in Chapter 11, and the management processes of its internal resources are presented in Chapter 12. The linking of the uniprocessor's internal activities to the external environment is expanded upon in Chapter 13. While other forms of processors, e.g., human/machine, operate on physical resources and/or their representation, computers operate solely on the representation of the resources by data; data management in computer organizations is of primary importance and its management processes are expanded upon in Chapter 14. Finally, multiprocessing systems for real time applications are presented in Chapter 15, pulling together the varied elements of computer systems.

Part IV, *Applications*, gives in-depth selective views of the management of an enterprise; it focuses the processes in Parts I, II, and III on systematic and purposeful activities in the generation of varied products and services as in manufacturing, program management, automation, finance, and radar-sonar processing. As the basis of most productive enterprises is an automation system, a generalized view of such a system is given in Chapter 16. In an enterprise, effective synthesis of any management system rests on decision making based on quality information. Such information is unfailingly obtained through extensive processing; generalized automated information processing under risky and uncertain conditions is presented within the context of radar–sonar systems in Chapter 17. As any enterprise can be viewed composed of programs or a program as a mini-enterprise, detailed program management implementation is presented in Chapter 18 applicable to the harnessing of resources toward reaching a goal, whether it is in the concept, design, development, operation and/or maintenance stage. Given any enterprise, its ultimate output is to provide a product and/or a service; a common approach to production and service operation is presented in Chapter 19. Underlying the viability of any enterprise is its financial health, and management of the enterprise as represented by its financial state is presented in Chapter 20.

In summary, improved work processes (decomposition and integration) and available resources (human-made and natural) determine the human economic condition. The foundation of human progress rests on creativity, where one seeks new and better work processes to accomplish better goals; creativity generates ideas, and innovation applies ideas leading to better technology; increased productivity in turn depends on fusing these four inclusive cornerstone elements: decomposition (specialization), integration (exchange), technology, and natural resources. This book weaves through these four elements within the context of varied enterprises. Part I, *Management*, explains the processes that effect judicious application of these elements in the accomplishment of desired goals. Part II, *Systems*, applies system processes to symplify work and work flow and delineates: a) the analysis processes to define the goals or purposes and to discover operations for their most effective realizations; b) the synthesis processes to compose and combine elements or parts thereof, however diverse, into a coherent whole. Part III, *Computers*, expounds on the facilitated conceptual realization of systems management through automated means to effect higher productivity. Part IV, *Applications*, focuses Parts I, II, and III on systematic and purposeful activities in the generation of varied products and services.

CONTENTS

PART I
MANAGEMENT AND SYSTEMS

CHAPTER 1 MANAGEMENT PROCESSES AND SYSTEMS 3
- I. Introduction 3
- II. Real-Time, Systems, Management, Computers: Definitions 4
- III. Mechanization, Automation, and Computer Systems 6
- IV. Management and Systems 9
- V. Management and Decision-Making 10
- VI. Discussions 11

CHAPTER 2 PROCESSES OF SYSTEM PLANNING 15
- I. Introduction 15
- II. The Planning Process 17
- III. Planning and Decision-Making 19
- IV. Classes of Problems in Systems Planning 21
- V. Discussions 22

CHAPTER 3 PROCESSES OF SYSTEM ORGANIZATION 25
- I. Introduction 25
- II. The Organization Process and Structure 26
- III. Decomposition or Breakdown of Work 29
- IV. Coordination and Integration of Work 30
- V. Span of Management and Decentralization 31
- VI. Design of Organizations 32
- VII. Discussions 33

CHAPTER 4 PROCESSES OF SYSTEM ACTIVATION 37
- I. Introduction 37
- II. The Activation Process 37
- III. Activation of Conflicting and Changing Activities 38
- IV. Activation of Human and Automated Systems 43
- V. Discussions 45

CHAPTER 5 PROCESSES OF SYSTEM CONTROL 49
 I.   Introduction 49
 II. The Control Process 50
      A. The Need for System Control 51
      B. Types of Control 51
      C. Levels of Control 52
 III. Budgeting 55
      A. Financial Budgets 55
      B. Time Budgets 58
      C. Quantity and Quality Budgets 61
      D. Effectiveness of Budgets 61
      E. Information and Budgets 61
 IV. Discussions 62

CHAPTER 6 PROCESSES OF SYSTEM COMMUNICATION 65
 I.   Introduction 65
 II. The Communication Process 66
 III. Communication Systems 71
 IV. Discussions 73

PART II
SYSTEMS SYNTHESIS AND AUTOMATION

CHAPTER 7 PROCESSES IN SYSTEM DESCRIPTION 77
 I.   Introduction 77
 II. System Representations and Their Implications 79
 III. System Analysis and Design 85
 IV. System Implementation 88
 V.   Discussions 96

CHAPTER 8 PROCESSES IN SYSTEM SIMPLIFICATION 97
 I.   Introduction 97
 II. System Simplification Processes 98
      A. System Top-Down, Elimination, and Grouping Processes 98
      B. System Search Processes 100
      C. System Simulation Processes 101
      D. System Test Processes 105
 III. Characteristics of Ineffective Systems and Their
      Improvements 106
      A. Ineffective System Characteristics 106
      B. System Improvement Characteristics 107
 IV. Discussions 108

CHAPTER 9 REAL-TIME SYSTEMS 109
   I.   Introduction 109
   II.   Basic Real-Time System Elements and Their Implementation 111
         A.   Real-Time System Elements 111
         B.   Real-Time System Implementation 113
   III.   Real-Time System Scheduling and Reliability 117
         A.   Real-Time System Scheduling 117
         B.   Real-Time System Reliability 118
   IV.   System Design Methodology 123
   V.   System Design Processes 124
         A.   Control System Process 125
         B.   Finite State System Process 126
         C.   Functional Decomposition Process 126
         D.   Data Decomposition Process 127
   VI.   Discussions 128

CHAPTER 10 MANAGEMENT PROCESSES IN AUTOMATED
SYSTEMS 131
   I.   Introduction 131
   II.   Processes and Data, Control, Logic 132
   III.   Processors, Data, and Instructions 135
   IV.   Architecture of a Computer System 139
   V.   Discussions 143

PART III
COMPUTER SYSTEMS AND AUTOMATION

CHAPTER 11 UNIPROCESSOR HARDWARE SYSTEM 149
   I.   Introduction 149
   II.   Basic Elements of a Uniprocessor 150
         A.   Digital Logic 151
         B.   Basic Units of a Computer 152
   III.   Uniprocessor Essential Organization and Activation 155
   IV.   Interfacing Uniprocessor Systems 160
         A.   Bussing Techniques 161
         B.   Input-Output Scheduling 163
         C.   Peripherals 164
         D.   Error Detection, Correction, Fail-Soft, and Maintenance 166
   V.   Discussions 168

CHAPTER 12 MANAGEMENT PROCESSES IN UNIPROCESSING
SYSTEMS 171
   I.   Introduction 171

II. Elements of High-Order Languages 173
 III. Human-Computer Communication: Compiler Organization 176
 IV. Operating System: Functions 182
 V. Operating System: Organization and Activation 184
 VI. Operating System: Portability and Efficiency 188
 VII. Fault-Tolerant Processing System 191
 VIII. Discussions 194

CHAPTER 13 INPUT-OUTPUT MANAGEMENT IN A
UNIPROCESSING SYSTEM 197
 I. Introduction 197
 II. Input-Output Processes in a Uniprocessing System 198
 III. Input-Output Data Organization 204
 IV. Input-Output Between Humans and Computer 210
 V. Input-Output Operating Characteristics 214
 VI. Discussions 215

CHAPTER 14 DATABASE MANAGEMENT SYSTEMS 217
 I. Introduction 217
 II. Database Storage: File Organization and Access Methods.............. 219
 III. Benefits of a Database System Manager 225
 IV. Database Management System Characteristics 225
 V. Database Management System Organization 229
  A. The Hierarchical Database 229
  B. The Network Database 230
  C. The Relational Database 230
  D. The Pseudo-Relational Database 232
 VI. Database System Architecture 233
 VII. Database System Design 236
 VIII. Database Management System and System Operation 243
 IX. Distributed Database Management System 245
 X. Discussions 250

CHAPTER 15 MULTIPROCESSING SYSTEMS FOR REAL-TIME
APPLICATIONS 255
 I. Introduction 255
 II. Multiprocessing Systems Architecture 258
 III. Software and Hardware Views of Multiprocessing Systems 264
 IV. Multiprocessing Communication Systems 267
 V. Multiprocessing Operating Systems 270
 VI. Multiprocessing Database Management Systems 274
 VII. Multiprocessing Reliability and Recovery Systems 275
 VIII. Discussions 277

PART IV
APPLICATION SYSTEMS MANAGEMENT

CHAPTER 16 AUTOMATED MANAGEMENT SYSTEMS AND
THEIR ARCHITECTURES 285
   I.   Introduction 285
   II.   Real-Time Control in Automated Systems................................. 286
   III.   Basic Elements in Electronic Automated Systems 289
       A.   Basic Constituents... 289
       B.   System Configurations 300
       C.   Automated Management Functions 302
   IV.   Automated System Architectures.. 304
       A.   Requirements.. 304
       B.   Implementation Approaches... 309
       C.   Management of Resources 313
   V.   Discussions... 317

CHAPTER 17 AUTOMATED INFORMATION PROCESSING
UNDER RISKY AND UNCERTAIN CONDITIONS............................... 327
   I.   Introduction 327
   II.   Signals, Noise, Channels, Processing, and Interrelationships.......... 329
       A.   Signal Noise and Channel Outputs..................................... 329
       B.   Information Processing Structures 333
       C.   Information Extraction Through Decomposition and
           Weighted Integration 338
           1.   Spatial Information Extraction.................................... 338
           2.   Spectral Information Extraction.................................. 343
           3.   Temporal Information Extraction................................ 344
   III.   Basic Signal Processing, Decomposition, and Weighted
       Integration 345
       A.   Source, Channel, Processor.. 347
       B.   Weighting Through Windowing and Gating...................... 352
   IV.   Actual and Lower Bound Localization Accuracies.................... 353
   V.   Basic Data Processing Decomposition and Weighted
       Integration 356
       A.   General Classes of Contact State Estimation Problems........... 359
       B.   Errors and Their Filtering 361
           1.   Causes of Errors.. 361
           2.   Characterization of Errors: Biased or Unbiased............. 361
           3.   Statistical Filtering of Errors 362
       C.   Elements in the Formulation and Solution of CLMA
           Problems 362
       D.   Expert System Concepts for Contact Localization and
           Tracking with Uncertain Conditions................................... 367
           1.   Expert Systems.. 368

| | 2. An Expert System Structural Elements .... 368 |
| | 3. Layered Integration .... 371 |

VI. Generalized Prediction, Processing, Wave Propagation, and Vibration Method 372
VII. Signal Prediction in the Presence of Multiplicative Noise 375
VIII. Discussions 377

CHAPTER 18 PROGRAM MANAGEMENT 379
I. Introduction 379
II. Basic Processing Flow in Program Management 381
III. Detailed Program Management Implementation 389
    A. Implementation of Work Breakdown and Organizational Allocation 391
    B. Technical Management Implementation 394
    C. Schedule Management Implementation 396
    D. Financial Management Implementation 398
IV. Discussions 401

CHAPTER 19 MANAGEMENT OF PRODUCTION AND SERVICE OPERATIONS 405
I. Introduction 405
II. Planning and Design of Production and Service Systems 406
    A. Capacity Planning and Design 406
    B. Location Planning 408
    C. Facilities Design 408
    D. Product Design 410
III. Activation and Control of Production and Service Systems 411
    A. Overall Operation Management 411
    B. Product Operation Management 412
    C. Scheduling Management 413
    D. Material Management 415
    E. Quality Management 419
IV. Discussions 420

CHAPTER 20 FINANCIAL MANAGEMENT 423
I. Introduction 423
II. Financial System 424
III. Financial Planning and Control 427
    A. Financial Analysis 427
    B. Capital Budgeting 428
    C. Financial Leverage 429
    D. Working Capital 429
    E. Short-Term Financing 429
    F. Intermediate Term Financing 430
    G. Long-Term Financing 430

	H. Mergers	430
IV.	Discussions	431

BIBLIOGRAPHY 433

INDEX 443

Part I:
Management Systems

1 MANAGEMENT PROCESSES AND SYSTEMS

I. INTRODUCTION

Managed organizations are more likely to succeed than unmanaged ones. A systems approach to management is presented here. An amplified examination of systems and their synthesis is developed in Part II.

A systems approach looks at the composite parts as together, interrelated, and amplified, and with a common output objective (Figure 1.1) and enables a critical view of the organization in action. Each system has a boundary that separates its parts from the rest of the world or the environment. The environment sets constraints or boundary conditions on the system as the system attempts to change it. When the system interacts with the environment, resources cross the system boundary, both ways, in and out. Resources crossing the boundary are processed by the system. As processing evolves, data on process evolution are fed back to human/computer elements to assess correctness and control the process if correction is necessary (Figure 1.2).

The world is composed of resources, human equipment, and material. To accomplish a set of objectives, the resources are selectively pooled together in a certain way. Without selective pooling, a multitude of combinations and connections is possible that lead to an undetermined end. To accomplish a stated set of objectives, plans need be set, resources need be organized, activities need be initiated, controls need be exercised, all in certain ways. Those pooled resources per set of plans, organizations, activities, and controls, we shall call the system. There are many such systems, big and small. The governing principles of operations remain similar independent of size. Size depends on what was pooled; the rest constitutes the outside environment. The more one understands the system or systems of resources resident in the outside environment, the more one is apt to understand their resultant forces on the goals one is trying to reach using the internal resources. The same follows for the resources pooled within the boundaries; they form subsystems, each with its derived sub-plan, sub-organization, sub-activities, sub-controls, all coordinated to support the system goals.

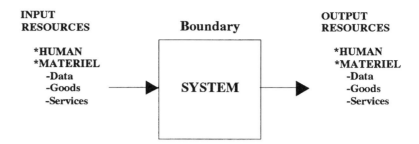

Figure 1.1 System.

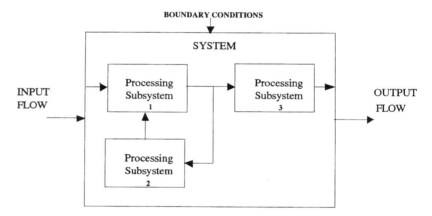

Figure 1.2 Decomposed system with forward (input to output) and feedback flows; people/computers judge progress against expectations.

Resources within and outside system boundaries seldom remain static. Their dynamic changes require constant system management where constraints are dependent on resource variability. In fact, static systems need no management.

II. REAL-TIME, SYSTEMS, MANAGEMENT, COMPUTERS: DEFINITIONS

Time is a valuable resource; however, it cannot be stored for future use, and once it lapses, it cannot be recovered. Effective time utilization is paramount to many successful system operations.

By real-time, we mean that essential activities happen without undue delays. We use it with an extended and contextual view of "real"; it means just-in-time or on-time. Here, real does not mean continuous at all times, but incremental. The increment is the time lapse for events of interest to happen. The size of increment, be it nanosecond, second, minute, hour, day, ..., depends on the management context. In general, the increment is longest at the higher levels of a system with decreasing values for the lower levels. In a corporation, for instance, real-time for its president may be monthly or quarterly, while real-time for one of its plant supervisors may be hourly or daily. Automated real-time systems are explained in Chapter 9. Real-time may have a hard deadline or a soft one; a hard deadline requires completion of tasks on time otherwise grave or fatal damage to the operation will occur; a soft deadline, though passed, permits completion of its tasks.

By resources, we mean what is available, acquired, or planned for and it may not be all that is needed to reach the objectives. It may be human, with specialties in engineering, finance, manufacturing, sale, etc. It may be material like minerals, parts ... and capital, like buildings, computers, machinery.

By system, we mean the resources within a boundary drawn together and being managed to reach a set of objectives. Its synthesis is explained in Part II. The outside resources to the boundary form the environment. The environment is outside direct management purview. It interacts with the system, influences and is influenced by the system through the flow of resources at the boundaries. Systems may be decomposed into subsystems and each may be further decomposed to one or more levels. The principal reason for decomposition is the facilitation of system synthesis evolution. Each decomposition will have its own internal boundary and its internal resources are, in turn, managed to reach an allocated set of objectives. All decompositions in a well-managed enterprise are, in turn, integrated to reach the parent system set of objectives.

By management, we mean the processes that resources enact and/or undergo to reach the objectives (Figure 1.3). Its processes are explained in Part I. Processes are the actions to execute the tasks pertinent to the objectives; tasks are made up of activities and events; events are those parts of the tasks that should be completed at a certain point in time; activity is the time and resources needed to go from one event to the next; this motion, when done with speed and in right direction, leads to expeditious execution of the activity. Managing is expeditious processing in a systematic way of doing things to reach the system's desired goals. The processes are those of planning, organizing, activating, controlling time and resources (human, capital, and material):

- Planning is thinking in advance of needed actions to achieve the goals.
- Organizing is setting in time/space the human and material resources within the system in preparation of performance of needed actions.
- Activating is coordinating the organizational activities to achieve the tasks leading to the goals.

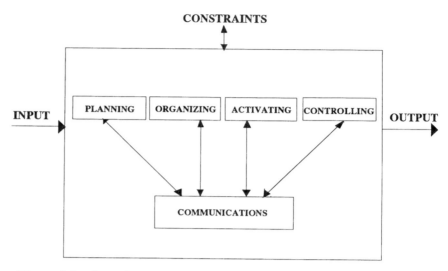

Figure 1.3a Generic management processes.

- Controlling is assuring that tasks are being performed by each part of the organization according to plans to achieve goals.
- Communicating is linking of the four processes to form a system.

These generic management processes exist at each level of a system. Their depth and extent vary. Humans only exercise ultimate management processes; computers execute delegated management processes; machines do only limited and elementary management processes.

By computers, we mean, among other things, machines that can perform or support in a prominent way the planning, organizing, activating, controlling, and communicating. Its processes are explained in Part III. With complete delegation, the system is said to be fully automated. Computers are found in many applications and their relative position to other resources in the management chain depends at which level of the system they are operating.

By applications, we mean slices across varied enterprises to view their operations using the concepts discussed previously.

III. MECHANIZATION, AUTOMATION, AND COMPUTER SYSTEMS

Manual operation is the activation of tools by humans to perform tasks. Mechanization is a coordinated set of tools that replaces manual operations by a machine. Automation is an increased mechanization; it insures that each machine jointly performs what they are supposed to. This is effected through feedback and correction. In that realm, computers replace human activities

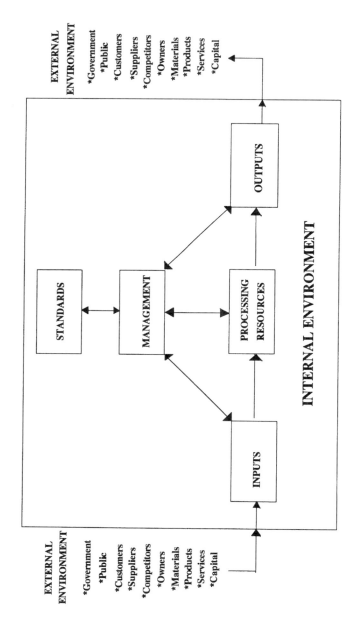

Figure 1.3b A managed system and its environment.

and thus realize further the automation process in advisory, associate, and/or autonomous roles. It is worth noting that one cannot automate old designs effectively unless they were conceived with automation in mind. Computers compute, do logical operations, search for relations, make decisions and thus relieve or simplify not only human manual effort, but also mental effort. The process of human-computer-machine work breakdown involves work reapportionment, (Figure 1.4). An amplified study of computer systems and automation is given in Part III.

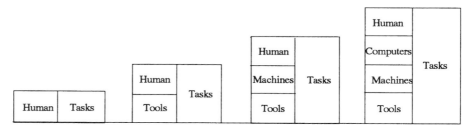

Figure 1.4 Enabling technologies and reapportionment of tasks along with increased productivity.

The computer is one machine that can be programmed to perform various functions. A program is a series of steps or instructions needed to complete a desired function or process. Those steps are stored in computer memory and direct the computer in its processing. A process may be simple, i.e., a division of two numbers or sophisticated, i.e., the control or landing and take-off of airplanes at an airport. The computer has enabled management to widen the boundaries that constrained concept definition, to extract relevant relationships, and to test new premises in the discovery of new ideas.

Automation is a continuous process of integration and optimization: integration of machines, computers, and people seeking the best possible way to realize some desired measures of output, like cost, reduced hazard, reliability, product information, accuracy, productivity, response, throughput, and shortened manufacturing time. The desired measures vary, but ultimately, they should, given the risk, yield the investor in some direct or indirect way the best return on his/her investment in the enterprise. The enterprise may be a service system, a processing system, a manufacturing system, an agricultural system, and, depending on the function within the enterprise, automation may be concerned with a single machine, a combination of machines, a process, people, their interaction, or that of the total enterprise. Within this framework, the computer can direct the gathering, processing, storage, communication, presentation of information and/or action. In this role, it is an intelligent system with the needed capabilities for perception, recognition, and action.

Mature and complex systems are capable of developing their own plans or changing those plans, of adjusting the allocation of resources, of setting

MANAGEMENT PROCESSES AND SYSTEMS

priorities of action, and of predicting system evolution. Such mature systems include humans as one of its components.

IV. MANAGEMENT AND SYSTEMS

The overall function of management is to utilize human and materiel resources to effect organizational goals. To that end, a system takes shape; made up of inputs, outputs, and processes. Each of the processes performs some portion of the overall function. For our purposes, the process type may be human, computer, and/or machine, each with its own characteristics and needs. Simply

- Human is the most flexible where people generate and/or carry out the instructions needed to perform the sub-function allocated to them.
- Computers are flexible enough where pre-stored or on-line instructions by humans are executed to carry out the allocated sub-functions.
- Machines are least flexible since only prescribed and fixed instructions are performed.

From a control viewpoint, machines are at the bottom of the hierarchy; machines, like a tractor, can literally move mountains, when guided, but cannot think through how to go about it.

Composition of systems varies. Large complex systems are a coordinated combination of the preceding three types of processes. Common and variant factors apply to the three types of processes. Those common factors pertain to technological sciences; the variant ones involve those from the behavioral sciences, such as leadership, motivation ... which are only relevant to humans. This broadened view of management will be elaborated on in Part I. Human organizations and computer organizations have something in common in the way they are managed; in some, computer organizations are set to mirror the human organizations they support. Society is full of such organizations, schools, businesses, governments, religions, military, commerce, civil, volunteer, manufacturing, finance. In sophisticated organizations, feedback is present to detect variations in expected performance and initiate control of the process. Too much control is an indicator of a poor system; in human systems, it shows lack of creative initiative; in computer systems it reveals itself in instability and poor performance.

The functions of management form a set of interdependent processes where:

- The planning process establishes what should be done before the start and continues as necessary with revisions during the evolutionary

process; plans contain the objectives, policies, programs, procedures, and methods.
- The organizing process breaks down the functions into process work to be done, defines their relationships, and assigns them to human/computer/machine to be done, and establishes the line of communication authority. This leads to an overlay of the human organization on the computer organization and machine organization. Unfailingly, the system organization is the selection of people/computers/machines to execute the work allocated to each position in the organization. For our purposes, organization has an extended meaning to include computer-machines. In that sense, computers, as they imitate human processes, require training as in a neuro-network; however, promotion, appraisal, and compensation in an organization remain human needs.
- The activating process sets the organization in motion to achieve the plans. Tangible and intangible factors come into play in the work environment. Common to three types of processes are the tangible ones which fulfill physical needs; i.e., heat, cooling, light. Singular to the human process are the intangible; i.e., morale, satisfaction, creativity.
- The controlling process insures that activated processes perform according to plans; corrective action to organizing, staffing and/or activating processes are applied to satisfy the planned/modified objectives. The primary tool of control is feedback. Too much control thwarts the organization's activities and denotes a bad mismatch between the system processes (plan/organization/activation) and system inputs/environments. Those mismatches lead to large stop-go behavior aimed at avoiding system breakdown/instability. Systems have centralized and/or decentralized operation.

In the evolution of a managed system from inputs to desired outputs within constraints at system boundaries, the management functions are interdependent, connected, and overlapping processes. To get a plan, one needs some organization to be activated and controlled. Once a plan is in place, the cycle of organizing, activating, and controlling repeats itself. The managing functions are not serial and static. The functions overlap, execute in parallel, and feed-back on themselves as they dynamically change.

V. MANAGEMENT AND DECISION-MAKING

Management, in its true sense, is choosing between alternatives; it is making sense out of choices or decisions to optimally attain a goal. The outcome of each choice is likely to be risky or uncertain. Thus, perfect and error-free decisions cannot be made so complex that plans be formulated and supervised without need for subsequent revision. Unfailingly, decisions made

MANAGEMENT PROCESSES AND SYSTEMS

at one point in time are based on current imperfect knowledge and state, and anticipated result at future points in time. As more facts are acquired, new decisions have to be made and new plans need to be formulated, and new courses of actions need to be taken to attain the stated goal. Thus management is more engaged in decision-making among alternatives than the supervision of the implementation of a certain course of action; the higher the uncertainty in the data, the higher is the risk and the higher is the level of the decision maker. Approaches to some applications in decision-making under risk and uncertainty are expanded upon in Chapter 17.

Decision-making is classified by a) *who is making the decision* and b) *the conditions under which the decision is made*. Elaborate decision-making may be made by humans/computers, singly or as a group as expanded upon in Parts II-III. The conditions or states of knowledge may be:

- Certain, where each alternative has a specific and known outcome.
- Risky, where each alternative has multiple outcomes, each with a known probability; risk is a quantitative measurement of an outcome such as loss or gain.

Decision making is a part of every management process. In managed decision-making, one needs to:

- Identify possible future conditions.
- List possible alternatives.
- Set pay-off associated with each alternative.
- Indicate likelihood of possible future condition.
- Set decision critera to select best alternative.

VI. DISCUSSIONS

Humans realized early that few major things can be accomplished alone; humans formed groups and coordinated their activities. While humans are unlimited in imagining, figuring, and inventing, they are limited in their ability to execute heavy physical work, to sustain extreme hardships, and to perform extensive repetitive mental work. To compensate, progressive groupings have been joined over time to include:

- Tools and animals, leading to the agricultural revolution.
- Machines, leading to the mechanical revolution.
- Computers, leading to the electronic revolution and the era of flexible automation.

While mechanical machines are adept at the batch processing type of automation, computers enable flexible automation to satisfy variances in orders, responsiveness to real-time events and controls of mechanical operation from order entry to product delivery. With automation, transfer of principal human efforts occur from physical work to mental work along with an increase in productivity. With higher productivity, higher profits ensue, which allow higher wages, which owners want for higher profits and workers want for higher compensation. Humans unfailingly seek a wider range of things, such as participation, communication, recognition, achievement, where compensation becomes an expression rather than the ultimate end.

A system is a grouping of resources with an integrated mapping of "why", "what", "who", "when", "where", "how" to each.

- The "why" expresses the goals.
- The "what" partitions the products and their descriptions by grouping work into segments and dividing the work into tasks, then reintegrating them into finished products.
- The "who" allocates/assigns products/sub-products to organizational units.
- The "when" sets the time for products/sub-products readiness.
- The "where" sets the space for products/sub-products execution.
- The "how" defines the tasks.

A system's manager plans, organizes, activates, and controls people, resources, and activities in order to move the system toward stated objectives (Figure 1.5). Planning sets the objectives and courses of action; organizing sets the authority and distributes the work; activating sets the organization into action to execute the work according to the plan; controlling insures that the activities progress toward the planned objectives. The management processes, set in a system structure, facilitate the decomposition and integration of objectives; major objectives are achievable through division into subsidiary objectives until we reach the elementary activities which are used as means, then integration of those activities upward to the parent objectives.

MANAGEMENT PROCESSES AND SYSTEMS

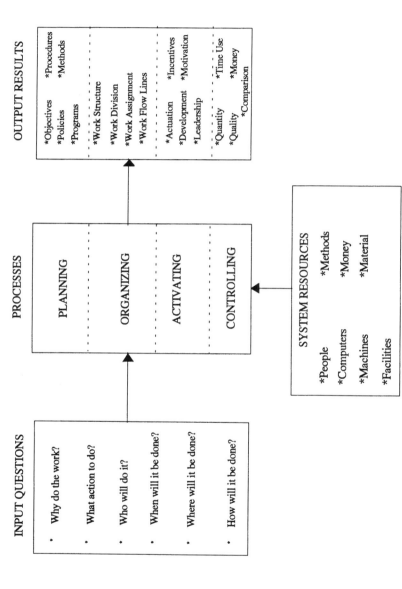

Figure 1.5 Management processes of human and non-human resources linked through communication.

2. PROCESSES OF SYSTEM PLANNING

I. INTRODUCTION

Planning determines what actions should be taken in the future. The path to the objective is the strategy which forms a bridge between present and future. When all things remain constant, the future becomes an image of the present and planning is almost non-essential. Planning becomes increasingly necessary when change is getting more extensive and/or is sudden. Change gives rise to future instability unless addressed, tracked, and responded to. Every enterprise, striving to grow in a dynamic environment, exercises planning to foresee the future and strives to coordinate all its resources into a more effective entity aimed at realizing its projection of the future. Planning enables an enterprise to progress as a coordinated system within long range forecasts made on projections of government policies, economic trends, population trends and/or technology trends (Figure 2.1).

The basic elements of planning produce conceptual answers to a sequence of questions that will be applied throughout the text in synthesizing systems:

- Why is it needed? It ties selected goals in a clear path to the objectives of the enterprise.
- What is needed to be done? It specifies the actions and sequence necessary to reach the goals.
- Where will it be done? It denotes the physical location and equipment assigned to execute given activities.
- Who will do it? It denotes the organizational units and the members assigned to execute given activities.
- When will it be done? It sets the timing for the start and end of activities; parallelism in activities is sought to reduce cycle time.
- How much will it cost? It sets the cost profile and value of the activities.

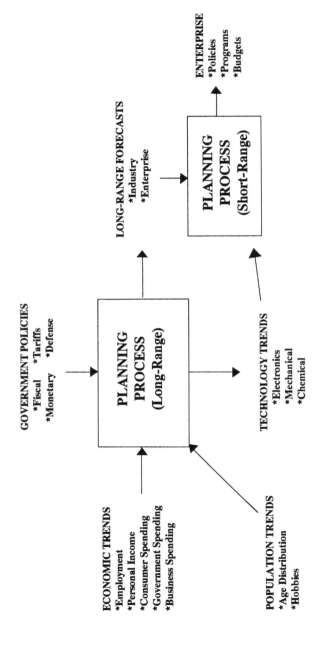

Figure 2.1 External premises to the planning process.

PROCESSES OF SYSTEM PLANNING

- How will it be done? It specifies the manner of getting the activities accomplished.

II. THE PLANNING PROCESS

Any successful enterprise integrates its multiple objectives and plans into a system of operation. Objectives are the drivers from which strategies are set or form the constraints against which the results of the strategies are evaluated. Though literal objectives and plans vary among enterprises, the governing principles are invariably common in the resulting system. An objective in a system may be:

- Profit as excess returns over expenditures of resources in a transaction or a series of transactions.
- Service as performance of any of the business functions, auxiliary to production or distribution.
- Social as in formation of cooperative and interdependent relationships among humans.
- Personal as in relation to individual characteristics, conduct, motives.
- Spiritual as relation to supernatural being or essence.
- Innovative as in introduction of new ideas, methods, or devices.

Plans express the delegation of authority to effect the objectives (Figure 2.2). Plans provide a predictive–corrective picture of system evolution over a definite period of time in the future. Picture details increase with proximity of events. The details are the result of foreseen problems with analysis as to their effect on the activity, and a plan of action that leads to stated objectives. Simply, a plan is a mental detailing of: why, what, when, how, who, how much, and where. Plans may be of multi-use or single-use. Examples of multi-use plans are:

- Policies, broad guides to thinking with room for judgement.
- Procedures, guides to action in a series of steps that yield an implementation of policy.
- Methods, ways to perform a step in a procedure.
- Rules, criteria to govern and regulate performance in a method.

Examples of single-use plans are:

- Programs, predetermined system under which actions may be taken toward stated objectives with an outline of the order to be followed, of the features to be developed, of the participating resources, and of the controls to be applied.

18 SYSTEMS MANAGEMENT: People, Computers, Machines, Materials

Figure 2.2 Decomposition and integration of objectives and their realizations through plans.

- Project, a mini-program with well-defined activities.
- Task, a mini-project with very well-defined undertakings.

Planning initiates the management function by setting objectives and control standards. The controlling function monitors the system's evolution and notes the discrepancies to plans. Those discrepancies could require a change in plans which might lead to modified organization and control functions. Controls are expressed by budgets on human/material resources; budgets may be:

- Purely time, such as time lapse.
- Quantity such as resource-time, man-hours, or kilowatt-hours.
- Common denominator to value, such as money.
- Quality, such as purity.

Planning occurs at all levels in an organization. The lower levels in a system are usually concerned with the immediate events and taking into account the local operating condition. The higher levels are concerned with long term system evolution, taking into account system operating conditions. The short term plans

PROCESSES OF SYSTEM PLANNING

should be made in a way that contributes to the goals set in the long term plan. They are based on information, premises, assumptions regarding future tasks and conditions. For an orderly and purposeful execution, plans should:

- Have definite, clear, and accurate objectives that tree down consistently to the lower levels.
- Be organized, e.g., hierarchical, when the various plans at the various levels fit into a correct and consistent pattern.
- Be feasible and economical where available resources and constraints are taken into account.
- Be flexible so that adjustment or adaptation does not stymie action.
- Be communicated to those entrusted with its execution.

When planning, there are usually many unknowns. To initiate the process, reasoned assumptions are made whenever facts are unavailable. As the process progresses, some facts become known and they replace the original educated guesses. For the remaining assumptions, provisions are made to adjust the plan accordingly or set upfront alternate plans if they turn out to be wrong.

III. PLANNING AND DECISION-MAKING

Decision-making occurs in the exercise of any management function. In fact, a decision is a mini-plan with all its elements. It implies action to solve a problem by an assigned individual/organization with a budget (funding schedule), resources, and control. The process leading to a decision is triggered by some deviation to a plan or expectation. The processes include:

- Diagnosis of the deviation and its source.
- Problem definition with identification of required changes, their criteria, external forcing functions, boundary/initial conditions, internally controllable and non-controllable variables.
- Collection and analysis of relevant data.
- Development of alternatives with conceptual, analytical, or prototype models.
- Evaluation of the alternatives and possible consequences of each under the constraints of the organizational elements, goals, and resources.
- Selection and implementation of an alternative.

The foregoing process may be executed totally by one individual unit, jointly with the organization or partly by individual and partly by the organization. Decisions aim to minimize the time and resource costs to the enterprise.

A decision or mini-plan is a choice from a set of available alternatives. There are two types of decisions:

- Programmed, when the decision is planned, based on system policy, procedure, or rule.
- Non-programmed, where the decision is planned, based on heuristics, personal judgement, or experience.

When a decision is made, its purpose is to reach an objective. The consequences of the decision may be:

- Certain, when the outcome relative to the objective can be fairly determined.
- Uncertain when the outcome is at risk.

At best, only probabilities of all or some possible outcomes are known. At worst, the probabilities of the possible alternatives are not known; then, intuition and insights form the basis of the right decision.

A decision is not usually made unconstrained. We seek to reach the optimum decision by minimizing the uncertainty or maximizing the certainty. At times, we are happy with a satisfactory one. When trying to reach an objective, multiple decisions are made. Thus, a total work breakdown–integration structure corresponds to a decision breakdown–integration structure to lead the work through decision points to a successful realization of objectives.

Setting up management functions is planning. Planning is the decomposition of the objectives of the enterprise into programs, the programs into projects, the projects into tasks, and the tasks into manageable sub-tasks. The decomposition is complete when all necessary work effort to reach the objectives is:

- Partitioned in a work breakdown tree structure.
- Described in a statement of work.
- Assigned in a responsibility matrix to an organization.
- Scheduled with start-end dates, dependencies, milestone events, or objectives of sub-tasks.
- Budgeted with cost/manpower/resource profiles linked to the milestone events.

Budget setting is planning while budget operation is controlling. Budgeting is the powerful tool to tie planning to controlling processes; budgeting continues this interaction among the processes through allocation and reallocation of resources to help those activities in need, needs brought to the surface in troubled areas through critical budgeting information, otherwise not accessible. Planning ties the overall system goals to program goals, program goals to key tasks, key tasks to key events and key events, to the budget dimensions, e.g., time, cost, quantity, and quality. Today's competition requires things to be developed concurrently with no pauses between them; this calls for coordinated parallel and

PROCESSES OF SYSTEM PLANNING

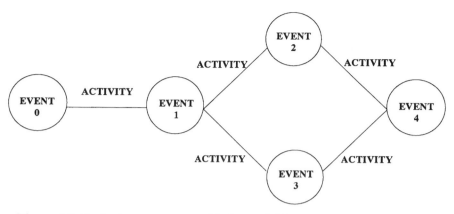

Figure 2.3 Tasks in a program with its activities and intervening events leading to task objectives.

sequential combinations of tasks; time monitoring is a critical harbinger of potential and/or actual problems. Monitoring of a task is done through discrete events (Figure 2.3); an event is the outcome of activities leading to it; activities are the needed work to complete that event; each activity takes time to perform; the time may be a single estimate, multiple estimates, weighted statistical estimate, e.g., mean and variance; the time length of an activity is to some degree in direct proportion to resources (people, facilities) assigned to it; thus in tracking the time constraint, other elements such as cost, quantity, and quality can be linked to form overall planning–controlling processes.

IV. CLASSES OF PROBLEMS IN SYSTEMS PLANNING

Many classes of management problems are addressed in planning when encountered prior to either initiation or at least completion of system operation; plans may overlap operations, but operations should not be completed before the plans' completion. Forecasting forms the basis of planning and is practiced in:

- Partitioning and allocation; in partitioning, system functions are divided into sub-functions to be performed by human and/or equipment the proportion of each and availability of each is at the basis of the planning process; allocation assigns a specific sub-function for performance by a certain human or certain equipment independently or jointly. Cost and/or scarcity of resources leads to their budgeting.
- Sequencing and scheduling; sequencing deals with the order in which work is to be carried out; scheduling indicates when work should be initiated and outputs should be completed. Those two processes balance the evolution of work within the system and avoid choke points, overloads, and idle time within the system. A system

balancing plan avoids large inventories and large delays and encourages parallel activities.
- Maintenance and replacement; with time, parts in the system deteriorate; maintenance is the replacement of those parts; failure of some parts can be damaging and costly; preventative maintenance occurs prior to failure to avoid greater and more costly damage following failure; the deterioration is predicted and planned for its correction.
- Search and estimation; here, uncertainties are filtered out, and incomplete information is filled in through some weighted interpolation–extrapolation. There are varied processes; on one end, brainstorming is an unconstrained mental speculation process to explain what happened and predict what will happen; on the other end, time ordered sequence of observation (time series) analyses are statistical or heuristic processes to recognize and extract patterns of information in gathered data over time, patterns which will continue with some confidence in the future; those future patterns form the basis for planning where the remainder of pattern uncertainty reflects on risk in the plan or decision reaching the desired goals. A general time series pattern can be composed of:
  - Trends which indicate a gradual change in the data.
  - Cycles which indicate wave-like variation in the data.
  - Irregularities which indicate sudden and severe changes in the data due to major changes in identified product, service, or situation.
  - Randomness which indicates the remaining unknown changes in the data.
- Simulation and prototyping; very complex systems do not lend their operation to only mental and statistical processes; then, simulation and prototyping are applied. Simulation is a computer model of the system; prototyping is a physical model. Models explore the action of an event and the system reaction to it, the effect of an event on other events, on system evolution, the dynamic interaction of events through the system within a developing scenario.

V. DISCUSSIONS

Effective planning balances two demands: securing high present returns while conserving future yields and growing present resources. Planning work follows certain basic steps:

- Problem definition; it describes the present system state and specifies the required improvements for a future state, for which planning is needed.
- Data collection; it brings data from internal and external sources and involves experience, past solutions, others' practices, technologies, opinions...

PROCESSES OF SYSTEM PLANNING

- Premises and constraints; it states the overriding assumptions, initial and boundary constraints used in developing the plan.
- Processing; it extracts information from the data within the stated premises and constraints, and involves detection, tracking, classification, identification, prediction, and targeting of relevant options.
- Chosen plans; it prescribes the course of operations and possesses simplicity, acceptability, controllable risk, progress check up, and flexibility to adjust to a changing environment.

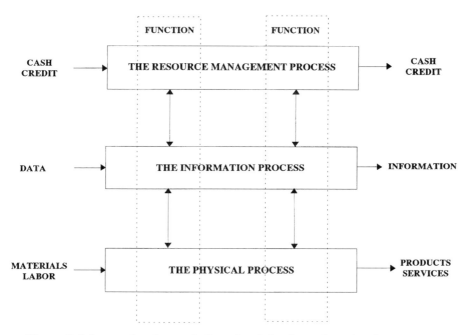

Figure 2.4 Improving for capital productivity through optimal conversion of system resources.

Success of an enterprise rests on the unity of purpose between the chosen plan and the functional operations. Operations carry out their activities through functions which form a convenient way of dividing responsibility for the system tasks. Without top down decomposition and integration, each of the functions sees the enterprise from its vantage point which leads to improving functional performance of the elements, not at getting the most out of the system as a whole. The plan aims to lead the system to the ultimate end of improving the productivity of capital through optimal employment of its resources (Figure 2.4). No plan is perfect; and events will develop differently from expectations, so assessment and revision based on feedback as well as making appropriate adjustments will be needed during the operational execution of the plan.

3   PROCESSES OF SYSTEM ORGANIZATION

I. INTRODUCTION

Organizing is the result of, or the process through which the total grouping of functions in a system is proportioned and allocated to reach the planned goals. At the root is the need for the ultimate decomposition of work to what can be accomplished by an individual unit within the organization. For our purposes, the unit may be a person, machine, and/or computer (software/hardware). Organizational principles are applied in Parts II, III, and IV. The result is a structure where grouping of similar and related activities occurs in what is called a department or component. The decomposition or division of work into an organizational structure requires, in turn, coordination or integration of individual work tasks for effective achievement of the overall goals. The work decomposition, its assignment, and its integration must yield a simple, not cumbersome, process that delivers the end product/service effectively. This reduces to the recognition, at the outset, that work by its nature is dynamic and should not be forced to fit in a static organization; the organization must be dynamically adapted to match, each time, the work at hand. Thus, organizational effectiveness is enhanced by two important features, the simplicity of the decomposition–integration processes and the speed with which the overall goals are achieved.

A system organization does not focus solely on people placement in a structural hierarchy; a person's wiring diagram does not display the essential process of how the work comes in, its flow through the organization, and how it goes out the door. System organizations map the people and their enabling technology in realistic structures (hierarchical/relational/network), and delineate the work flow in a simple correspondence between people, enabling technology, work, and work flow; complexity in this correspondence is detrimental to productivity.

Simply, there is the work, there are the people who know how to do it, the resources that enable it to get done, and there is the traditional wiring diagram. Whenever the wiring diagram is not directly and simply tied to the tasks in the work process, a re-wiring is in order to avoid impediments to the work.

II. THE ORGANIZATION PROCESS AND STRUCTURE

The organizing process involves:

- Listing and description of all the products and services to meet the organizational objectives. This forms the work requirements.
- Partitioning the total work to be performed by type of workers, i.e., human, computer (hardware, software), and machine. Each has capabilities and limitations. Each may perform some of what the others can do, but within certain limits. Basically, a machine does physical work; a computer does analytical-synthesis work, where the software contains pre-stored instructions and the hardware does the physical execution. Humans do both analytical synthesis and physical work.
- Allocation of each work type to an individual organizational unit or part through breakdown of the work load into tasks small enough to be reasonably performed by that specific unit.
- Setting of a structure (arrangement of parts) with formal mechanisms to integrate the parts; the mechanism includes the lines of authority and control as well as the transport (material/data/instruction) that flows through these lines.

Thus, an organization is a depiction of an arrangement of parts, along with the established mechanisms through which activation coordinates each part's work contribution into a unified whole. The parts may be data, computers (software/hardware), machines, people as depicted in Parts II-IV. The basic structures may be (Figure 3.1):

- Hierarchical; it is depicted by a "tree" where each node, as a work-center with its subsidiary parts, is linked to one superior, and all pre-set links among the nodes form no cycles; a cycle is a series of links that begin and end at the same node, without need for backtracking on any link; connectivity from node to node follows the links to integrate the functions of the organization; connectivity may be physical or logical through address pointers. Hierarchical charts may be drawn top-down, left to right, or circular; a circular chart has its highest position at the center, equal positions on the same circle, and lower levels away from the center.
- Network; it is depicted by a tree with added cycles to it and/or multiple superiors at any node.

PROCESSES OF SYSTEM ORGANIZATION

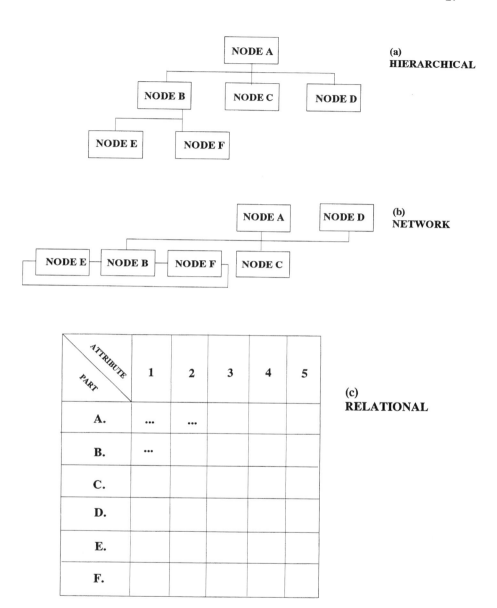

Figure 3.1 Examples of organizational depiction: a) hierarchical, b) network, c) relational.

- Relational; it is depicted by a table where the attributes of the parts are listed and where no pre-set link pointers are established; each part has at least one unique attribute which forms the key (pointer) to it; connectivity of the parts is done on-line through dynamically linking unique attribute(s) that fulfill the sought after organizational function.

Thus, the hierarchical and network structures are based on a pre-set and tightly pointer-connected concept, while relational structure is based on attributes related on-line to satisfy a request for an organizational function. The time evolution of the organization may proceed in parallel or in series.

There is no single best way to structure organizations. Generally, they start hierarchically (centralized) and grow decentralized, distributed, and semi-autonomous as they grow large and diversified. A hierarchical chart conveys the static shape of the organization. It gives:

- The hierarchy with levels of management indicated by the vertical number of boxes with a manager attached to each box.
- The division of work with organization work load indicated by labels in each manager's box (node).
- The chain of command indicated by solid up-down flow lines.

A hierarchical human organization gives only the broad outline of the organization, the different areas of responsibility, and the grouping of work segments. By itself, it is not totally descriptive, specifically of human organizations:

- The levels of management are not necessarily indicative of the actual influence in organizations.
- The labels are not indicative of the scope and depth of responsibilities and activities undertaken.
- The horizontal and diagonal channels of communication (data/instructions) are not depicted.

Other descriptive documents provide the definition for static elements and their dynamic behavior in response to external and internal events in the organization. Unfailingly, varied overlays exist that describe more realistically how an organization functions. These overlays form the web that could either restrict work agility or propel it to new heights.

Comparatively, the organizational description is precise and complete for computer-machine systems; where it is lacking, failure occurs and recovery is incomplete. For human systems, there is imprecision and incompleteness; it comes about, basically, from our inability to define the power within and among humans; human units defy exact definition. On the contrary, computers, machines and their peripherals are well defined up-front and work activities are defined not to exceed those powers; if it occurs, failure occurs. Thus computer-machine systems are predefined. Their behavior in response to all

PROCESSES OF SYSTEM ORGANIZATION

types of events is prescribed. Artificial intelligence systems are an attempt at mimicking human versatility.

III. DECOMPOSITION OR BREAKDOWN OF WORK

At the root of any organization structure is the ultimate goal of reducing the total organizational work load down to unit work loads that can be performed by individual resource units. Formal breakdown of work to the unit may be done by:

- Function, where one activity type or related activities are brought together under single management, e.g., engineering, manufacturing, marketing, finance.
- Product, where those activities using same or similar technology and manufacturing are brought together under single management, e.g., commercial product, military product.
- Geography, where those performed in a certain region are brought together under single management, e.g., Northeast, South.
- Customer, where those activities performed for a given customer type are brought together under single management, e.g., children's toys.
- Time, where the activities' operational sequence dictates the grouping under single management, e.g., basic research, advanced development, development, production.
- Process where the activities belonging to a certain type of operation are brought together under single management, e.g., data processing.
- Equipment, where each type is brought together under single management, e.g., build type, such as electronics, mechanical, chemical.

Actual division of work is usually a combination or matrix across the preceding formal division of work. For any structure to execute any project, allocation of the resources among the units has to happen and coordination of those resources are required, as they respond to events and execute the needed activities; if the structure is cumbersome and/or time consuming, re-structuring is in order.

At the root of work division is the experienced benefit, which ensues from specialization. Since the beginning of time, greater productivity has resulted from job specialization, i.e., art, science, medicine, engineering; then, within each, further specialization occurs. Job scope is thus reduced while job depth is increased. Depth and scope are always in the balance to effect greater productivity.

IV. COORDINATION AND INTEGRATION OF WORK

Integration of a given organization to perform a job requires coordination. Coordination provides the integration of the activities and objectives of the separate units into the top organizational goals (Figure 3.2). The greater the specialization, the greater is the need for coordination. The need springs from differences among the specialties based on:

- Differences in standards for progress applied in each specialty type, e.g., art vs. science.
- Differences in inter-communication modes among the specialty types, e.g., finance vs. marketing.
- Differences in goal perception, e.g., program management vs. engineering.
- Differences in time scale, e.g., basic research vs. manufacturing.

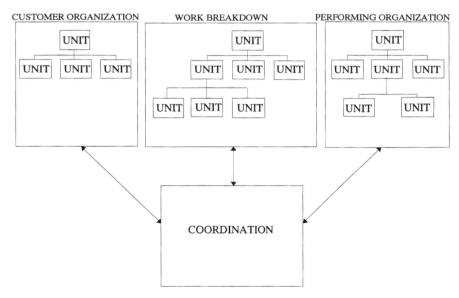

Figure 3.2 Coordination provides for integration of work units across organizational units to satisfy customer units.

As the whole is divided into parts small enough to be executed by an individual unit, several mechanisms are available to integrate the diverse activities into the desired whole. Those mechanisms include:

- Organizational structure, where work conflicts are resolved through chain of command setting/interpreting of policy.

PROCESSES OF SYSTEM ORGANIZATION

- Horizontal communication across the organization, where: a) policy is voted on, or b) policy is carried out rather than formulated.
- Product or project manager, where all activities relating to the product or project are coordinated and where each specific role is explained.
- Round-robin, where integration is rotated among a set of units.

The type and strength of the mechanism depends upon the mapping of work breakdown on an organization. Where it is one to one, the dominant mechanism is the organizational hierarchy and the execution is simple and direct. An interleaving occurs with many to many mapping and the coordinating mechanisms become involved in a dominant way.

V. SPAN OF MANAGEMENT AND DECENTRALIZATION

The span of management is influenced by several factors. Among them are:

- Speed required to meet the deadlines or milestones with concurrency and dependency among those milestones.
- Variety or similarity and complexity.
- Stability of plans.
- Direction and control needed by subordinates and availability of predictive tools of status.
- Coordination required by the various units.
- Ease of communication.

As the span of management is reduced because of one factor or another, delegation and decentralization grow. In delegation, some of the management functions are completed at some lower level in the chain. With delegation, decentralization grows and introduces speed in the organization. For efficiency, integration or centralization has to happen at some level again. Otherwise, the units operate independently from the common goal of the organization. The right balance has to be struck between centralization and delegation where three to five organizational layers between producers and consumers form good design guidance. Several factors enter into the delegation process:

- Rapid response decision-making in a rapidly changing environment.
- Informed decision-making closer to the scene of action.
- More responsibility and initiative at the lower levels.
- Cost and risk associated with decision at varied levels.

With delegation of work goes delegation of authority; there are five bases of authority: expert, referent, legitimate, coercive, reward, where one or multiple of those bases are available for successful execution of the delegated work.

VI. DESIGN OF ORGANIZATIONS

There is no universal organization structure to suit all purposes. A structure depends on a combination of:

- The environment in which it operates: when the environment is stable with only minor fluctuations, few product changes occur, and a mechanistic system is best suited, along with a well-defined chain of command. When the environment is changing or turbulent where rapid and radical products happen, an organic system is best suited with communication across all levels, and command held at the local level.
- The characteristics of its members, where people with their preferences mark the structure as managers, and their reaction to it as subordinates.
- The technology used to carry its activities, where complexity requires automated management services of its resources, be it hardware, instructions, and/or data.
- The strategy adopted to reach its goals, which triggers the activities that individual units have to carry, and the structure best suited to carry them in the environment in which the organization finds itself.

In designing or re-designing organizations, some critical questions must be examined:

- Are there clusters of people/computers/machines/materials dedicated that contribute to the product, rather than a team dedicated to work customer product?
- Are the processes to deliver what the customer wants complicated by the mapping to an existing structure?
- Is the work pulverized to suit functional structure of what should be a simple process?
- Is there extensive communication to cope with organizationally created boundaries?
- Are things done right the first time without a lot of rework and repetition?
- Is there excessive system reserve (people, time, money) to accommodate uncertainties?
- Is there excessive overruns in ongoing activities?
- Are there excessive controls to prevent a basically unstable organizational system from going awry?

If the answers are "no", the organization is perfect; don't touch it. If the answers are "yes", the organization is broken; working harder will only increase productivity by 5-10%, not enough to meet the nineties and beyond competitive need of 50-70% improvement.

The message is simple and direct. We don't allocate a simple task across multiple computers/machines with undue interfaces and communications when a single computer can execute the process. On the other hand, human functional organizations can dictate break-up of a simple task, executable by a single person, in order to adapt it to the functional structure:

- This results in complex execution of otherwise simple tasks.
- This involves a lot more resources.
- This requires a longer time.
- This costs a lot more funds.
- This is prone to lower quality.

Organizations, however they originate, are not static. They grow and change with time. With change in management focus, organizational structures evolve and change, as well as the control system and the return (reward) emphasis. All changes may not synchronize when organizational focus, structure, control, and reward may be at different phases of the change process, Table 3.1.

VII. DISCUSSIONS

Organizing produces a whole structure with its parts so integrated that their respective relation is guided by their relation to the whole. Organizing has two elements, parts and relationships; the parts can be considered as the work units composed of humans, computers, machines or parts thereof; the relationships bind the participants to work together through a connection or dependence: splintering of the work processes to satisfy a given organizational structure rather than the setting of an organizational structure to satisfy simply the work processes leads to compounded difficulties in the integration of the parts to make a whole due to unnecessarily complex interfaces.

Organizing is necessary since most jobs require the work of more than one single unit, be it human, computer, or machine. With such jobs, the managers deal with the collective actions of many minds and hands, and computers and machines where the individual units reach the overall goals only due to their synergy and their simultaneous actions. Organizing harnesses the resources (human, equipment, material) in an effective manner so they can perform the required activities by the plans. It is intended to enable people–computers–machines to work together toward attainment of the goals at hand. The organizing activities include:

- Work decomposition or breakdown into manageable and related packages.

Table 3.1 Organizational Characteristics and Variable Growth

Focus	Make and Sell	Operational Efficiency	Market Expansion	Consolidation	Problem Solving
Structure	Informal, individualistic	Centralized, directive, functional	Decentralized, delegative	One staff, product watchdog	Teams participation
Control	Market results	Standard cost center	Reports and profit center	Plans and investment	Mutual goals
Reward	Ownership	Salary increase	Individual bonus	Profit sharing and stock options	Team bonus

PROCESSES OF SYSTEM ORGANIZATION

- Resources decomposition or breakdown into manageable and related units.
- Assignment of work packages to compatible organizational units.
- Flow chart of work in relation to organization chart.
- Work integration and fusion of activities now divided for effective performance.
- Adjustment/modification of work/organization in light of control results.

This view of the organizing process falls into system description process. A system view reveals more precisely the various parts and their relationships, how those parts act and react to trigger designated actions. Consideration of an activity usually reveals that it is the result of many other sub-activities which in turn are due to many other sub-sub-activities. An effective vehicle for framing the activities, sub-activities or sub-sub-activities is the system representation in terms of discrete units, each having an input, process, and output. A system view frames and simplifies the conception, development, the decomposition into discrete parts and their integration according to their relationships into an organized whole with a unity of action toward high order system goals.

Systems are endemic to most activities. We encounter natural and human-made systems. For instance, the activities in the human body can be decomposed into subsystems, e.g., nervous, digestive, circulatory, sensing and those in turn into subsystems; a firm is a human-made system whose internal parts act and react together to achieve set objectives and whose external parts act and react to the outside environment including customers, suppliers, stockholders, and government (Figure 3.3). Viewing the organizing process in its generality, it includes as parts and relationships (Figure 3.4): a) human organization, b) computers organization, c) machines and devices organization, d) material (physical, chemical, nuclear...) organization; those types or organizations must function as an integrated system. Organizing is dynamic; organizations change due to: a) fluctuations in demand of products/services; b) obsolescence of parts and/or functions; and/or c) technological innovation. The ultimate judge on the fitness of an organization is not how well each organizational unit did its work, but how well the integrated process satisfies the stated objectives.

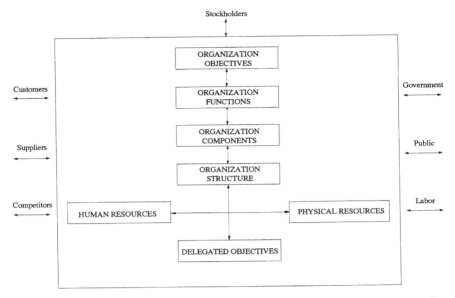

Figure 3.3 The organization and its environments, internal and external.

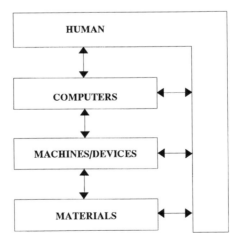

Figure 3.4 Generic elements of system organization.

4   PROCESSES OF SYSTEM ACTIVATION

I. INTRODUCTION

Activation puts the organization into action according to plans and monitors, and adjusts and/or corrects the performance per established control processes to reach the set goals as detailed in Parts II-IV. The performance rests on behavioral characteristics of the organizational elements, be they human, computer, or machine.

Predictability of outcome depends on the complexity of the situation, the elements' characteristics, and the mutual interaction of both. Predictability rests on depth of understanding. Our understanding of systems or elements which are linear, deterministic, time invariant, and spatially homogeneous is well founded; predictability of their behavior under activation is well assured. On the other extreme, our understanding of systems or elements which are nonlinear, stochastic, nonstationary, time varying, and spatially inhomogeneous with complex initial and boundary conditions is hazy, and predictability of their behavior under activation is uncertain. While computers/machines fall predominantly within the first category, people fall invariably within the latter category; people unlike computers/machines respond more to a total situation rather than highlighted specifics within it; here lies the enigma when engaging the human mind to create and innovate.

II. THE ACTIVATION PROCESS

Activation is the process of directing the task-related activities of group members. The processes, as appropriate to the situation, task requirements, and group members may be:

- Authoritative, where specific orders and decisions are issued from a superior with execution carried out by group members.

- Consultative, where general orders and high level decisions are issued by a superior regarding set goals but following discussions with group members.
- Participative, where goals and decisions are reached by the group members within limits defined by superiors.

To succeed, flexible application of the processes is required to blend together the task requirements, the situation at hand, and group members. Not one process is conducive at all times to the desired performance.

In general, performance is a function of knowledge, ability, and motivation. Knowledge is the sum of information at hand. Ability is the power to perform. Motivation is the stimulus to action and is a function of needs, rewards, and/or consequences of action.

An integrated description is characterized in Figure 4.1. The job characteristics are the input function. The performance characteristic yields the output function. The work situation defines the boundary conditions. The activity function is summarized by three sub-functions, gain, state transition, and feedback. The gain is defined by the system needs or criteria for stimulus or motivation. The state transition summarizes the knowledge and ability to respond to input, boundary, and output characteristics. The feedback denotes the consequences of the response. In human systems, for instance, each individual has resources around his evolving needs (Figure 4.2). His initial needs are physical and evolve into security needs, social needs, esteem needs, and cap at self-actualization. In automated systems, the needs remain physical (cooling, heating, power...); widespread criteria are used to satisfy those needs (deterministic, statistical).

There are basic learning procedures during system activation:

- Identification specifies the desired/unwanted activities.
- Measurement collects the frequency over time and circumstance of those activities.
- Evaluation determines the factors causing those activities.
- Control implements the mechanism for extracting the desired activities and rejecting the unwanted ones.

III. ACTIVATION OF CONFLICTING AND CHANGING ACTIVITIES

Conflict originates from varied sources; many have their roots in the way the organization is set up. Conflict occurs in varied forms (Table 4.1). Regardless of its form or coverage, sources of conflict include (Figure 4.3):

- Sharing of resources, where their limited availability can create conflicts. Shared resources may be manpower, materials, equipment,

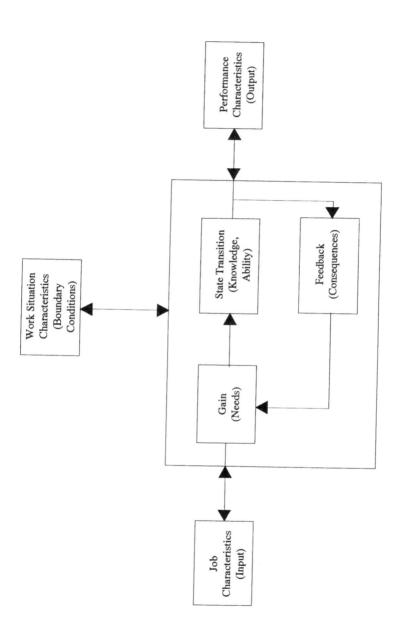

Figure 4.1 Characterizations of activities at the system or unit level and their interaction with the external environment.

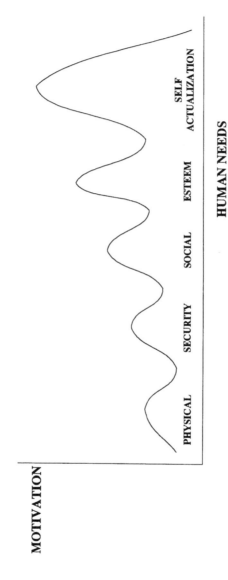

Figure 4.2 Increased resonance in motivation with evolution of human needs.

PROCESSES OF SYSTEM ACTIVATION

Table 4.1 Examples of Conflicts

Conflict	Computer System	Human System
Within an individual unit	Single computer	Single person
Between individuals	Two computers	Two people
Between individuals and groups	Application and resource manager	Cost analyst and proposal team
Between organizational units	Data manager and resource manager	Engineering and manufacturing

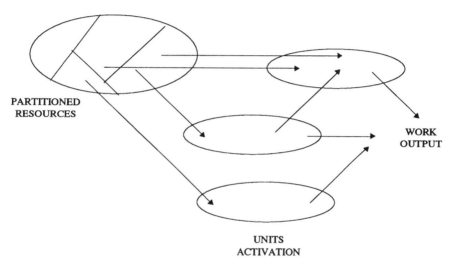

Figure 4.3 Conflict due to sharing, interdependencies, ambiguities, and differences in goals.

facilities, money. Conflict arises when their allocation in amount and/or time may be less than what is needed (real or perceived).
- Interdependencies of activities, where the work on one task cannot be continued or completed until an input is received from another unit. Work is partitioned, then allocated to the units of an organization with the expectation of completion of each by a certain time. Conflicts arise when the input is not what is expected (quality, time).
- Ambiguities of work, where the expected results are understood differently by the organization units. The work definition is not communicated so that sender and receiver both agree on meaning; computer systems have standardized protocol where the involved meaning is strictly specified and strictly followed by the parts.

- Differences in goals, where organizational units, depending on specialties, have different accounting processes for the results of their efforts. Conflicts arise when the metrics among the specialties differ, i.e., quality vs. quantity, short term vs. long term.

To minimize conflict, activities include:

- Planning the design to avoid sources of conflict through segmentation and tiering to support priority demands especially in time of reduced capacity.
- Organizing so as to avoid splintering of tasks with resulting complex and multiple interfaces just to satisfy an existing functional organization.
- Action to create cooperation, where the parties work together to attain mutual goals.

When conflict arises, short of re-planning and/or re-organizing, procedures for negotiation and solution include:

- Rotation of the decision maker.
- Finding a third choice to the conflicting two choices.
- Assignment of scores to all choices and selecting the highest scored choice.
- Testing to some criteria of the choices.
- Setting prices on choices.
- Using arbitration.

Change is often brought to a system during activation. Change is brought by external forces and/or internal forces. Within the system, change is brought to the structure, the technology, and/or the people. These changes include:

- Its structure modifies such items as the management hierarchy/levels/spans, the flow lines of work, and/or communications (command/information).
- Its technology modifies such things as research, engineering, production, methods, equipment.
- Its people modify roles, selection, training.

Generally, the types of changes are interdependent. Adaptation of the activities must occur, otherwise, conflict arises. A technology change introduces automation to a variety of small tasks and supplants people in the structure. Thus implementation of a change in an element (structure, technology, people) entices changes to the other elements. Changes occur at the individual level, group level, or system level.

IV. ACTIVATION OF HUMAN AND AUTOMATED SYSTEMS

Since the beginning, units have come or have been brought together into groups to undertake some sort of action. Singly, that action could not have been undertaken successfully. Groups come in different sizes and cover varied endeavors from family to business, religious, political, military, trade, educational, social, or civic groups. Human group activation rests on the group concept of "leadership" which puts authority in the hands of one, few, or distributes this authority among a wider circle. Through its ideas and actions, the "leadership" influences the ideas and actions of others.

Influencing calls for knowledge of the individual goals, their synergy with group goals, e.g., their respective motives for their navigation pathways around obstacles to those goals, and for the operating environment around the group.

Motives are the internal human stimulus to action and explain why/how people respond to organizational influences toward organizational goals; a well-motivated organization works in a created environment where the individual motives and the group motives are aligned so that jobs are efficiently and effectively carried out leading to high productivity. Contrary to human systems, automated systems have their stimulus to action comparatively well understood and their productivity well bounded and predictable; this is elaborated upon in Part III. On the other hand, the potential of human systems is practically unbounded and is regenerative through a continuously creative process where motives, environment, input, processing, and output remain mysteriously linked.

System productivities rest on physical and mental factors:

- Physical factors influence variably human, computer, and machinery units in a system; they pertain largely to the environment where the units find themselves; the factors include working conditions, unit interactions, policies and practices, security, status, pay where the latter factors are meaningful to humans.
- Mental factors influence mostly humans in a system; the factors include the job itself, challenging work, recognition, feeling achievement, increased responsibility, opportunities for advancement; the job factor could pertain to the job computers will undertake.

System effectiveness rests on an integrated approach of the factors where the characteristics of the activators or leaders, of the activated followers, and of the situation are merged to determine the "best" system. The activators' characteristics include: a) *their capabilities and traits*; b) *behaviors* marked by task-oriented and/or people-oriented approach; and c) *experience/expectations*. The characteristics of those activated or followers include: a) *their capabilities and traits*; b) *task relevance*; and c) *experience/expectations*. The situation includes: a) *external environment*; b) *objectives of the group*; c) *technology involved*; and d) *adopted structure*.

While machines have preset actions and reactions, humans and computers have the ability to deal with abstraction, to adapt to many environments, to respond to many stimuli in different ways. Additionally, humans exhibit a singularly natural ability to learn and create; the resulting range of capabilities spans from the fairly primitive, instinctive, or conditional reflex to the rational or thinking response where the interpretive processing of the same stimulus could result in different responses. While computers and machines have prescribed, though complex, behavior and respond to a stimulus directly, humans respond to the interpretation of the stimulus within the environment it is received; a human may receive a stimulus in one environment and respond favorably to it and receive the same stimulus in a different environment and reject it out of hand; on the other hand, computers or machines with different behavior from group norms are considered errant and are removed from the group for maintenance or replacement. People, unlike machines, respond to a total situation.

Humans process stimuli based on their past experiences, evaluations, and expectancies; though some mechanisms are understood, the interpretative and creative processes are not; the description of the human processing operator is by no means predictable:

- Non-verbal information input, superimposed on a message changes the message content; only humans interpret and process facial expression (wink, smile), hand movement, tone of voice, eye contact.
- Ability to ferret out information with vague unclear rules while tolerating ambiguity is only established in humans; computers and, to a much greater extent, machines insist on clear, specific, and directive procedures.
- Quantitative ability to formulate algorithms and ease of processing of qualitative attributes are human characteristics; computers and, to a much lesser extent, machines only execute formulated algorithms or heuristic processes.
- Ability to recognize needs and weaknesses and create new mechanisms and processes to fulfill those needs and compensate for those weaknesses is only a human characteristic; humans create computers and machines and not the other way around, and through them a civilization was built in science, technology, engineering, law, medicine, commerce... .
- Ability to learn, adapt, associate and infer is a human characteristic; computers are designed at best to mimic that human behavior.

With such diversity in the internal dynamics of the human processor, it is no wonder that the external dynamics within human groups are further

PROCESSES OF SYSTEM ACTIVATION

diverse. An understanding of the group as individuals, each with a basic drive, need, and goal helps in estimating areas of common agreement, in explaining behavior, in motivating group members. While understanding of the individual is helpful, it is also helpful to recognize that the individual is subjected in a group to resonances, damping, distortions, dispersions, and nonlinearity that alter, detract, or enhance the individual's behavior. After all, the minimum essential in group formation is the belief by the individuals that the group represents the best potential for fulfilling their individual interests. Otherwise, considerable expenditure of effort is made to keep a group together rather than to engage it toward the goals that triggered its formation in the first place. By comparison, understanding computers and machines as a group is a more tenable proposition and will be treated in Part III.

Group productivity rests on the processes applied to secure group action. Processes fall into general categories of information giving stimulation to action and/or evaluation. The varied techniques range from dialogue to small group exchange, interview of an expert panel, discussion among the most concerned group members, lecture by an expert, brainstorming, symposium, conference, and workshop or institute. Depending on the applicable format used, the chosen technique provides mutual interchange of data, information, ideas, opinions between members of a group, exploration of issues of mutual interest, crystallization of own thinking, formation of group consensus on a solution, activation and execution of that solution. The processes are moderated or free-flowing, organized by topic, constrained by subject and/or time, limited to certain participants or all inclusive, all for the principal purpose of insuring that group members are enabled to work together cooperatively to reach chosen goals. Cooperation rests on the potentially positive categories of individuals in the group; groups are composed of three categories of individuals:

- Group builders: those include the types that create harmony, compromise, encourage, and facilitate.
- Group detractors: those include the types that block progress, dominate the group, seek recognition, and negate.
- Group actuators: those include the types that initiate work, seek and/or give data/information, elaborate, summarize, and/or orient and keep focus.

V. DISCUSSIONS

Activation is characterized by action rather than contemplation or speculation. Planning and organizing realize no product or service until activation is initiated. Activating sets into action the organization to realize the plans as a function of time. To every action, a reaction is experienced; a resistive reaction dissipates some of the action; the result brings about change and motion due to its resultant force; the force stems from energy exerted by physical, intellectual, or moral means; a force applied toward a desired direction of motion yields productive work where one expands physical or mental efforts to overcome resistance/obstacles and achieves an objective.

Activation requires coordinated action by the units in the system based on influence and power; influence is the process where one unit follows another unit's direction, suggestion, or advice; influence is based on power which results from a unit's inherent or positional attribute. The type of applied power determines the activating mechanism which overlays the organization and does not necessarily map to the formal organizational structure (Figure 4.4). The bases of power are:

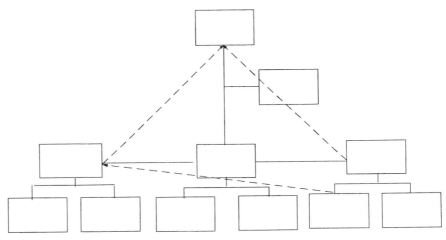

Figure 4.4 Overlap of a non-formal structure over a formal organization structure.

- Coercion, where the influencer may punish the follower through withholding or frustration of a need.
- Reward, where the influencer may grant a reward and satisfy a follower need.
- Legitimacy, where the influencer is followed by virtue of his organization position.
- Referral, where the influencer is followed based on blind faith.
- Expertise, where the influencer has special knowledge that can satisfy the follower's needs.
- Representation, where the influencer is followed based on reflection of the follower's needs and interests.

Coercion and control over rewards are having decreasing influence. Legitimate power is on the decline since the formal hierarchical organization is being increasingly flattened into a more network or relational one. Expert and, to a lesser degree, representative powers are increasingly the major sources of influence and are both dependent on information; some information is internally generated and others are externally collected through communication; the external communication flow and centralization of information contribute to the base of expert power. In most present endeavors, there is an essential need for information to make decisions and take action.

While stimulus, action, and reaction of computer-machine systems, though complex, are inherently predictable, human systems can be quite unpredictable; unpredictability is not usually a design feature in automated systems. Identifying human wants or needs puts to the test experience and proficiency in management; the task is extremely difficult; human wants or needs are dynamic and may be obscure and/or conflicting, and this emphasizes the challenge of human relation in management; besides their identification, their satisfaction may not be simple or feasible. In controllable automated systems, the activating process is strictly prescribed at the outset; the workstations are connected to each other by work flow and control flow plans, all pre-set by pre-programmed instructions with decision points, schedules, and controls. In human systems, their activation processes may be strictly or loosely prescribed; they are subject to review, variability, and surprise since human performance is highly dependent on identification of human motives and their marshalling, an information processing area with large uncertainty.

The trend is to have computers and machines do more and more of human functions. Computers managing computers and machines have become widespread; there, the series of activities are integrated with little or no human effort required in the development and production of products and services. As computers and machines displace human operators and elevate them to do increasingly mental functions, humans are the agents of this displacement. Humans conceive computers and machines to mimic, however clumsily at times, their sensing, thinking, and action where transducers and processors mimic: a) *sight* with radar, infrared, photonics receivers, etc.; b) *hearing* with sonar, audio systems, etc.; c) *smell* with gas detecting, gas analyzer devices, etc.; d) *touch* with pressure, temperature, capacitor gauges, etc.; e) *taste* with acid, alkaline, pH meters, etc.; f) *action* with motors, valves, solenoids, etc.; and g) *thinking* with computers.

5 PROCESSES OF SYSTEM CONTROL

I. INTRODUCTION

There is no single or unified method of control. There are series of controls, each different, each used to deal with different problems or system elements in the organization. These controls need, however, to be interrelated to produce a managed system.

The control process is the means and not the end. Through it, actual activities are made to correspond to planned activities. Not every activity is observed continuously in time; this is not productive. In control, progress of key observables toward planned goals is monitored at key positions in time/space; detection, classification, identification, and tracking of deviations from expectations are noted, and corrective actions are undertaken. Controls based on selected observables are prevalent and are applied, for example, to:

- Standardize quality, i.e., deterministic, like blueprints, inspections... or statistical, like minimum square-errors, maximum likelihood.
- Measure productivity, i.e., output per employee, output per machine, output per computer.
- Motivate performance, i.e., promotions, bonuses in human organizations.
- Plan and program operations, i.e., budgets, forecasts.
- Limit span of authority, i.e., management layers.

Control cannot be static; it has to be adaptive to account for changing situations. Control does not guarantee success; it facilitates it. Only the right control of the right plan can facilitate success. Even with control, a system can become unstable and break up. Applying excessive control fails to harness the resources to achieve the set goals.

II. THE CONTROL PROCESS

Whether considering the system in its totality, a component or sub-component in that system, the representation in Figure 5.1 applies; only the size, multiplicity, and/or complexity of the blocks changes; the flow lines between the blocks are of two types, information and corrective action. The control process is enjoined to the basic building blocks (input, processing, output) of the system (Figure 5.2).

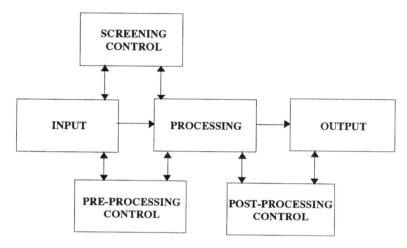

Figure 5.1 General process control.

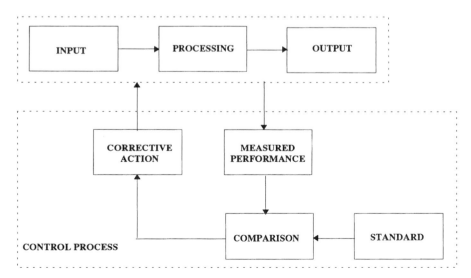

Figure 5.2 Overlay of control process on operation.

PROCESSES OF SYSTEM CONTROL

A. The Need for System Control

Many factors enter that require control in a system. Systems do not naturally function to expectations:

- Partitioning and allocation of functions: this delegation or decomposition of functions diminish the organizational responsibility to execute a plan. Without control, managers have no way of telling that progress is or is not satisfying the plan. With control, checkups take place at well-positioned milestones and corrective actions are taken where necessary before a critical failure occurs.
- Faults and failures: people and/or computers/machines make errors and fail. Without proper control, systems progressively deteriorate and devalue until bankruptcy. With control, reconfiguration, repair, or replacement take place to reset the system and to reconfirm its original purpose.
- Size and complexity: as a system grows and its functions become complex and diversified, coordination of its elements toward the common goals is needed. Without control, contribution of the elements will grow independent and most probably at variance from the stated system goals. With control, the system is constrained to perform against standards; otherwise an alert flag is set.
- Change: when plans are formulated and controls are set, premises about the environment are made. Each environmental change or mismatch goes unnoticed, and progress against the wrong plan continues until instability and failure set in. Adaptive control is needed to detect, identify, and self-correct to an evolving new plan. There is a strong or weak dependency among the systems in an environment; a variation in one may require a corresponding adaptation in another; those systems may be individual, family, town, country, corporation, program, industry, university, etc... .

B. Types of Control

Three types of control processes are encountered:

- Screening control, where specific conditions are met before operation may continue; those controls are yes–no. Types are found in system management by exception, e.g., safety checks, quality inspection to catch discrepancies.
- Predictive real-time controls where deviations are detected as soon as they happen, and corrective action is inserted before a particular sequence of operations is completed; those controls require timely information about changes in the environment in order to institute timely corrective action.

- Post-processing controls, where deviations are detected after a sequence of operations is completed; corrective action is applied on similar activities in the future.

Whatever the type of control might be, the steps in each are basically:

- Defining desired results where they are linked down to the individual human/computer/machine assigned to deliver them.
- Setting standards and methods for gauging performance: the standards must be relevant to the goals and, most importantly, measurable. The source of standards may be historical, imposed by external sources, or predetermined by higher authority, or sometimes subjective. Most importantly, they should be flexible to adapt to a changing environment.
- Measuring performance: the more sophisticated and complex the system is, the more the goals and performance measures are not directly related; most likely, the measures are or appear to be components of the goals or correlates of the goals. The actual frequency depends on the rate of activity change and the need for immediate corrective action. Frequency should focus on the major determinants so as to avoid over-control.
- Evaluating the deviation of performance from standards: this involves an assessment of: a) *the importance of the deviation so as not to track minor ones*, b) *the source of the deviation as to plan or process*. For simplicity of control, feedback should be directed down to the unit performing the work for correction and upward for management by exception.
- Applying corrective action: this brings performance back up to standards when deviation warrants it.

Positioning of control systems should be set at the strategic points determined to be critical to the overall success of the activity at hand. Too much control is costly and detrimental. Focus should be on the deviations that cause the biggest harm.

C. Levels of Control

Control should be designed to be exercised where performance is taking place. Otherwise the system becomes imprisoned by the control system. Highly centralized control loses the simplicity needed to have speedy response.

System standards vary depending on the system goals and the levels where they are applied within the system. The system goals are partitioned and allocated to its varied components and down to its individual entities; broad

PROCESSES OF SYSTEM CONTROL 53

controls at the upper management levels are increasingly detailed as they are treed down to the lower levels.

No single control is applicable to all system activities. Each type of activity requires its own form of control. At a corporate level, key goals are set in multiple areas. A variety of standards are applied to suit the specific organization; then the series of controls are interrelated to focus the individual activities toward the upper system goals. All along, the control parameters give slices and views of the system, where a common denominator is developed, such as currency, to integrate widespread budgets, such as a) *quality budgets*, e.g., signal-to-noise ratios or errors at the activity level, b) *time budgets*, e.g., response time, schedules, cost... at the task or program level, eventually to c) *the financial statement* at the corporate level. Chapter 17 discusses automated process control amid risk and uncertainty.

At the corporate level, there are varied standards which are eventually mapped down to the program and individual level for realization (Figure 5.3):

- Profitability: profit measurement methods have included the rate of return on investment, percent earnings to sales, or profit minus cost of capital investment; those are important stockholders/bankers measures, as to the viability of the corporation.
- Productivity: many factors influence productivity, i.e., automation, process layout, human efficiency. Indicators on the efficient use of resources have been: depreciation dollar cost as a metric of utilization of equipment; payroll dollar cost as a metric of utilization of humans.
- Market share: it is a measure of available orders each product or service has in a company; a decline in the measure would signal a customer dissatisfaction or aggressive market penetration.
- Organization development: organizational structures can lead to excessive delays; work flow design can yield unacceptable product defects. Comparative measures with industry performance would flag such problems.
- Human development: like material resources, human resources need to be planned and developed; measures on the staffing process include internal promotion, recruiting, education (degrees earned in prior year, internal courses); low numbers can indicate failure to develop ability and skills of organization membership to increase their productivity. Employee satisfactions are monitored through surveys, continuous communication; participation in decision-making to insure preventative maintenance prior to failure (loss of employee, low production).
- Product development: products need to be planned and developed; measures include internal research expenditures, enabling technology, contracts acquired, support of important customer metrics (lower cost, high reliability, simplicity in operation).

54   SYSTEMS MANAGEMENT: People, Computers, Machines, Materials

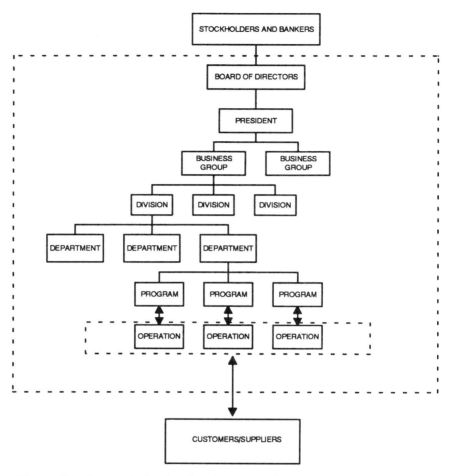

Figure 5.3 Corporate levels of system control.

When corporate entities are viewed from these standards' perspective, they show a common thread of similarity of principles down to the actual operation level. At all levels, transformations are taking place. Each transformation function involves many variables (processes and data) pre-set, reset, and changed to satisfy the goals and constraints of the system (Figure 5.4).

Measurement and control are included to insure that pre-set variables lead to desired goals. The measured as well as controlled variables may be directly or indirectly tied to the system goals. When the system controls are done by computers, transduction and signal conditioning tie the computer to non-compatible system elements. Generalized feedback measurements and controls are depicted in Figure 5.5. A sample system is depicted in Figure 5.6.

PROCESSES OF SYSTEM CONTROL

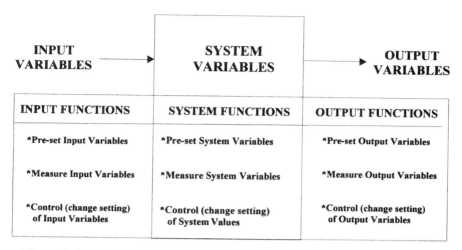

Figure 5.4 Input–system–output transformation functions with implied feedback and control.

III. BUDGETING

Budgets are the system resources allocated in support of each planned activity within the system during a projected period of time. Meeting or exceeding the budget denotes system effectiveness. Budgets as control devices are used to coordinate the multiple activities toward the common objective. They put the plan into effect and communicate it; during activation, budgets signal deviations from plan, but identification of the cause and remedy need to be investigated separately. In reviews, actual performances against the budgets are compared, deviations are explained, and corrective actions are set. Chapter 18 expands on their utility in program management.

Budgets may have a common denominator and thus have a direct relation to the overall system objectives. A common denominator may be dollars, and an overall objective may be profit. In such a system, dollar budgets are communicated to all subsystems and units and the system has an integrated control. More often, budgets have multiple denominators, directly related to the overall objective. When not translated and integrated into a common currency, the management system operates with varied budgets where multiple control devices are exercised. The variety is likely to be found more at the operational level. Some common budgets include financial, time, quantity, and quality budgets.

A. Financial Budgets

Financial budgets denote the money flow in the system, where it comes from, where it is expensed, and how much is left at the end of a period. From a budget point of view, systems can be divided into:

56 SYSTEMS MANAGEMENT: People, Computers, Machines, Materials

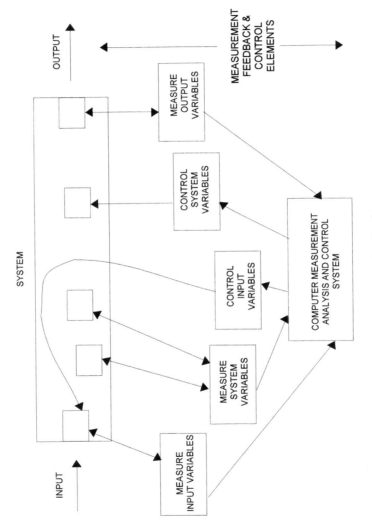

Figure 5.5 Automated system control through feedback.

PROCESSES OF SYSTEM CONTROL 57

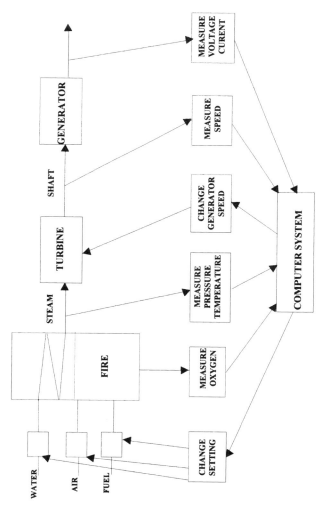

Figure 5.6 Sample feedback control.

- Revenue components, where orders and sales are generated; those components are judged on the basis of orders' capture, dollar size against the budget; the orders for the year are made up of back orders and projected new orders. Revenue components include sales.
- Expense components, where the actual costs of direct and indirect labor and raw material are compared to budgeted ones at review time. Overruns denote inefficiencies in such components as engineering/manufacturing.
- Profit components, where the expense components are totaled and substracted from the revenue components. The net in contrast to the budget provides one control measure where profit is assessed against the assets used, where depreciation of buildings and equipment and cost of investment are subtracted.

Varied statements are prepared and are useful in detecting positive and negative trends and thus instituting corrective actions if necessary. The balance sheet budget (Table 5.1) integrates all budgets at the end of the period to denote if actual results correspond to planned results in assets, liabilities, and net worth. Analysis of financial statements is done through ratios of figures within the statement itself. The ratio is either observed over time for performance and/or compared to others in the industry for performance. Those ratios are derived to indicate:

- Liquidity, where the ability of the corporation to meet its immediate financial needs is assessed.
- Net worth, where the general health of the corporation to face the future is assessed.
- Profitability, where viability of the business to continue is assessed.

B. Time Budgets

While routing establishes the activities to be executed, the sequence to be followed, and the resources to be used, scheduling assigns the calendar time to execute the activities. A schedule is a management control of time as a resource. Time is used as a coordinator to synchronize the activities. Activities are mini-events scheduled to complete an event. Scheduling of activities for all events needed to complete all the tasks in a project/program is a planning process, (Figure 5.7); securing reports on progress against the plan and acting on deviations is a controlling process. The desired characteristics of an event are not explicitly indicated in the schedule and are defined elsewhere. A milestone is a controlling decision, event, product, phase that has to occur by a certain time. Milestones are used as control parameters at review time of progress, of cost, and the need for schedule modifications. The sequence of activities that take the longest time is on a critical path and is tracked closely

Table 5.1 Financial Statements

	Indicators	Formula	Comments
Statement	Balance sheet	Net worth = Assets − Liabilities	*Net Worth: Residual value if company dissolves *Assets = Cash + Security + Account Receivables + Inventory + Land + Plant + Equipment + Patent + Good Will *Liability = Accounts Payable + Expenditures + Mortgages
	Income statement I	I = Sales − Costs	*Income available for stock or capital reinvestment *Cost: Labor, Material, Taxes, Operations
Ratio of Figures from Financial Statement	Debt ratio	Total Debt/Net Worth Total Debt/Total Assets	*Measures the company's short term financial condition
	Coverage ratio	Earnings/Total Interest	*Ability to meet interest expenses *Earnings before interest and tax
	Liquidity ratio	Current Assets/Current Liability	*Ability to meet short term obligations *Current assets = Cash or Easily Convertible Resources
	Profitability ratio	Net Profit/Net Worth Profit/Total Sales Profit/Total Assets	*Measures efficiency of operation *Profit: Net Profit after Tax
	Operating ratio	Sales Force Cost/Total Sales Sales/Inventory Sales/Total Assets	*Measures functional efficiency of operation
	Return on interest	Profit/investment	*Measures investment viability

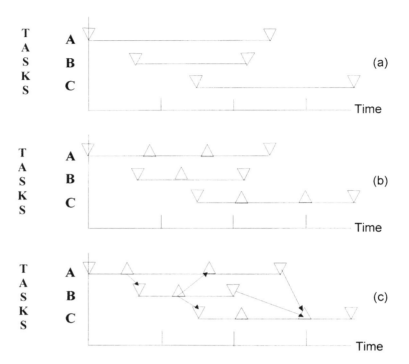

Figure 5.7 Time control of tasks in a program; (a) scheduled time for start and stop, ▽, of tasks; (b) milestones, △, within each task; (c) milestone dependencies across tasks.

as a control parameter. The activities in a critical path change as the plan is realized since scheduled activities may take longer or shorter than expected.

The length of the time interval between tracked events in the control process depends on the rate of information variability and this is usually linked to the level in the system:

- For an operator, it is short and may be in microseconds, seconds, minutes.
- For his manager, it is hours, days, weeks.
- For the higher levels, it is in months, quarters, year.

Time management control parameters are used in varied ways:

- Fixing time and event, e.g., quarterly report.
- Fixing time, e.g., weekly, monthly... and track events.
- Fixing event and track it in time.

PROCESSES OF SYSTEM CONTROL

C. Quantity and Quality Budgets

Quantity budgets are concerned with the amount or size of products/services generated. Seldom is a quantity set without a quality measure attached to it, otherwise unsatisfactory products/services would be provided just to satisfy the stated quantity quota while satisfaction of customers is not considered, and as a result the desirability of the product/service is weakened.

Quantity reflects on the speed of production; it yields a measure of productivity. Highly productive systems have their flow from input through process to output unhindered by delays in storage; when utilizing effectively the full system capacity, the quantity of products/services is increased. Quantity can be used as a variable to regulate the flow at input, within the process, at output; for instance, quantity of orders, or orders potential at the output are used to control the quantity of raw material purchased to make the product, and the quantity of processors engaged to produce the products.

Quality reflects on product/service satisfaction; it yields a measure of reliability, dependability, consistency. Quality is assured through inspection of the product/service against a set standard; some tolerance to deviation from the standard is usually allowed within an upper and lower control limits; acceptance is predicated on deviation within the set limits.

D. Effectiveness of Budgets

For budgets to be effective, their control parameters need to be:

- Pertinent to the activities at the level applied. For instance, while top managers are concerned with budgets and financial results on a greater basis, lower managers are concerned with day-to-day operational activities and controls relating to overtime hours, down time, milestones accomplished.
- Few in number so as not to overconstrain the activities, delay the detection–correction process to a deviation complicating the system with minor objectives. Controls should focus on the primary objectives and provide when and where a corrective action must be taken. In fully automated computer systems, the how is added.
- Adaptive so as to reflect shifts not only in standards, but in activities, in importance, in plans, in environments.

E. Information and Budgets

To control the activities in a system, information must be embedded within the different budgets. Information as separate from data implies decision and action relative to the activities in a system. Effective control systems require:

- Timely information as to the time an event has occurred or information on a given event at a pre-set time interval.
- Accurate information where the quality or closeness to reality is provided.
- Complete information where all relevant information is included.

Computer-based data systems enable the processing and delivery of such information. Chapter 17 expands on information processing under risky and uncertain conditions. This facilitates the real-time decision-making process. Decision about events as they are taking place constitute a real-time control system.

IV. DISCUSSIONS

The real value of managing is to achieve results. To realize this value, planning, organizing, and actuating are applied but seldom do their activities occur flawlessly thus triggering the need for controlling. Control overlays the managing process and:

- Measures performance.
- Compares performance against standards and establishes differences.
- Adjusts for differences through corrective action.

Any activity can be controlled through observation of its control variables including: a) *quantity*, b) *quality*, c) *time use*, and/or d) *cost*. Neither all activities nor all elements within those activities are controlled. It is not necessary to control each activity, nor use all control variables. The contribution to the control process through the control variables depends on the management function being performed; for instance, cost is controlled in production through expenditures with suppliers on material, while it is controlled in sales through expenditures on advertising, in finance through expenditures of interest on loans, in personnel through expenditures on salaries.

Controlling is streamlined by tracking the important variations, the key milestones, and the key activities which lead to control by exception. One streamlining approach is overall control. Overall performance control provides a top-down approach to assessing the health of an operation against the plan. While some activities in need of corrective action may go unnoticed, this approach concentrates on the overall objective of viewing the entire enterprise with its varied and complex components together as a whole system. Some overall control variables include a) *profitability*, b) *market standing*, c) *capital resources*; while those variables concentrate on the consolidated condition of a large, diversified, and complex enterprise, they bypass the activities within a function of a component in the system which may be in need of corrective action; such an approach leaves local activities to local control and leads to local autonomous or independent control within the confines of broad system objectives. Centralized control of local activities fails to yield desired results and is better left to local or regional control.

6   PROCESSES OF SYSTEM COMMUNICATION

I. INTRODUCTION

Communication provides the exchange of pertinent data, information, and decisions in an organization and among organizations; it represents a transport of data on products/services rather than the actual transport of physical products/services; physical transport is a time consuming and costly process best undertaken, if at all, in the final delivery of a product/service. This strategy is facilitated by the growing maturity of electronic technology which is revamping the way we have done things; those things rested on the industrial technology with heavy reliance on physical transport for exchange and integration. Communication links the management processes in Chapters 2–5 by:

- Providing data/information held by one process or sub-process to another process or sub-process.
- Providing decision taken by one process or sub-process to another process or sub-process.

While a system is a realization of planning, organizing, activating, and controlling processes, communication makes the system workable as it provides the facility to exchange information and effect integration of specialized activities among the processes. However, the exchange of information must be pertinent to the sender/receiver; a receiver omission is not necessarily an exclusion when the message is not relevant. Otherwise, the communication system becomes deficient as it transmits desired and undesired (reverberation/noise) messages about/to each of the management processes causing their saturation, deterioration, and eventual breakdown.

Selective transmission channels are important to fulfill the goals of intended and pertinent communication and include broadcast, multicast, and point-to-point. Message coding is used to optimize the information transfer between sender and the receiver and to prevent its unintended reception. Communication management is necessary, otherwise information overload arises where pertinent information is either missed, processed late, or processed incorrectly. Part III expands on those concepts within the context of automation and computer systems.

II. THE COMMUNICATION PROCESS

The communication process includes (Figure 6.1):

- A sender with a need and purpose for communication of data-information-decision to a receiver(s).
- Encoding of the communication in a symbolic representation (words, numbers, gestures, pulse...) of the real communication.
- Message as the physical form of the encoded communication.
- Transmission as the emitter of the message into the channel.
- Channel or the mode of transmission (paper, air, water, space, wires...).
- Receiver as the collector of the message from the channel.
- Decoding as the interpretation of the message and its translation into information.
- Noise, distortion, dispersion, interference as the factors that confuse the information and may lead to a different interpretation of the communication.
- Feedback, as the response of the receiver to the sender; it repeats the above steps with a reversal in role of sender and receiver. The degree of feedback establishes a two-way communication and allows for a reduction in confusion and ambiguity.

Transportation is the movement of people and products from one place to another (Table 6.1). Communication is the movement of information (relevant data about people and products) from one place to another (Table 6.2). Transportation or communication provides the distribution mechanism to link producers to consumers, servers to clients, of physical products or information; it began on waterways, followed by railways, highways, airways, and now information on electronic-ways.

By its nature, communication simplifies the work process by avoiding the need for intermediate and time consuming processes in physical transport of products or sub-products over space. Physical transport calls for the exertion of a force proportional to the product's mass and the required acceleration needed to reach the desired location in due time; the expanded energy can be costly and is proportional to the force applied, multiplied by the distance travelled and the time taken. Communication expands much less energy, easily handles overlapping processes, and enables real-time system operation with ensued improvement in productivity.

PROCESSES OF SYSTEM COMMUNICATION 67

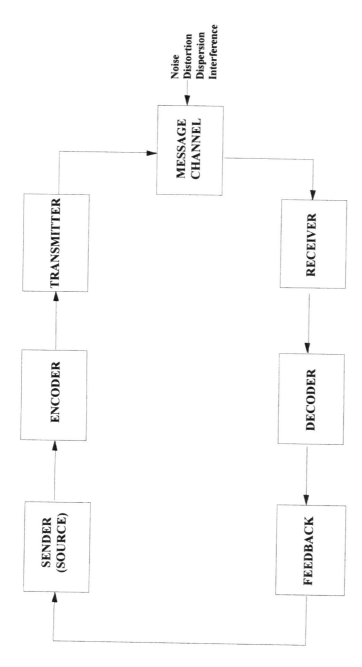

Figure 6.1 Communication process where feedback restarts the loop.

Table 6.1 Transportation Highlights

Date	Mode
Land	
Beginning	Human backs
4000 BC	Animals
3500 BC	Wheel in Mesopotamia
300 BC to 400 AD	Roman Network of Roads
Air	
1760	Cugnot of France Steam Powered Self-propelled Vehicle
1783	Mongolfier of France Hot Air Balloon
1825	Great Britain Public Steam Railroad
1885	Daimler-Benz of Germany Gasoline-Powered Automobile
1903	Ford of USA Mass Produced Automobile
1903	Wright Brothers of US Piloted Air Machine
1939	German Jet Aircraft
1957	Soviet Union Artificial Satellite
1969	US Landing on Moon
Water	
2000 BC	Phoenician Sailing Ships
1807	Fulton of US Commercial Steamboat

Table 6.2 Communication Highlights

Date	Event
20000 BC	Paintings on cave wall
8000 BC	Signaling with drums, fires, smoke
4000 BC	Writing with shapes representing symbols
1500 BC	Phoenician alphabet with one sound consonants and vowels
AD 150	Trai Lun, China paper
1800	Gutenberg, Germany book printing
1830	Daguerre, France quick photography
1844	Morse, US message sender over telegraph line
1867	Sholes, US typewriter
1876	Bell, US telephone
1901	Marconi, Italy transmitter of morse code over radio signal
1920	Station KDKA radio broadcasts
1936	Public TV, UK
1962	Telstar, US Communication Satellite
1980	Computers, in telecommunication of messages and documents

PROCESSES OF SYSTEM COMMUNICATION

Communication fulfills a business end as well as human need. It is the process of sending and receiving information messages so that understanding is effected and intended results are achieved. It occurs between human–human, human–computer–machine, computer–computer, computer–machine, and machine–machine. These are formal and informal channels of communication. Like formal channels, informal channels carry information in four directions, but usually meander like a grapevine; this type exists in human organizations. The formal channels are set by the formal system and reflect the authority, control and assigned functionality; the channels carry information down, up, laterally, and diagonally and thus fulfill the communication process within any organizational structure:

- Downward channels pass directives (information) down the hierarchy from one level to the next for the usual purpose of getting action or coordination on a goal.
- Upward channels pass up information on activities and actions relating to what has happened on their assigned work; in its feedback process it carries control data on the system state to the higher levels.
- Lateral channels pass information among units at the same level with the purpose of functional coordination of activities.
- Diagonal channels pass information among units across an organization chain to gain advice or exercise a function.

Not unlike any other organization, communication may be executed hierarchically, relationally, and/or through a network; communication flows are effected through some patterns. Basic patterns are depicted in Table 6.3 where:

- The ring units communicate past or through their neighbors in a progressive transmission of information.
- The star or tree is a centralized pattern and the units are subordinates and communicate through the focal unit; it maps a hierarchical organization.
- The chain units communicate past or through their neighbors.

The patterns depict a physical flow or logical flow. A system includes a multiplicity and combination of those patterns. Usually, simple tasks may be performed effectively with a centralized communication pattern while complicated tasks call for decentralized patterns.

Means of communication can be words, numbers, pictures, actions... . Communication breaks down due to faulty transmission through complex organizational channels aggravated by a multiplicity of units that may:

- Conduct selective screening.
- Add symbolic meaning of words to listener different from sender.

Table 6.3 Basic Communication Patterns

TYPE	PATTERN
RING	Pentagonal ring with nodes 1, 2, 3, 4, 5
STAR	X-shape with nodes 2, 3, 4, 5 connected through center
CHAIN	Linear: 1 — 2 — 3 — 4 — 5

- Provide resistance to change.
- Introduce unwanted physical conditions, e.g., noise, heat.

Communication is improved through:

- Redundancy, where information is more easily enhanced through filtering and estimation processes in noisy, distortional, dispersion, or interfering channels.
- Feedback, where reports, actions, and results from the receiver are observed and tested for compliance with the communication.
- Usage of a language the receiver actually will understand.
- Matching of the communication to the receiver to facilitate the filtering of the message.

Communication relays the data/information characterizing the decisions made by system processes. Decision making involves the setting and communication of:

PROCESSES OF SYSTEM COMMUNICATION

- Objectives with a set measure of effectiveness.
- Alternative processing actions.
- Outcome of each alternative.
- A choice of a course of action.

Depending on the level, the decision may impact an individual unit, an area, or a total system. As each course of action increasingly yields multiple outcomes, the choice has risk due to the involved uncertainty. Uncertainty exists because of variables not under control of the decision maker(s). Communication provides the links to exchange data and reduce their uncertainty, and to enact decisions and realize them.

III. COMMUNICATION SYSTEMS

Communication systems exist wherever information is transported from one point to another, among human-to-human, human-to-computer, from computer-to-computer and from human-to-computer-to-machine, Parts II–III. Text, voice, and images are conveyed to all distances through myriad types of systems, e.g., radio, television, telephone, radar, sonar, telemetry, infrared, laser, and x-rays.

Communication systems transmit information from one point in space and time to another point. The information transmission to the receiver relates to signals changing with time in an unpredictable way. Chapter 17 elaborates on the varied types of signal transmission channels. A signal or message is the physical realization of information.

Intuitively, the higher the signal changes are, the higher is the possible information content. Increasing the information content means changing the signal as rapidly in time as possible and over as many amplitude values as possible. What is possible in time is limited by the time response of the signal storage devices or their inverse, the system bandwidth. What is possible in amplitude subdivision is limited by undesired noise fluctuations. Thus there are two fundamental limitations to communication, system bandwidth, and signal to noise. There is a minimum time required for energy change and a minimum detectable signal-amplitude change required by noise. Since the unpredictable part of the signal amplitude needs be transmitted, it is not necessary to transmit the continuous signal, but rather samplings of it quantized at a rate equal to at least twice its maximum frequency of change.

The elements of a communication system attempt to produce an acceptable replica of the message at the receiver and include:

- Transduction which converts the message from one form to another that is more amenable for processing and conveyance.
- Transmitter which matches the message to the transmission medium; modulation is one process designed to match the message

to the medium through the choice of frequency carrier; coupling devices include modulators, filters, amplifiers, and antennas.
- Channel which connects the transmitter to the receiver through a transmission medium; the medium may be the atmosphere, water, empty space, cable, or wave guide. Depending on the medium, signal, or message contamination occurs. Besides reduction in signal strength and time delay in reception, more serious signal alterations may be encountered that cause uncertainty and confusion in the communication; distortion and dispersion alter respectively the amplitude-frequency content and phase-frequency content of the signal and require compensations to separate from the intended modulation; interference adds an extraneous signal of a form similar to the desired one and requires its separation; noise adds random variations and requires filtering so as not to mask the signal bearing the information.
- Receiver which extracts the desired signal and restores it to its original form; it includes demodulators, filters, and amplifiers.

Modulation is the alteration of a carrier wave in accordance with the message; coding may be added to prevent intrusion by the medium or an unintended receiver. There are two types of modulation; a) *continuous wave modulation* where the carrier is a sinusoidal wave form, and b) *pulse modulation* where the carrier is a periodic pulse train. Continuous wave modulation is a continuous process and is well adaptable to a continuously varying message where the carrier is at a much higher frequency from the message, and where the message modulates its amplitude, its phase, or its frequency. Pulse modulation is a discrete process and is well adaptable to a discretely varying message. Notwithstanding, quantization may transform the modulation into a digital signal; coding converts further the digital signal from one symbolic language to another. Modulation is essential to communication. Modulation:

- Facilitates radiation of the signal through a smaller antenna whose physical dimensions need be approximately one-tenth the carrier short wavelength.
- Enables reduction of contaminants through a larger transmission bandwidth or selection of further out operating carrier frequency.
- Allows concurrent operation of communication stations through assignment of different carrier frequencies or channels, through multiplexing of multiple message transmissions on the same channel.
- Enables reception of different transmissions with the same equipment by positioning the signal within its fixed frequency spectrum.

While the communication process is straightforward, there are many barriers to effective communication:

- Failed emission where screened or distorted information is encoded or an incorrect encoding takes place.
- Notch filtering where reception is done selectively because of an environmental influence or a constraint at a certain point in time.
- Discounted message where source credibility by receiver is low because of a predisposition to accept different messages or of its low standing or unreliability.
- Mismatch where encoder–decoder attach different meaning to the words, their context, their position, or their timing.
- Contamination where a channel adds noise to the message and/or distorts it and interferes with its understanding.

IV. DISCUSSIONS

Managing large systems involves a work decomposition process to the individual units followed by an integration process of their collective efforts into the desired whole. Without communication among the units, the decomposition and integration processes could not be realized. By the same token, communication could not compensate for complicated organizational interfaces and diverging goals. Efficient communication enables better job performance, acceptance of policies, rules, and regulations, thus winning cooperation and fostering smooth operation.

Communication is at least two-way: for humans, it is not only talking, it's also listening; it is not only writing, it's also reading; it is not only viewing, it's also touching. For automated systems, this two-way mechanism is realized through sensing and feedback control.

Effective communication rests on some basic tenets:

- Information: the sender can't pass information, the basic purpose of communication, if the sender possesses incomplete or incorrect data to transmit; next, the sender must transmit the information of interest to the receiver.
- Medium: the interfacing of the sender to the receiver impacts the clarity, speed, and coverage; complexity grows with distance between sender and receiver and influences in turn the choice of technology and transmission medium; dissemination media vary from the elementary but most effective face-to-face for humans, to memoranda, to higher technology such as television, video teleconferencing, radio, telecommunication, and computer networking.
- Integrity: information is fused as it flows through the organization going from senders to receivers; however, considerable change may

occur and its distribution represents a challenge to conveyance of a relevant but still representative replica of the original information.
- Recognition: the symbols used to represent information must be commonly understandable by the sender and receiver; these include symbols representing objective reality such as words for places, objects, materials; figurative symbols, on the other hand, may connote different meanings and should be used among senders–receivers of similar background. Often, figurative symbols are used purposefully to prevent unwanted interception of a communication by a third party, and/or to facilitate communication among the parties; for instance, the combination of two symbols, 0-1, has enabled the realization of the computer revolution.

Part II:
Systems Synthesis and Automation

7 PROCESSES IN SYSTEM DESCRIPTION

I. INTRODUCTION

Work processes (decomposition and integration) and resources (human-made and natural) are joined through processes in system description. The speed and complexity of modern management systems is increasing, prompted by the increasing shift from material transactions to more information or data transactions. Early societies operated with material transactions; later, a transaction value was set in terms of currency or credit, and the exchange went from material–material to material–data and data–data, where data are symbolic of the material value. Initially, humans worked on the data and machines; the machines, manipulated by humans, worked on the material. Societal progress created a system of transactions that is predominantly information- or databased. This trend has been accelerated by computers which enable operation on large databased systems and their visualization in space and in time.

A system is part of an environment whose resources have been bounded for management of their activities (Figure 7.1a and 7.1b); it possesses four basic elements:

- Boundaries that limit its content from the rest.
- Resources that make up the major entities or objects within the boundaries and refer to its productive capacity (people, computers, machines, material, currency...).
- Management processes (Figure 7.2) that: a) *define its objectives* (what the resources are collectively to accomplish as a result), b) *develop its plans* (what it must do and what activities and when they need to be executed), c) *organize its resources* (what structures and interactions are set), d) *activate the organizations* to execute the plan, e) *evaluate and control* the results against the objectives. There are basically three management layers: a) deciding what

78   SYSTEMS MANAGEMENT: People, Computers, Machines, Materials

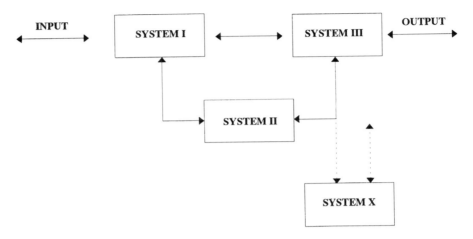

Figure 7.1a World made up of systems.

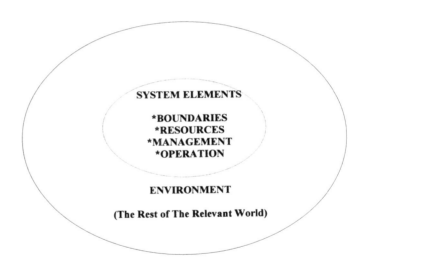

Figure 7.1b System elements.

should be done; b) deciding who should do it; c) seeing that it is done.
- Operation that executes the activities with the state depicting the system resources (content, characteristics...) at the desired moment of time against the objectives.

Simply, a system can be viewed as made up of objects; objects are carriers of observable properties; properties may be quantitative, e.g., size, and/or

PROCESSES IN SYSTEM DESCRIPTION

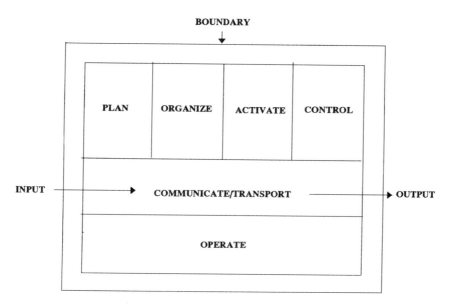

Figure 7.2 A generalized system; major processes.

qualitative, e.g., nice; observables in physics include time, space, energy, velocity, force, mass, pressure, frequency, and charge; any construct of objects which carries observable properties will be called a system; from the list of observables, a set is taken to explain certain behavior of the system; this set defines the state of the system and its evolution in terms of independent state variables e.g., time/space. Localizability in time and/or space of objects turns out to be a major pursuit in human/computer/machine activities; for instance, Chapters 14, 17, 18, 19, 20 deal, respectively, with localizability of data objects in storage, package objects in program, target objects in environment, material objects in inventory, and monetary objects in finance. The search for localizability of objects rests on a search for invariant properties through:

- Inductive logic, where one seeks to establish comparison (correlation) between objects; closeness of correlation may be quantitative or qualitative, deterministic or statistical as in Chapter 17.
- Deductive logic, where facts are derived from theoretical hypotheses.

II. SYSTEM REPRESENTATIONS AND THEIR IMPLICATIONS

A system or its subdivision may be represented in varied ways:

- Equations, e.g., algebraic, differential, integral, logical.

- Physical model, e.g., scaled version.
- Written description, e.g., functional write-up.
- Drawing, e.g., a cross-section.
- Visualization, e.g., a three-dimensional view in space/time.

Those representations are models or abstractions of reality where they may be classified as mathematical, schematic, or physical.

Usually the representation is a combination of the above and is applied to the elements in the system and their relationships. The elements may be physical or logical and include from the system such things as work centers, computer centers, personnel, machinery... . The element size depends on the detail desired. The signals, interactions, connections, or transactions among the elements denote the states of the system; the flows may be material or symbolic (data)... . Thus an input state of a set of variables or attributes is transformed according to the element properties into another set of variables or attributes. Prediction of transformation from one state to another state depends on the complexity of the elements and their interactions.

There are varied tools to represent a system. At the root of many of those representations are the elements and their relationships (Figure 7.4):

- In system block diagram, the elements are represented by rectangles and the states of the input-output (I/O) by arrows; each rectangle represents a grouping of details where further details can be represented by lower level rectangles with their attendant I/O.
    - The organization chart is a set of black boxes arranged in a tree-down fashion; it depicts where the system functions and sub-functions reside; those organizations are static and may pertain to human or material, e.g., corporate organization (Figures 7.3 and 7.4), work breakdown structure, software planning diagram, hardware planning diagram... .
    - The flow chart (Figure 7.3) is a set of black boxes, and it depicts through successive decomposition the sequence and detail of transformation through the organization and shows the activation process. An object-oriented system treats its black boxes as blocks fully specified by their external functional and physical specifications; then blocks with different internal compositions but analogous external characteristics can be substituted for each other; similar (methods, variables) objects form a class; this object-oriented approach allows products to be offered in a variety of internal configurations and permits product reuse/replacement.
- In a system flow graph, the elements are represented by arrows and the states by circles; block diagrams and flow graphs are equal to each other.

PROCESSES IN SYSTEM DESCRIPTION

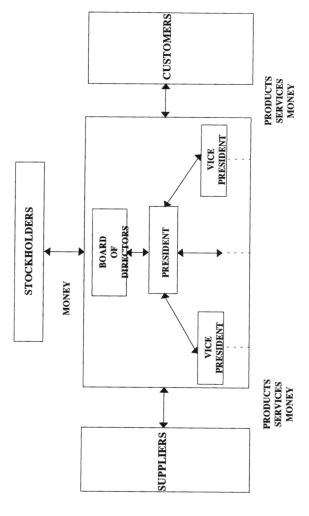

Figure 7.3 Corporate chart and its environment.

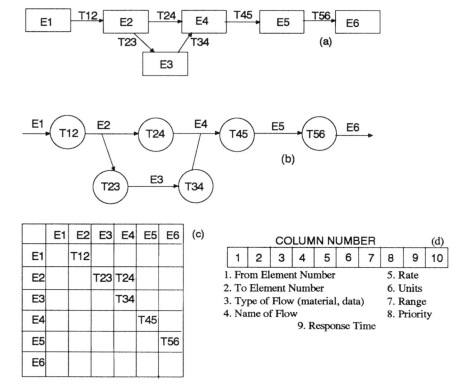

Figure 7.4 Varied schemes for system representations: (a) block diagram, (b) flow graph, (c) matrix, (d) record.

- As a matrix or table, the elements are lined along the vertical columns and horizontal rows; the interactions are indicated at the intersection of rows and columns. Details of the interactions vary from indication of existence to full description:
  - The input-output matrix shows the sources of data/products and users of data/products along the rows and columns with the transformation at the intersection.
  - The decision matrix is a summary expression of decision rules in a multiple action situation that identify future conditions or states of nature, possible alternatives, pay-off associated with each, a decision criterion to select the best alternative. Test combinations may be set along the columns while tests and actions are set along the rows with yes–no decisions at the intersection.
- As a record, the information on elements and their interactions are set in fields and ordered by column numbers; the character of the transaction is detailed across the allocated columns; the fields may

PROCESSES IN SYSTEM DESCRIPTION

be coded by a number or letter. Multiple records form a table. Multiple tables form a database description.

These varied representations give sectional views of elements and transactions. Cross-sectional views may also be developed. Samples include:

- Which elements deal only with material or information flows? Typically, information elements are usually involved in decision-making.
- Which elements receive more flows than they send? Typically, these elements are decision makers or resource managers.
- Which elements have the greatest variety of inputs and outputs? Typically, higher level managers deal with a greater variety of interfaces.
- Where are the major information/material delays in periodic/aperiodic reporting? Typically, delays entail system cost.

Elements fall into three building block types:

- A transformation or conversion operation block; it is a transformation of an input state into an output state.
- A decision or logical operation block; it is a transformation operation which is dependent on the input/output.
- A feedback or correction operation block; it is a transformation operation which is dependent on the output.

Each of the elements or transactions in a system is a grouping of details about the system. Groupings are of the same class when they behave similarly in system operation, share the same needs, display common attributes; objects of the same class possess the same methods and variables but may contain different data; classification is a powerful tool in system analysis, synthesis, development, and operation. Hierarchy of the classes denotes the flow of authority or control among the elements of the system and thus is also critical to understanding the system and predicting its interactions. A system hierarchy may be static, transient, or dynamic. When changing, it may be described deterministically, statistically, or may be random. Hierarchy seldom stays static with time in complex dynamic systems. A system may involve different sets of hierarchies all existing at the same time. One may encounter hierarchy of objectives, hierarchy of plans, hierarchy of organization, hierarchy of work assignments and thus an individual entity may belong to different classes depending on the hierarchy under consideration.

In defining a system, one defines all possible transactions among the elements as well as each of the entities within an element. An event is said to occur when a change in the system occurs. An event may be internal or external to the system; samples of events are:

- Modification of an attribute in an entity.
- Production or elimination of an entity.
- Modification to the class.

Rules of change must be defined, i.e., predetermined, conditional, random. The schedule of events defines the evolution of the system and gives its status.

System complexity increases with increase in information content. Information requires memory for storage. Memory stores the diverse system options and actions in response to variable input conditions. Planning, organizing, activating, and controlling such a system become proportionately more complex. The order of the system denotes that complexity. A first-order system responds without memory, while a second-order one utilizes memory in taking action. Naturally, action needs be taken before conditions change again; so too much memory search and assessment slows system response and limits its use. Thus, symbols in the memory must be unique, non-redundant, organized, and accessible for quick storage, retrieval, manipulation, and action. Symbols have been set in stone, on paper, in sand, in screens, and in semiconductors, etc. The major types of symbols include:

- Data to represent the system goods and services.
- Instructions to manipulate the data toward given objectives.
- Instructions to revise the structure, the manipulations and their order, or to create new ones to satisfy changing environments and objectives.

The presence of symbols' type and amount denote a system hierarchy where:

- The simplest system is zeroth order; it has only a transformation with no memory or feedback.
- The first-order system has transformation and feedback but still no memory.
- The second- and higher-order systems have transformation, feedback and memory to assess and act on a variety of inputs, to predict future needs, to select from set plans of actions, and to account for past history.
- The learning system has new goals, plans, organization, activation, and/or control applied to satisfy evolving internal/external objective-setting environments.

Those four levels depict the hierarchy in a system where the lower orders are at the device levels, and higher orders at the management levels.

The preceding representations are applied during the system development phases:

- System definition phase states the overall requirements in terms of system functionality behaviors which are described through some system models. From this, the number of components or elements required to realize this overall system behavior are identified.
- System design phase defines the components making up the system in terms of stated functionalities of the interface or protocol requirements; from this, each component or element is designed to satisfy the specified behavior and protocol.
- System implementation realizes the components through re-use, purchase, or building from scratch; as a component includes increased functionality, the potential for re-use decreases.
- System integration assembles the components into the overall system and readies it for use.

A strict waterfall approach to the phases is unadvisable; concurrency in the development process allows feedback loops of knowledge resident or required within the phases and across the phases.

III. SYSTEM ANALYSIS AND DESIGN

Analysis involves an existing system. Design involves development of a new system; in confirming the design, analysis is performed. Analysis and design go hand in hand. When the results of an analysis show a shortfall in an existing system, new design follows. In analyzing an existing system or designing a new system, the management and operation functions in Figure 7.5 are encountered. Those functions may be executed singly or by combination, such as:

- Manually and/or mentally by humans.
- Through devices operated by humans.
- Through machines controlled by humans.
- Through computer automated devices/machines.
- Through selective human control of computers.

In analysis or design, the following descriptors (Figures 7.6 and 7.7) are established:

- System objective where the purpose of the total system is set with recognition of its full potential over time.

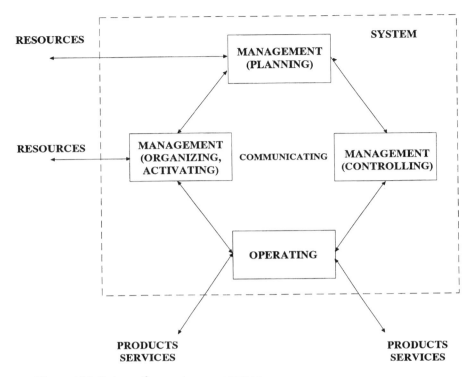

Figure 7.5 Interactive system processes.

- System boundaries and interfaces with other systems where they present points of complications with the possibilities of interactions, transfers, and/or sharing of resources.
- System elements or components where each is a grouping of major productive resources in the system and contributes in a significant way to the system objectives; the components set what the system does, how the work is accomplished, the basic system flow paths and major deviations or exceptions.
- System sub-components or work centers where each performs one major sub-function through a grouping of interrelated activities.
- System procedures where actual processing is described for normal operation as well as control of procedural exceptions (error prevention, lost data/material prevention, phony data/material prevention...) through counts checks, limits checks, totals checks, sequence checks, structural checks, as well as dual processing.
- System timing where how soon the varied data/products must be processed and transported for use in the system operation or its control. Usually the closer to the physical operation the user is, the faster is the need for the detailed data. When the user is at the

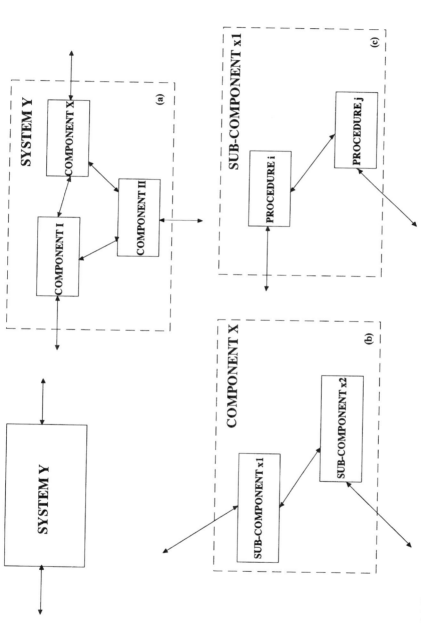

Figure 7.6 Decomposition of processes: (a) system to components, (b) components to sub-components, (c) sub-components to procedures.

88 SYSTEMS MANAGEMENT: People, Computers, Machines, Materials

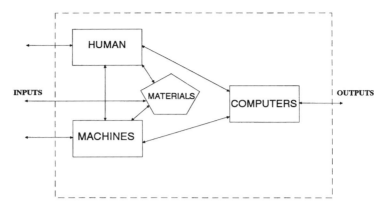

Figure 7.7 Decomposition through partitioning and allocation.

executive level, the data are usually stripped of details, the frequency of reports is low, and summaries and trends are more pertinent.
- System capacity where volume (number of transactions) per type per period of time over time is tracked.
- System availability where how long the system is doing what it supposed to is set.
- System effectiveness where information/product value vice cost, complexity of required processing, or size of storage is made.

Humans, computers, and machines join to change the state of materials. Machines handle primarily materials. Computers handle data (material descriptions), and in that sense are universal machines; humans handle both material and data. Their combination leads to efficient and effective handling of each other and material. Ideally, machines would perform the physical process, computers the information process, and humans the resource allocation process (Figure 7.8).

IV. SYSTEM IMPLEMENTATION

The development cycle begins with system requirement, includes system analysis and design, and ends with its operation. When implementing a system, one approach determines (Figure 7.9) in serial or interactive cycles:

- First, the required behavior of the system as viewed at its boundary; this is called the essential model.
- Second, the system functions needed to support such an external behavior.

PROCESSES IN SYSTEM DESCRIPTION

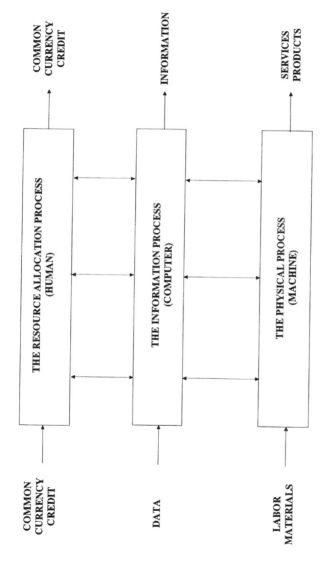

Figure 7.8 Optimized view of human–computer–machine functions.

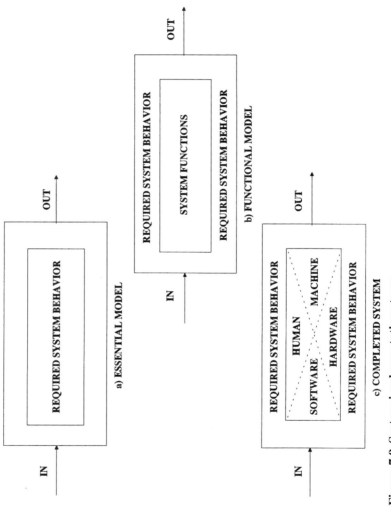

Figure 7.9 System implementation stages.

PROCESSES IN SYSTEM DESCRIPTION

- Third, the partitioning of the functions among the needed resource types, e.g., human, computer (hardware and software), machines, followed by their allocation to specific resources or entities.
- Fourth, acquisition and/or development of the entities and their integration into a working system.

Throughout the implementation, five elements are tracked within the system entities, from requirement through development and operation. For each entity, be it the system itself or lower level components of it, also be it human/hardware/software realization of it, the elements are:

- Interfaces; their content is set in diagrams or tables where a depiction is given of their purpose, information passed, direction of transfer, protocol, format, periodic/aperiodic conditions of transfer, time response to transfer, operator screen layout and function keys; consistency and completeness within the entities and across the entities are validated.
- Functionality; its content is set in terms of input–output and processing; for each function or sub-function, the input–output are set through a description of their class, purpose, source/destination, frequency, checks, units, and operator actions; processing is set through a description of purpose, parameters involved, timing sequence of events, allocated timing, mathematical equations, process control, timing and undesired events restrictions, validity checks, and responses to abnormal situations.
- Performance; its content is set through a description of memory size, memory location, size and number of files, size and number of tables, number of concurrent users, processing time, number of transactions per second, availability, recovery times, and constraints on implementation.
- Services; their content is set through a description of availability, databases, adaptation, and programming; availability is set through a description of check points, back up, recovery, restart, error handling; databases are set through a description of their location (local, remote), access, sharing mechanism, frequency of use, organization, collection mechanism; adaptation is described through re-configuration mechanism; and programming is described through language, compiler, and timing.
- Qualification; its content is set in terms of methods used and traceability; methods are set through demonstration, test, and analysis; traceability is set through a linkage matrix from present entity to parent's entity.

A system, be it human, controlled, automated, or both, receives data/products from the outside world then processes that data/product before sending it back out. Whether considerations are at the system level, component level, or lower levels, the same constituents are encountered; that is events, processes, connectivity (Figure 7.10):

- Events are usually based on knowledge of the system's purpose or functions, and the event list triggers the processes; events are either time triggered or flow triggered; events contain all necessary information needed to do work on that event or based on it, and rules for re-submittals. Events are needs responded to because of their bearing on the purpose of the system; building an events' list stems from primary events delineated from functions/purpose/expectation of system components, i.e., its people, hardware, software, machines. To find out about more events, one looks for variations, opposite, preceding, succeeding, and/or missing events.
- Process is the activity, modification, transformation, or whatever happens to the flow on their way through the system. A process with interface flows to the outside world is the system context description; further detailed description is leveling, that is breaking up the process to lower level processes to give more details; what is involved in leveling includes:
  - Elements (data/material) transformation rules.
  - Operation on data/material elements to produce output flow.
  - Putting elements in stores.
  - Fetching elements from stores.
  - Connecting elements in one store to ones in another store.
  - Showing what sets of elements are processed, where they come from, what they are, and where they go.
  - Showing range, unit, and precision of input/output elements.

  Many processes are time dependent; the meaning of a function key depends on the mode of operation at the time; so output is dependent on the time for the same input; data dependent output is dependent on data input and data in store. Process organization can clarify connectivity through process categories and connectivity between categories.
- Connectivity is the flow and/or store between processes; the stores are delays; the flows are the pathways the events flow through; a direct flow from one process to another has a tight causal effect; a store allows the next process to get to it when it has time. There are three kinds of flows: descriptive data on material control data of the material or the material itself. What is involved in data description includes:
  - Graph of what is connected to what.
  - Names of sets of data used by processes.

PROCESSES IN SYSTEM DESCRIPTION

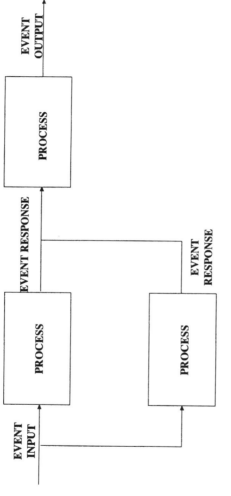

Figure 7.10 Events, processes, connectivity, and responses.

- Names of sets of data produced by processes.
- Names of sets of data stored by processes.
- Names of sets of data in a category.
- Names of connections between categories.
- Names of unit, precision, and permitted values of data elements in a set.
- Relationship between data in two stores.

Control flows switch to turn processes on/off thus making a process time-dependent. Control flows are triggered by control processes such as a human pressing a function key. What a control process does is graphically described by state transition diagrams which include (Figure 7.11):

- The states the process may be in; there is always at least one state, the initial state.
- The condition that causes transitions to one of the states.
- The action taken when a specific transition takes place.

Control flows act like switches on the processing system; this flow is not manipulated into anything by the system. Sometimes the control itself is a process that controls other processes in the system (Figure 7.12); for instance, in human–machine interface, function keys are reserved for controls.

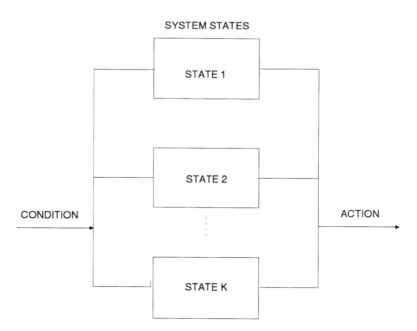

Figure 7.11 Control description.

PROCESSES IN SYSTEM DESCRIPTION

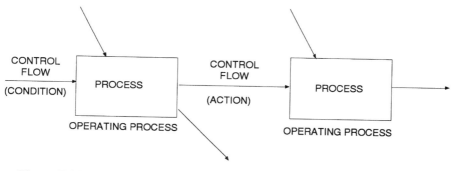

Figure 7.12 Control processes.

In automating a system, a sequence of instructions are carried out by the computer without human intervention. Computer capability is denoted by:

- How fast can instructions be executed with a measure as millions of instructions per second.
- How fast can data be stored for retrieval with a measure of megabytes per second.
- How much data can be stored with a measure of megabytes.
- How fast can data/messages be transferred from one storage to another with a measure of bytes/messages per seconds.

Instructions either perform work related directly to an application or support of that application. A given application is usually decomposed into execution units called tasks and storage units called data sets; those tasks and data sets are assigned to processor and storage. Sharing of data happens in one of two ways:

- Two processors access a single storage device.
- The data are duplicated with a copy in each processor; here a process is added to copy and update from one processor to another.

At times a process may need to be split across processors, then the process is not kept whole but is decomposed into two processes and flows established between the two processes. In evolving a system one usually designs:

- The essential model in which the required behavior of the system is described.
- The implementation of the processes and their connectivity to support the required system behavior. Care in process allocation is needed; direct data flows between processes add more complexity

than when a delay or store is available; flows with multiple data add more complexity than flows with a single piece of data.

The boundary that separates all the processes from the environment delineates the system within which a set of rules and patterns are followed in response to external events.

V. DISCUSSIONS

Decomposition or breakdown of the whole into parts and their subsequent integration are essential processes to definition, design, implementation, and operation of systems and/or their management. Such systematic decomposition and integration processes flow down and are applied during system definition, preliminary design, detailed design, coding for software, fabrication for hardware, unit test, integration test, system test, installation and test, and demonstration and customer training. At any subdivision/integration level, the basic elements involve:

- Input or state of things prior to transformation; some state characteristics may vary with time and/or space; some may be subject to limits or constraints, such as magnitude.
- Transformation, transition, or conversion from one state to another state; some transformation variables may change with time and/or space; some may be subject to restrictions or constraints; the alternative transformations are judged according to a set criterion.
- Output or state of things following a transformation; the resulting state characteristics may vary with time/space; some may be subject to limits or constraints.

The varied system representations facilitate problem definition, visualization and solution, its communication, prediction of operation in control system, development, and training. Whatever is the purpose of the representation, reconciliation or compromise among conflicting criteria leads to a search for a "best" transformation among the alternative solutions. While the search leads to an expansion of alternative solutions, the decision process provides reduction and/or integration into an optimum solution. Underneath it all, cost is a principal driver, that is:

- Cost of arriving at the specification of a solution.
- Cost of making or developing the solution.
- Cost associated with using it.

8   PROCESSES IN SYSTEM SIMPLIFICATION

I. INTRODUCTION

The underlying purpose of any system is to do work using the available resources. Work simplification deals with conceiving and applying the effective use of humans, equipment, material, time, and space. System simplification processes lead to higher system productivity where waste is trimmed and essential operations are instituted.

The objective of a large system rests on complex and variable sets of actions and counter-actions. Underneath the system's apparent complexity and variety, a simplified view leads always to common treatment of classes of elements rather than individual treatment of each element in either analysis and/or synthesis. Simplification is reached through three major approaches which constrain and filter the set of possible elements and their relationships, through top-down, elimination, and grouping approaches:

- Top-down approach formulates the total system goals, objectives, major constraints, measures of effectiveness, priority of actions, the internal allocation of resources, and overall changes or modifications to all. The top-down approach is applied in complex and changing systems to state the guiding principles under which elimination and grouping operate.
- Elimination approach deletes from consideration details considered unimportant to the view at hand. Elimination is applied when looking for optimization and detailed action.
- Grouping approach clusters together details into blocks considered descriptive of the subject at hand. Grouping is applied when looking for comparison between blocks of details and estimation.

While the top-down approach concentrates on the underlying goals that guide and integrate the system elements, elimination and grouping work focus on system blocks to improve a system's procedure, process, or element. These approaches are interwoven in the consideration of three basic system activities: system search, system stimulation, and system test.

II. SYSTEM SIMPLIFICATION PROCESSES

System simplification rests eventually on the methods applied to work out non-essential activities. It begins by considering:

- The system policies which set the boundaries within which program activities will take place.
- The program which integrates the diverse procedures into a unified whole that satisfies the objectives.
- The system procedures which define chronologically the series of tasks.
- The methods which define the actions to be taken to perform a task; tasks should not be fragmented across organizational units with resulting unnecessary interfaces.

A. System Top-Down, Elimination, and Grouping Processes

Some general top-down methods include (Figure 8.1):

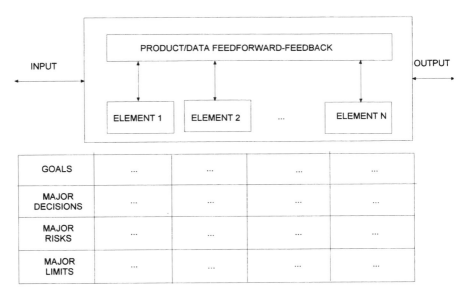

Figure 8.1 Top-down framing of system.

PROCESSES IN SYSTEM SIMPLIFICATION

- Framing overall goals, i.e., ultimate outputs and/or desired states of the system rather than the detailed structure of its block; those overall goals are usually less time variable and must be free of self-conflicting goals.
- Defining the meaning of the goals by setting measures of effectiveness to gauge how close an outcome of one alternative is to a desired goal; this allows a trade-off of alternatives toward the optimum application of resources and decision-making.
- Avoiding conflicts among the allocated goals of individual elements and/or group of elements within the system; system conflicts follow when the outcome of an operation can be valued differently by the elements within the system; each individual element may not result in an effective operation of the system. Thus goal-directed activities of each element should be assessed to insure its positive contribution to the total system operation. Organization has a hand in reducing conflict through avoidance of product fragmentation across organizations that results in overburdening interfaces during the integration process.
- Weighting the contributions of element and group to the system through some coordinated view of system goals, major decisions and their basis or measures of effectiveness, major risks, major constraints, their partitioning and allocation down to the element system level (Figure 8.1).
- Ferreting the major risks attached to major decisions and connected to the amount of system resources committed to the effort.
- Identifying major system constraints set to reduce the conflict of goals, physical constraints, material, or information in origin.
- Assessing the ratio at which the system and its resources can change as goals, measures of effectiveness and/or constraints grow or change in time. Goals are the desired states or results while measures of effectiveness gauge progress toward the goals, and constraints filter the alternatives applied to reach the goals. The system must reorganize its resources in response to changes. The evolution of the change is based on projection processes.

Some general methods for eliminating and grouping include:

- Bounding the domain of system variables to the area of interest (Figure 8.2) usually leading to a differential range, area of highest peak, and area of most likely occurrence; details outside these area boundaries are eliminated from consideration.
- Bounding the dimensionality of the system, i.e., the patterns or modes of operations, the set of possible conditions, the extremes in operation; this reduces the set of system elements or their relationships

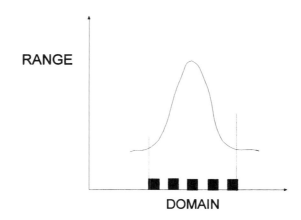

Figure 8.2 Bounding the domain of a system.

by considering a smaller set extracted on the basis of cluster in a given space, time, or attribute.
- Bounding the range of system variables by statistical procedures, by thresholds on system measures, i.e., size, amplitude, frequency, and/or attributes, to eliminate consideration of data and factors not significant.
- Bounding by statistical methods into smaller sets the data on system elements or their relationships pertaining to place, type, time of events, i.e., average, variance, total.
- Bounding by conversion from continuous variables to discrete variables, from variable to attribute, i.e., sampling, contour mapping per attribute.
- Bounding by prototyping the typical block of elements and their relationships, i.e., typical U.S. families, typical U.S. city.
- Bounding by substitution or transformation to bring unrecognized elements or relationships into familiar form, i.e., Fourier transform, coordinate transformation.
- Bounding by symmetry where total system blocks and relationships may be depicted by a smaller set of each, i.e., bilateral symmetry as even–odd.
- Bounding by partitions where system blocks and relationships are thresholded so the interactions may be studied through studies of independent sub-systems.

B. System Search Processes

Search is a predominant objective of many systems; its simplification is essential; basically, it is the selection of objects with specified characteristics within or remote from the system. Unfailingly, search involves detection, classification, location, and tracking functions. Searches vary from selection of an errant component in a system, to selection of a book in a library, to selection of a stock from the exchange, to selection of a target from multiple returns, or to selection of an item from a file. Search methods can be simply categorized into:

- Search with full information through a) *logical sequencing*; b) *item–characteristic data organization table*; c) *characteristic–item data organization table*; d) *characteristic screening tables*; e) *weighted matches*; f) *indexing or assignment of key words to documents*; this type of search is expanded upon in Chapter 14, "Data Base Management Systems".
- Search with partial information through a) *random sampling*; b) *extreme values*; c) *tests on a group of possibilities*; d) *unimodal optimum*; or e) *hill climbing*; this is elaborated upon in Chapter 17, "Automated Information Processing Under Risky And Uncertain Conditions".
- Search for sources of failures through: a) *metered probes*; b) *continuity test*; c) *creating independence*; d) *signal tracing*; e) *binary split*; f) *stress methods*; or g) *block substitution*; this is elaborated upon in multiple places throughout the book.

Searches involve decision-making. Because of uncertainty, risk lies between the two extremes of full and partial information.

C. System Simulation Processes

Simulation leads to simplification of reality. Simulation is the creation of reality in a few but relevant system particulars. It allows the system designers to analyze the run time behavior at the design stage and verify that the processes and their interactions satisfy the system requirements The advantages of simulation include:

- Ability to predict performance and influence the design prior to development.
- Ability to repeat experiences, rare or costly, in reality.
- Ability to shorten cycle time from concept to market.

Principal elements of simulation are:

- Conceptual simulation where mental precision is relied on to imagine what would happen under sets of varying circumstances.

- Scale model simulation, such as factory layout, war games on maps where system properties and behaviors are learned from manipulating scale models.
- Symbolic simulation where system definition is represented by mathematical relations whose solutions are obtained manually or by a computer. It proceeds in a downward decomposition strategy followed by integration to verify the system up front, from concept to production. Simulation begins by insuring the correctness of system requirements; this high-level simulation is increasingly detailed in a downward migration until the whole is verified. In computer systems, the parts, e.g., boards and integrated circuits, are individually simulated for functional verification, followed by their concurrent simulation for inter-operability verification; next, microcode and diagnostic routines are simulated for combined hardware/software verification; finally, simulation of system software as well as application software are re-executed to identify new functional design problems due to the parts.
- Human simulation where human participation is needed because system definition involves characteristics and judgements not reducible to quantitative relations or logical statements.

Each of the preceding elements may singly, but usually together, form the system simulation. A sampling of the needed descriptors to set up a simulation includes (Figure 8.3 and Figure 8.4):

- Work centers which are the breakdown of system resources into sets with indicated functions, content (human/computer/devices), and characteristics. The information is tabulated with items such as queue set-up time, operation time relatable to unit of work, storage... .
- Work-loads which are the breakdown of system products/services into jobs. The table includes allocations to each of the work centers described in work center characteristics.
- System routes which are the connectivities of work centers; the table includes transport mode, time, cost... of communication/control data and products/services.
- Work routes which are the system integration routes used by the work flows among/within the work centers. The table items include, for each in-process job, a routing of where it has been, where it is going, and the time required at each work center.
- Schedule which is the tracking in time of the start and end times of all work loads at work centers and on route; the table includes items on the next event list, time of transport from one work center to another, event arrival, event set up, run completion, batching, and rejection.

PROCESSES IN SYSTEM SIMPLIFICATION

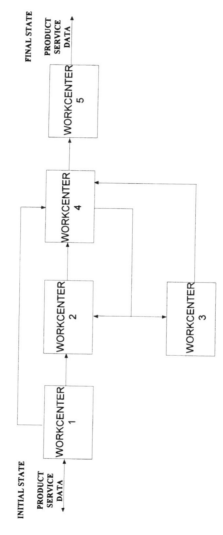

Figure 8.3 Work centers and needed connectivity routes.

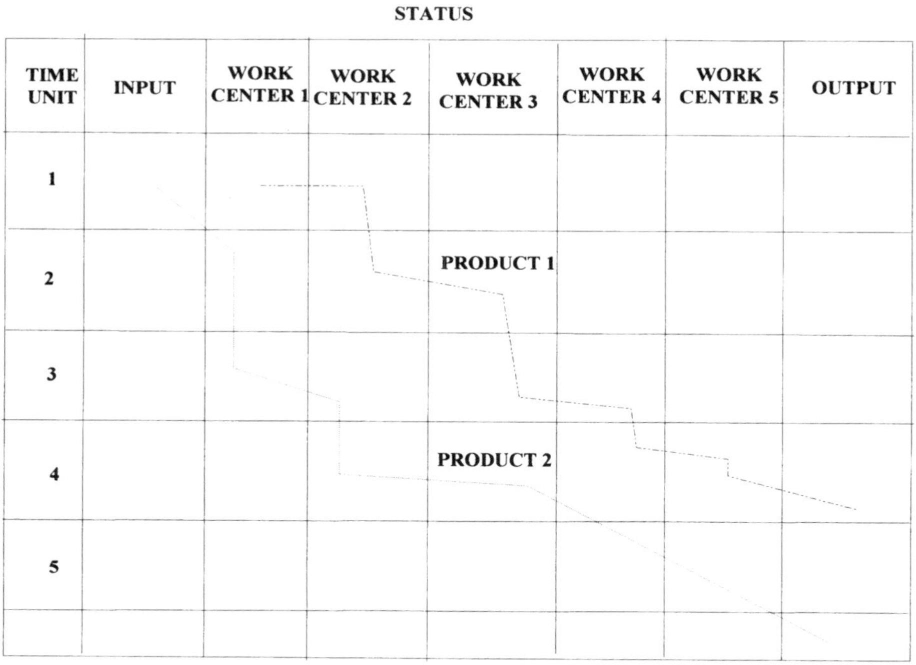

STATUS

TIME UNIT	INPUT	WORK CENTER 1	WORK CENTER 2	WORK CENTER 3	WORK CENTER 4	WORK CENTER 5	OUTPUT
1							
2				PRODUCT 1			
3							
4				PRODUCT 2			
5							

Figure 8.4 Depiction of job movement with granularity controlled by sample time.

- Conditions which give state circumstance, bounding, or scheduled action at a given time. There are multiple bounding types on center/work/routes: a) *time condition*, i.e., initial/final condition or state; b) *spatial condition*, i.e., boundary conditions; c) *priority condition*, i.e., setting the order of precedence in case of conflict over shared resources; conditions for interrupt at any time of ongoing work. The table includes all logical possibilities with rules of operation under all possible contingencies; the yes–no responses to possibilities have simple, associated, control procedures compared to the "maybe" response.
- Monitoring which is inspection of operation centers, work, and routes against a set of standards. The table includes items on milestones, quality, quantity, and rate. A cross-sectional depiction of system status is given in Figure 8.4. At a given time, this tabular structure gives a system status of jobs being performed across all work centers. With time evolution, it shows a given job's movement across work centers as it is integrated.

In simulating a physical system, a model or symbolic representation of the real system has to be developed; this involves:

PROCESSES IN SYSTEM SIMPLIFICATION

- Making a list of key words to isolate the major objects in the system or its components.
- Making a list of simple concepts.
- Constructing relationships that involve multiple objects and concepts.
- Setting quantification.
- Setting cause and effect among the key words with arrows showing relationships; you end up with a hierarchical network or relational structure.
- Examining a static model to obtain an explanation/prediction of an object in the model.
- Firing up the dynamic model where the objects are initialized and their evolution in time/space is observed.

The process of system verification, be it real or simulated, is through testing. Tests are conducted on the total system or by partitioning it into independent divisions, each validated separately.

D. System Test Processes

Many reasons exist for system tests: simplified understanding of system operation; simplified verification of expected/predicted system performance; simplified diagnosis of system failures; and simplified system synthesis. There are two types of tests, external tests and internal tests. External tests are less intrusive and are less disruptive of system operation:

- Verification of input conditions to assure conformance to specifications; this is specially important when the system lacks the ability to reject inappropriate inputs and operates on them equally as appropriate ones.
- Determination of output response characteristics (type, time, frequency, quality) against set or to be set specifications for synthesis or assurance of appropriate system operation.
- Conservation of or balance between input and output as in conservation of physical quantities (energy, momentum) or in a balance sheet in accounting.

Internal tests are intrusive, but are more revealing of operation:

- Probing is multivaried; it begins with visual inspection; it reads internal checkpoints; it tests for flow and connectivity on selected sequences of elements. Test re-check points, when installed, facilitate maintenance.

Both external and internal tests combine to extract information about the system. Direct and indirect testing may be conducted. Direct testing yields the desired variables. Indirect testing means that the measured variables are transformed, e.g., mathematically, into the intended variables.

Indirect testing is more preponderant; it is less intrusive or destructive; it enables measurement at a long distance from the test object; many indirect testing procedures are based on the interaction of waves with tested objects (radar, sonar, x-rays...).

Prior to testing, some partial or complete information is known about the system; the information is used to isolate and check separately the system parts. The more complete information about system function and structure allows system blocks to be preset with diagnostic routines simply or in combination, to check on operation and/or results.

Tests enable many things:

- Comparative assessment of candidate concepts/designs for the performance of a given objective.
- Sensitivity of system performance to parametric variation of input data and/or system settings.
- Logical correctness of the system, i.e., control rules such as FIFO (first in, first out) as to effect on some criterion (loading, processing, time, minimum cost, maximum cost, maximum utilization...).
- Test of the actual system, rather than its logical operations, in the real world. Variations in testing between logical and actual grow with heavy humanly controlled systems where emergency conditions (breakdown, loss...) arise to interrupt the sequencing of work and where response processes are not consistently defined and/or followed.

III. CHARACTERISTICS OF INEFFECTIVE SYSTEMS AND THEIR IMPROVEMENTS

A. Ineffective System Characteristics

Many factors enter to make a system ineffective, complex, and certainly not simple. Among them are:

- The goals and constraints of the organization are not conceived and/or allocated to the individual element(s), and operations proceed according to perceptions.
- The effectiveness measures are not appropriate or correctly oriented, and, though met, do not support system goals.
- The operating standards for the system are not set and deviations cannot be controlled.

- The vital processes are not protected or insured in case of failure and either act inappropriately or fail to act in the face of imminent danger.
- The growth and change of the system are not planned for and the system structure and/or its settings become inappropriate with evolution of time.
- The system is not protected against intrusive disruptions and does not recognize their presence.
- The boundaries among system elements are not sufficiently linked, and coordinated flow is disrupted.
- The required resources are underestimated, and system goals go unfulfilled.

B. System Improvement Characteristics

Many simplifying characteristics render systems amenable to enhancements. Among them are:

- Modular construction where a limited number of well-defined building blocks or objects are variably combined to satisfy system requirements. Module size, variety, and interconnection form the elements of trade-off in planned modular construction. The resulting benefits are many in design, development, production, operation, and maintenance. Modularity yields volume; volume justifies design costs; it facilitates assembly and testing; vs. custom; it simplifies production; it leads to automation; it reduces differences in training; it allows prompt evaluation of alternate designs; it eases maintenance; it reduces special module types; and it facilitates information transfer.
- Information handling where the process takes place in representative patterns rather than in physical storage and movement of material; it replaces costly inventory with information. First, process type involves data management, where data combination and restructuring are carried on data files organized by modular construction; this organization and the computer's ability to reach and manipulate the data in it allow the varied summaries of recombined data in response to different queries. Second, process type involves skills management where the procedures leading to a desired result are executed by computer to control device manipulations. Third, process type involves human–computer interactive processing where alternatives are promptly assessed based on data and procedures' selection and application.
- Goal and constraint honing where alternate plans are reviewed and adapted to remove defective elements within them.

IV. DISCUSSIONS

As a system grows in size, variety, rate, and complexity, its definition, design, development, operation and/or maintenance become unwieldy unless it is viewed through simplifying processes. Simplicity of systems is an essential complement to systems' speed. Notwithstanding the computer's procedural capabilities to deal with size, variety, rate, and complexity, the simplifying processes remain a multiplier in:

- Design and development of those computers and their processes.
- Creative information processing where non-procedural elements require human intervention.

The prescriptive processes to simplify treatment of general systems' problems apply to the functioning of any system and yield internal views and external views of essential system properties. These essential properties yield meaningful relationships and reveal how the various parts act and react in response to designated action.

Simplification processes sort out what could be a tangle of details into recognizable patterns and allow the manager, be it human executive or computer program executive, to be concerned with broader system control issues such as allocation of shared space and time, setting of priorities and values, and assignment of human, computer and/or machine resources to deal with changing input conditions.

Such processes provide the connecting links to understand, design, or improve a general system that could consist of electrical, electronic, mechanical, fluid, optical, and chemical components; they enable the investigator to cross the traditional boundaries of specialized disciplines, follow the work flow, information flow, or material flow in a system and manage the diverse operations which, by necessity, cut across the historical disciplines to lead toward definite system objectives.

9   REAL-TIME SYSTEMS

I. INTRODUCTION

Time is considered a valuable and unrecoverable resource in all kinds of endeavors, be they commercial, military, manufacturing, or service systems. Real-time systems are developed in Part III, and their applications in Part IV. Terms have been coined to express this urgent or rapid utilization of time, e.g., real time, just-in-time, short cycle time, all to imply its maximum and effective utilization as it is decomposed and integrated in productive work. The basic enabler of a real-time system begins from a concrete and simplified system description. A real-time system is not a speeded-up system which may contain essential and non-essential work activities. A real-time system implies a speeded-up system which is simplified to contain only the essential activities for execution of the work at hand in the most economical way.

Since the beginning, time has been an integral part of systems' operation and control. In their evolution, systems have gone through several distinctive stages (Figure 9.1):

- In the first, systems were elementary and characterized by humans who were tool users (hammers, wheels, needles, swords, ploughs...) and the pace of individual work varied.
- In the second, time rules were added where the clocks controlled labor and its resulting cost.
- In the third, machines, as integrated sets of tools, were added to do combined and rapid work controlled by humans and clocks; machines do physical work and have a single dedicated use.
- In the fourth, computers were added to replace human functions in controlling machines; computers controlled by clocks rapidly manipulate representation of physical work and in that sense have universal uses through their separately imparted instructions.

Computers have enabled the wide realization of real-time systems. A real-time system completes each of its activities rapidly, just before or just in the

110 SYSTEMS MANAGEMENT: People, Computers, Machines, Materials

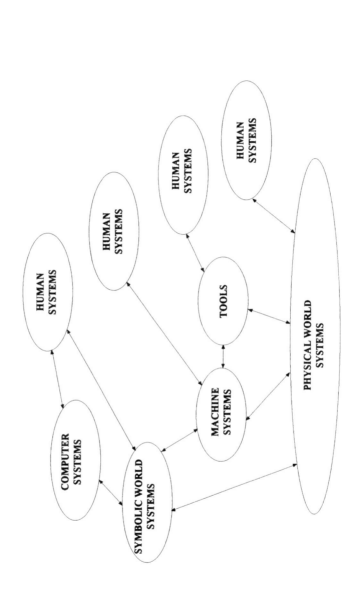

Figure 9.1 Human progress beginning with tools, then machines (automated tools), then computers (automated machines) used in transforming the physical world and increasing human productivity.

REAL-TIME SYSTEMS

time required. The speed with which work can be accomplished unattended by humans now increased productivity and reduced cost; time is the critical parameter in this system operation. Then, the system is made up of a controlling part and a controlled part (Figure 9.2). As such, it reacts to inputs from the environment in which it is operating in time to respond and control its behavior. That control may be direct as in automated computer controlled devices such as valves or switches or plants or missile systems. It may be indirect when personnel receive the data and act, in time, accordingly. The control is described by instructions called the software program. The software provides levels of abstractions between the problem to be solved and the hardware set to solve it.

II. BASIC REAL-TIME SYSTEM ELEMENTS AND IMPLEMENTATION

A. Real-Time System Elements

A real-time system is composed of three basic elements:

- Processes; the function of a real-time system is to execute a set of activities or tasks within time constraints. Of the set of possible activities, some interact to respond to a specific external event. To implement one activity, the suitable set of instructions and processing hardware are needed, which in total constitute the process. The simplest process is one that carries an explicit interaction with the outside world (Figures 9.3 and 9.4). When the processes depend on each other, they must cooperate thus requiring communication of information and synchronization (activation, suspension) among each other and interaction with the devices they control (Figure 9.5).
- Communication; information is communicated in two ways (Figure 9.6); a) *a direct transfer from one process to another* through what is called a channel, and b) *access to a shared repository* through what is called a path.
- Synchronization; in addition to communication processes need to manage each other's activities, that is to wait for, start, or stop other processes. This is called synchronization, and interrupts are used to signal the needed event.

In setting a system with those basic elements, an approach describes:

- First, the system functions.
- Second, the activities those functions will actually execute.
- Third, the processes to execute those activities.
- Fourth, the access mechanism from one activity to the next.

112   SYSTEMS MANAGEMENT: People, Computers, Machines, Materials

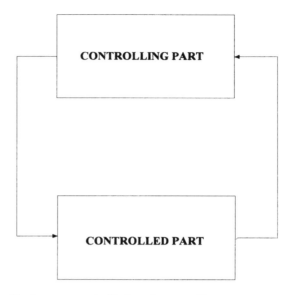

Figure 9.2a System as controlled and controlling parts.

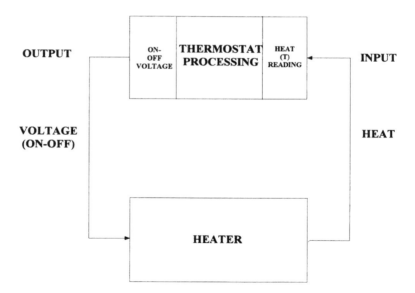

Figure 9.2b Single and simple process or activity with control.

- Fifth, the synchronization or coordination of the unfolding of activities.

Figure 9.3 Single process of activity.

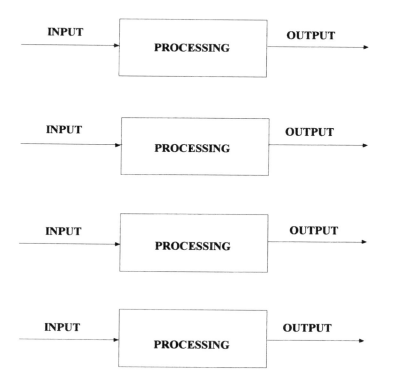

Figure 9.4 Multiple independent processes or activities, e.g., heating, washing, drying.

- Sixth, the control to report on progress of activities against set standards.

B. Real-Time System Implementation

Depending on the system, single or multiple processes implement its functions. The constituents for process implementation include the following resources:

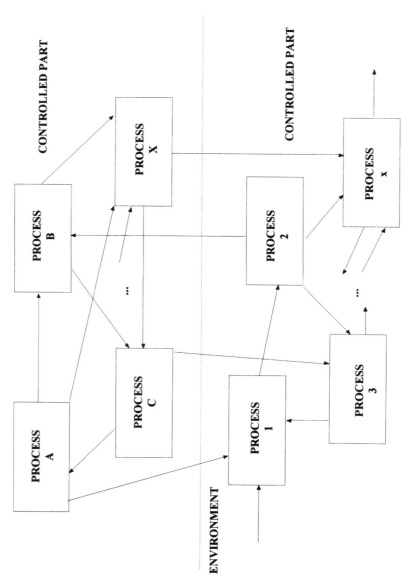

Figure 9.5 Sample of dependent or coupled system processes.

REAL-TIME SYSTEMS

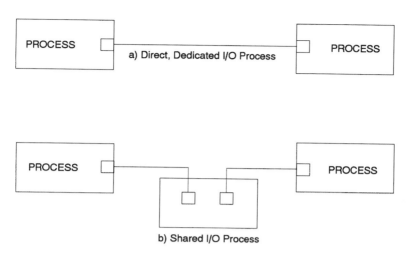

Figure 9.6 Interconnection between any two processes to exchange data done through a) dedicated b) shared storage constituting an I/O process of its own.

- Hardware to execute the process, e.g., central processor and memory for computational work and machines for mechanical work.
- Programs for processes to instruct the central processor on what to do.
- Communication among processes.
- Synchronization (cooperation) among processes.

Implementation options include (Figure 9.7):

- Dedicated resources for each process.
- Shared resources among the processes.

Sharing triggers the need for management; otherwise, cooperation (synchronization) between the processes will be elusive. When sharing, facilities must be provided to determine:

- Which events trigger a change in process execution from one to another.
- The needed mechanisms to support a change.
- The timing for a change or schedule of changes.

To execute a process, hardware (processor plus memory) and program (instructions plus data) have to come together. Unfailingly, processor sharing occurs with:

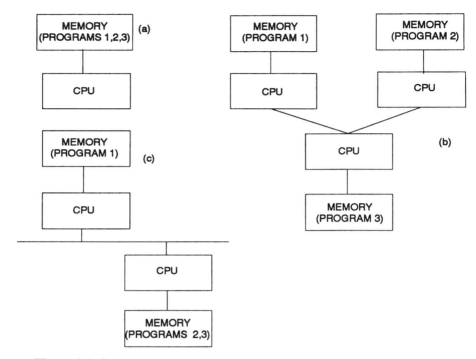

Figure 9.7 Options in resource ownership: (a) centralized and shared; (b) distributed and dedicated; (c) distributed and shared.

- Serial execution where a process starts after the preceding process has completed. A simple filter performs that way. Unless the process is very short, response to environmental changes is lost in the controlled system.
- Pseudoparallel execution where the process instructions and re-initiating data are saved in some volatile memory (registers) to allow it to restart from where it left off when it was interrupted. Interrupts are handled by hardware when it saves first the current environment of a process and then brings into activation the process that is supposed to handle that interrupt. An interrupt may cause such a change in the environment that no return to the interrupted process occurs.

Memory sharing occurs with:

- Sharing instructions residing in one location of memory; for instructions with data that may be modified in parallel by two processes, a lock–unlock mechanism must be set at the start and end of instructions for protection.

- Sharing of memory itself when the instructions are copied from large storage as they are needed for operation upon detection of an event.

Communication allows data accessible by one process to be accessible by another process. The data are copied to another memory through a channel or made available to the process needing it at a shared storage area:

- A channel carries the information through a message format which contains ID (sender, receiver) data or a pointer to data. Either of two mechanisms are widely used:
  - The queue where first-in from sender is first-out by receiver or where priority order is given to the message. In a distributed processor, we have a sender buffer (if awaiting transmission) in-transit memory and a receiver memory. When memory space is occupied, the system may send data to secondary storage or refuse the transaction.
  - Shared storage to hold system tables, files, data, where a monitor is used to protect the shared data from errant processes.
  - Circular buffer with a pointer for loading and a pointer for unloading.

Synchronization can be implemented through two facilities, wait and signal. Semaphores are used as the synchronizing agent. Initialization sets the value of the semaphores; a signal increments the value by one; a wait decreases the value by one but not lower than zero. Executive processes include those services used to support the existence and monitor the activities of the remaining processes, i.e., interrupt handling process, swapping, memory allocation, scheduling, file system control, I/O device driver, fault protection.

III. REAL-TIME SYSTEM SCHEDULING AND RELIABILITY

A. Real-Time System Scheduling

At the heart of real-time systems is the scheduling function. Scheduling is a decomposition and integration process of time and forms the plan for resource activation; it sets the coordinated time and sequence of each activity and of each operation within an activity. Where resources are shared, competition is created and competition can cause deadlock. Which process is selected and when and how to execute depend on the relative significance of that process. This significance is expressed in the immediacy of attention conveyed to it. That significance is expressed in three levels:

- Level 1 is assignment on demand basis; those processes require nearly instant service. They preempt level 2B and level 3 processes;

Level 2A process is allowed to run to completion before transfer of control. Within level 1, a first-come first-served approach is used. Highly urgent level 1 processes may have dedicated signal lines to interrupt the processor.
- Level 2 is assignment at prescribed time basis; those processes are polled at prescribed intervals of time given by a real-time clock when the process may interrupt the processor; a level 2A process is allowed to complete its function uninterrupted; a level 2B process may be interrupted by a level 1 process or level 2A process.
- Level 3 is assignment on standby basis; those processes are assigned resources when they become available and interrupted when a higher level request is submitted; the assignment is on a round robin basis with preset time slices.

The designation of which process deserves prior attention may be fixed at design or may remain variable. A schedule manager, who has overall system status, is pressed into service to re-designate process priorities; the manager has to take into account the deadlines of all the processes in the system and their dependencies and satisfy the most time critical process string. Complex deadlock-avoidance algorithms are used to provide the efficient sharing of resources. System saturation is usually caused by excessive level 1 interrupts which are by nature unscheduled–scheduled, and the system has no control over their judicious timing.

B. Real-Time System Reliability

Real-time system implies continuous availability of operational resources to deliver the controls and services as required. Systems do not operate continuously without failure. Generally, hardware fails because of aging and software because of residual errors. Real-time system failures are critical since time then lapses without completion of system activities because of system unavailability. System availability is a function of a) *how long the system operates without failures* expressed by the parameter function of mean time between failures, and b) *how long it takes to repair the failure* expressed by the parameter mean time to repair the system once it has failed.

Stress testing is the main avenue to:

- Ferret out and expose faults at system nodes (hardware, software).
- Expose faults in certain pathways of the code or hardware design.
- Trigger the occurrence of timing conditions disruptive to the execution of software/hardware.

Real-time systems experience race conditions, deadlocks, performance bottlenecks, and missed deadlines. Logic analysis is needed to monitor and

REAL-TIME SYSTEMS

capture, from applications running in real-time, the sequence and timing of all events including interactions between tasks, interrupts, message queues, semaphores, and application level events.

Regardless of system testing amount, residual faults and weaknesses in the system remain. This is illustrated by the various combinations of pathways and nodes contributing to output, O, from input, I, operation, Figure 9.8. Some remaining faults are encountered under some circumstances and have to be dealt with automatically or manually. Two principal mechanisms are used to capture progress made prior to failure:

- Rolling back to a point (checkpoint) where the system is known to have operated correctly and restart the operation from there.
- Sharing of critical modules where the system is switched over from the faulty one to the healthy one. Otherwise, the faulty part is taken off line for maintenance.

To uncover faults, checks (tests) are placed at various levels in the system's hardware and software; the response is checked against the reference; variance causes a declaration of a fault; when the response is triggered by a specific module, then detection and localization happen concurrently. Otherwise, the tests need to be set to isolate the fault to a single module. This isolation is done automatically or manually; following isolation, repair or replacement is carried out. The checks' levels against expected response and their order follow (Figure 9.9):

- Hardware level to include exercise of CPU instructions, level tests, writing bit patterns to memory, check sum tests; those tests should be short and thorough; to insure occurrence, resetting of the real-time clock interrupt is used as a watchdog.
- Software module level to include a) *range checks* on data falling within expected range; b) *state checks* on sets of data against expected values at a certain time condition; c) *outlier's checks* on data falling beyond reasonable values.
- Process level to include checks on expected behavior at the process boundary: a) *checks on outgoing data for correctness;* b) *lack of output data beyond expected time*; c) *over-filled inputs buffer beyond expected delay;* d) *different versions of the same information.*
- Audit (supervisory/executive) level to include checks on the consistency of redundant information in the system databases: a) *checks on system state* implied by one database against those by another; b) *checks on processes* resetting their timers at given points in their execution cycle; c) *checks on data consistency* between controlled system data and controlling system data.

120 SYSTEMS MANAGEMENT: People, Computers, Machines, Materials

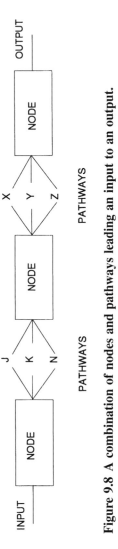

Figure 9.8 A combination of nodes and pathways leading an input to an output.

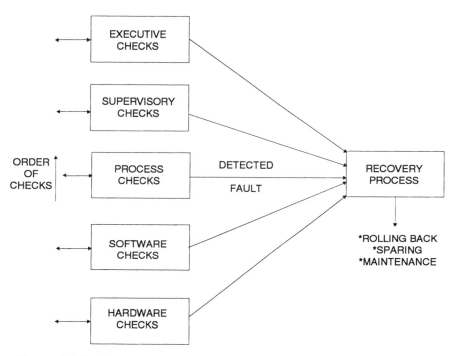

Figure 9.9 Performance monitoring, fault detection and recovery overlaid on regular system operation.

- System level to include checks on expected behavior at the system boundary: a) *checks on expected activity or lack of it at the boundary,* e.g., electrical activity, communication activity among neighbors; b) *checks on destructive behavior of systems in terms of fail safe devices* (relief valves, fuses).

The checks and recovery processes monitor the normal operation processes; some assurance is needed that the monitor is performing its functions also. In general, watch-dog timers are used at the various monitoring levels that are reset once the checks are performed.

Once a fault is detected at a certain level and localized (Figure 9.10), recovery from that fault may use one or multiple mechanisms to avoid initialization and restart at time zero which can take minutes to effect:

- Exception handling is the procedure for local error recovery because of temporary effects, such as overflow or failed handshake.
- Rolling back is a restart from a checkpoint, a summary of the most recent state of the processing cycle; this mechanism succeeds if the condition causing failure is transient and will not repeat itself on

122 SYSTEMS MANAGEMENT: People, Computers, Machines, Materials

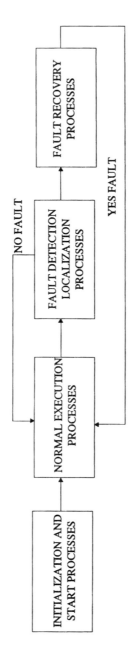

Figure 9.10 Normal and abnormal process flow.

another try. Rollback of a failed process often implies rollback of interactive processes with the failed one to purge the contamination that occurred during the time between actual failure and detection rollback.
- Sparing is a provision for redundant hardware/software used as back-up in case of failure; spare hardware is a replica of the operating one; spare software is at best independently coded to the same design specification as the operating one.

Fault detection, localization, and recovery present an overhead to the system timing. Not withstanding, faults can propagate very rapidly and can cause breakdown of the system unless attended to on a priority basis. The seriousness of the fault dictates the level of priority assigned to the recovery. Unfailingly during recovery, work load increases with backlog of work.

In design choices, reliability provides a cumulative value to product quality; in it rests the true cost, the initial cost of acquisition, and the ongoing cost of maintenance; to it is linked the cost of failure; from it springs the design strategy. When the repercussions of a failure are minimal or non-existent, a cheap throwaway product may be the design strategy.

When failure triggers a crash to a critical operation with costly repercussions, design for maximum reliability is the desired strategy. Throughout, the viable product should be inexpensive relative to the cost of delay or failure of the operation.

When the initial cost is high, the on-going cost should be low, first through maximum reliability, then self-diagnostic processes, simplicity of in-field repair, segmentation in customer service and adaptation to focus most importantly on customer service.

When short on capacity, a more reliable system matches its reduced capacity with high tier and most critical customers to continue service to them. Since service drops in quality when a system reaches 75% of its capacity this policy focuses valuable operations on the most valued customer.

IV. SYSTEM DESIGN METHODOLOGY

In conceiving a system within real-time constraints, a design methodology needs to be followed that delivers within those critical constraints. It starts, first with customer needs, objectives, and constraints; second, it identifies the requirements and then refines the design constraints; third, it decomposes and allocates the requirements to lower levels, refines the interfaces and integrates the whole; fourth, it transforms the functional architecture into a physical architecture and verifies it. This process follows, basically, three nested iterative loops between requirements and their functional architecture, the functional architecture and its physical architecture, its physical architecture and the requirements. This design process delivers:

- What the system is projected to do and how timely once developed, i.e., its objectives and goals.
- Description of the fundamental parameters bounding the system, its functional and physical dimensions, its response time and capacity.
- Derived basic limits of operation resulting from the set dimensions.
- Enunciation of the system control strategies to satisfy the goals within the stated boundaries and limits.
- Resulting system organization, its basic block diagram.
- Initial system verification through simulation and prototyping.
- Fixing of failed system parts by iterating on the preceding process.
- Description of the physical nature and characteristics of the system to be controlled through probes of the system and their output characteristics.
- Derived requirements on the processing elements, their signal conditioning, spatial processing, spectral processing, temporal processing, local control strategies for graceful rather than catastrophic degradation.
- Derived system requirements on hardware and their capacity.
- Derived system requirements on software.
- Derived system requirements needed for distributed processing and character of distribution, their hardware, software, data and people distribution.
- Derived system provision for a back-up facility for continuous reliable operation during failure.
- Development process, host, support, and target environment relationships.

V. SYSTEM DESIGN PROCESSES

In effecting the design methodology, varied processes are applicable to facilitate the conception, decomposition, implementation, and integration stages. These processes include:

- A control system process.
- A finite state system process.
- A functional decomposition process.
- A data decomposition process.

In general, the processes are interchangeable and their ease of application depends on the problem at hand. In large systems, a combination of the above processes are variably applied.

A. Control System Process

Systems are found interconnected in varied ways and types. A control system is a type of elements interconnection set to clearly drive the system in real-time toward the desired objectives and goals. Many control methods provide system responses to input commands through:

- Adjustment of system parameters to compensate for deficiencies.
- Alteration of the processes themselves, their structure, components, and parameters.

Optimal control systems are those designed under a minimum performance index which denotes the quality of the system, i.e., how well it meets the objectives it was designed for. A basic control system structure adds three fundamental operations to the process (Figure 9.11):

- Feedback operation to connect resulting system output to incoming input.
- Difference operation to compare where the input is and where the system perceives it to be.
- Controlling process combines the output of the difference operation with some control process (human/computer) to drive the controlled process toward desired output. A control system can naturally be reduced to the standard input–process–output format.

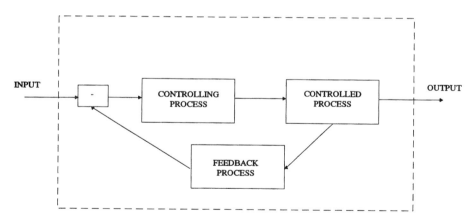

Figure 9.11 Input, process and output in a control process.

In general, characteristics of large control systems include:

- Many of those control loops nested within each other; the outer loops deal with the lower levels of controls while the inner loops deal with the higher levels of controls.
- Nonlinear operations.
- Time varying operations.
- Operations function of multivariables.

The combination of the above characteristics increases the complications.

B. Finite State System Process

There are varied system design methods; one tracks the state of the system. A system is in one of a finite set of states that it can be in. While in one state, it performs the set of activities specific to that state. While transitioning, it carries out the sequence of actions that move it from one state to another. The simplest system has only on–off states, i.e., light bulb.

The finite state system may be depicted using:

- State tables containing a row for each state in the system; the row contains the state number together with pointers to the beginning and end rows in the action table.
- Action tables containing a row for the action in response to an event and the state the system will be in at the end of the action. In certain situations more than one next event/action is possible; an auxilliary action table is set up to depict possible sets of actions and next states that could result.
- Event table containing the sequence of events the system will encounter.

C. Functional Decomposition Process

Functional decomposition is one approach for breaking down system functions and their interfaces into a hierarchy of smaller and smaller functions and their resulting interfaces. The level of break down is continued until the representative individual in the implementation/development group can muster understanding/operation of those lower level sub-functions. Those sub-functions are defined in terms of the events that trigger them, and the attributes and behavior they follow. Basically, those sub-functions are described in terms of software/human instructions and data to be executed by the hardware. It is not unlikely that participants in the decomposition process lack the general experience to effect implementation without decomposition. A software/hardware designer may not understand the system he is asked to control nor the mathematics he is asked to implement, but has the requisite expertise in software/hardware design and implementation.

The resulting decomposition in software–hardware–people is the system processes that execute the total set of functions. Not all system functions execute at all times; a thread is an execution of elemental parts of the functions from system input to output; a set of related threads from all the system's threads can form a mode of operation.

The decomposition process yields different system structures depending on the designers involved. General groupings of characteristics are judged desirable in the system breakdown:

- Tasking, where a grouping of system elements (hardware/software/people) is brought together to execute a single task.
- Temporal, where a grouping of system elements is brought together to execute things occurring together in time.
- Data, where a grouping of system elements is brought together to operate on a single data type.
- Procedure, where a grouping of system elements is brought together to execute a single procedure.

Some interface characteristics are judged desirable in holding together the preceding elements to make a system:

- Full decoupling, where a system element or sub-element may be modified or replaced without perturbing other elements or interconnections.
- Partial decoupling, where a system element or sub-element may be modified while partially perturbing other elements/sub-elements, i.e., channels, common storage control mechanisms, data transfer mechanisms, hardware, software, operation, maintenance.

In activating the system processes, the basic operational structures are:

- Sequencing, order in which tasks and sub-tasks are executed, one after the other.
- Iteration, repetition of a sequence until the desired refinement is reached.
- Decision, selection of sequencing/iteration is initiated under a criterion test and its progress path is selected.

D. Data Decomposition Process

This process defines the data flow through a system. Its depiction gives the movement of data/information within and among system functions, the sources and destinations of data, and the places where they are stored; in a message processing system, it describes what happens to a message as it traverses the system from input to output. This representation shows the dependencies among the components of the system. It begins with the given data items at the input and desired data at the output as initially defined entities; definition of intervening data and activities represent the data decomposition process. Compared to functional decomposition, this process automatically yields a close decomposition to the physical problem, where what is given and what is desired are preserved.

The data structure from the physical problem drives successively intervening data structures and functions and then processes to transform input data into desired output data.

The basic operational structure in data decomposition corresponds to that of functional decomposition where sequence, iteration, and decision are now done on data items rather than on process.

VI. DISCUSSIONS

System concepts have proliferated in industry, commerce, and government. Generalized system definitions, descriptions, and their predictive behaviors have been developed. Real-time systems stress the critical time aspect of their on-going processes and set a time constraint on the use of system resources. After all, time means money, since labor and equipment are at the source of cost, and reduction in cycle time means higher productivity and lower cost of both.

In real-time systems, time is the lowest common denominator and serves as the central parameter in the architecture of such systems. Responsiveness or response time of the system to a change, a request, or an event must take place promptly; promptness may be milliseconds to seconds in response time as in aerospace systems aimed to control missile systems, and minutes to hours, days, or months as in the case of commercial or industrial systems aimed to get products to market. Objects in extended waiting or storage, regardless of what they may be, present a vexing problem to productivity; objects may be bits, bytes, or messages in computer systems, or products on store shelves.

Whatever the field of application, a real-time system interacts with its environment and controls its behavior. The system is composed of a managing part and a managed part. The managing part contains the plan, the organization, and activation and control processes; it is constituted of human/computer elements since they have the flexibility and the capacity to execute such instructions. The managed part contains the physical mechanisms to carry out the instructions from the managing part; it is constituted of material, devices, and machines.

REAL-TIME SYSTEMS

The speed of the computer in the execution of management instructions and its flexibility in dealing with complex processes make real-time system implementation a reality. Computers can:

- Operate on data in execution of varied mathematical and logical instructions.
- Summarize data, e.g., mean, median, variance.
- Filter data to yield information needed for decision-making.
- Evaluate alternatives and make decisions.
- Find relationships in associations or common agreement among existing data.
- Find relationships between existing data and new data as in prediction.
- Make new patterns to accommodate new experiences.
- Store the actions to be taken in response to every event or class of events.

Besides speed, a real-time system must be reliable with emphasis on lack of failure vs. recovery from failure. A reliable system displays an extended time between failure and a minimum time to repair, enables continuous operation in some degraded mode during failure, and provides correct and complete answers regardless of the situation at hand. Thus, a real-time system is out of service, if at all, for a short time. In ensuring a reliable system, the computer manages the resources and provides soft failures and rapid recovery from hard failures through rapid fault detection, localization, and reconfiguration.

10 MANAGEMENT PROCESSES IN AUTOMATED SYSTEMS

I. INTRODUCTION

Processing is execution of real-life tasks. It is performed principally on the physical environment by humans and machines and on the symbolic environment by humans and computers, Figure 10.1. Automation occurs when tasks are performed by computers managing machines. While machines perform a variety of complicated physical processes, computers, as explained in Part III, perform two comparatively simple physical operations, on–off switching. On the other hand, the computer's powerful impact is in symbolic processing devised ingeniously around the two simplistic physical processes.

While humans can perform in both physical and symbolic environments, their real asset is in providing the binding processes in areas where there is a lack of definition combined with variety, multiplicity, uncertainty, and high risk in the process. It must be stated at the outset that not everything can be automated nor is it worthwhile to do so; the automobile, as a familiar example, has machines to produce motion, computers and devices to monitor machine performance, the human to drive the automobile; to get a computer to perform the driving would be a complicated, difficult, and economically unappealing prospect considering the infinity of situations the instructions within the computer would have to account for.

Notwithstanding, the road to much higher productivity and product/service quality is enabled through judicious application of the computer to the right processes. The computer provides the engine of the information revolution where more of the tasks are performed in the symbolic environment. How computers can manage themselves, machines, and humans is delineated in the following sections.

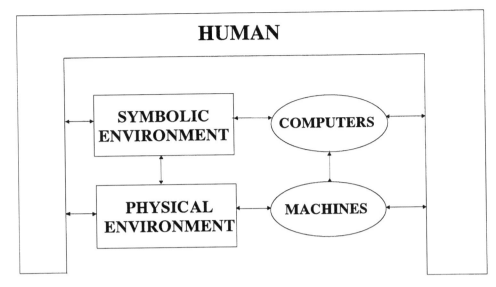

Figure 10.1 Humans–computers–machines and their environments.

II. PROCESSES AND DATA, CONTROL, LOGIC

This section sets the transition of processes from the physical and manual domain to the symbolic domain through data, control, and logic as enablers of automation. The issues may be presented through well familiar processes.

A process is a real-life task, i.e., washing clothes. Abstraction of a process may be represented by a simple example (Figure 10.2); a processor is what executes the process, i.e., clothes washer (human/machine); a second processor may be connected with the first, i.e., drying clothes. There, cooperating processes are needed; linking of two processes is called interfacing; the procedures (rules) for sending and receiving control signals or instructions, data, or objects are called protocol.

Any two processes may be done concurrently if multiprocessors are available or in sequence. A third process may be connected with the first two, i.e., ironing done by human/machine. Then more cooperating processes are needed and so on... . An interrupt to one of the processors may be caused by a higher priority process, such as a telephone ringing. The interrupt may disrupt processing depending on its time of occurrence. A computing system is a set of automatically interacting processes; it enables the stringing of processes to execute work without human interaction. Messages between processes contain actions. In response, processes perform activities. They utilize resources, e.g., CPU time, memory space, peripherals, data. Processes can be grouped according to the function they perform. Data are a resource of principal interest in database processes. Processes initiate operations which manipulate data. The state of the database is the set of values defining the content of the database.

MANAGEMENT PROCESSES IN AUTOMATED SYSTEMS

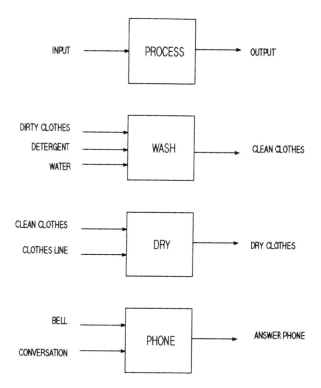

Figure 10.2 Samples of processes.

Actions/operations at the user level are called transactions. A transaction is a logical unit of work. Application programs/people issue calls for actions/transactions. Within an application, a task is broken down into a series of steps; the steps are made into a sequence of program instructions; an instruction cycle is the amount of time a computer takes to do an instruction. Objects or data are real or a representation of facts that is manipulated by a process. When data are interpreted in a well defined context, we call it information. Storage is needed whenever the process is not available for operation on object/data, Figure 10.3.

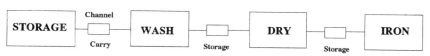

Figure 10.3 Familiar processes and object/data flows.

Control provides for coordination of the processes toward their successful conclusion. There are two strategies:

- Synchronous control requires time dependence of the execution rate of processes.
- Asynchronous control allows for time independence.

Synchronous control relies on timing by clock signal, to promptly initiate processes so that storage is not needed or available. The speed of the slowest process execution determines the system speed processing. Faster processes wait while slower ones are completing their tasks. It implies a strict centralization control. Asynchronous control requires buffering (storage) to accommodate the different speeds with which processes are executed. The minimum speed is a system constraint dictated by buffer capacity. Larger buffers lead to speed independence. Monitoring of resources yields control signals, e.g., "OK to proceed", "buffer full", "hold". System constraints are provided by such monitors. Control performs various functions. Among them, it:

- Initializes processes and readies them for execution.
- Starts processes.
- Reacts when a process has gone wrong, e.g., reset.
- Indicates when something should happen.

This is realized through signaling. The state of the processor is denoted by indicators such as lights and dials.

Logic gives the internal structure of a process. It is the plan to specify execution of a process. The process state is a report on its progress. The changes in the reports are the states' transition. The possible states and reasons for varying from one state to another are usually depicted in a state transition diagram. The number of states is finite. Some state indicators include "idle", "ready", "in process", "error", "buffer full", "OK to proceed", "buffer empty". The basic elements of a logic plan include:

- Sequencing to do this, then this, then that, and may be done singly or in parallel.
- Decision or selection to execute alternative sequences depending on the case at hand, i.e., in case of "a", then do sequence "1"; in case of "b", do sequence "2".
- Refinement through repetitive sequencing while the condition for initiating the sequence remains.

The basic aspects of logic planning are used in programming, Figure 10.4:

- First, declarations (enumerations) of all the ports (I/O) and state of the process are made.

MANAGEMENT PROCESSES IN AUTOMATED SYSTEMS

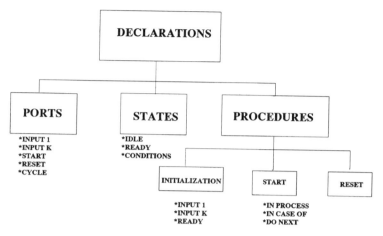

Figure 10.4 A process logic representation.

- Second, the procedures to execute the process logic are described; the procedures are started by the control function, e.g., initialization, start, reset... .

III. PROCESSORS, DATA, AND INSTRUCTIONS

There are three types of processors: human, computer, and machine. Machines predominantly process materials while computers and increasingly humans process data representation of materials. For each, a different data representation is pertinent. For the human, the data are represented so that the user can understand, visualize, and make interpretations. For the computer, the data are highly encoded. High level tasks, e.g., arithmetic, are broken down into low level tasks (on–off) so that digital circuits can understand and operate. An important aspect of data processing is how the data are organized; that is data structure or record. In a computer, each data item is referenced to a scalar value; all data of the same type, for instance, are collected into an array; data of different types form records; the motivation is to facilitate processing. For the human, much more of a variety is available to enhance processing.

To perform automated cooperative processes (Figure 10.5), the basic physical processor building blocks that support processes involved with data, control, and logic include:

- Local storage (Figure 10.6) called memories or registers when in CPU, or buffers when in I/O, with the difference in those storages not so much in structure, but in size and access time. Each element or cell in the storage holds an address and data or instruction. The cell size is an indicator of precision.

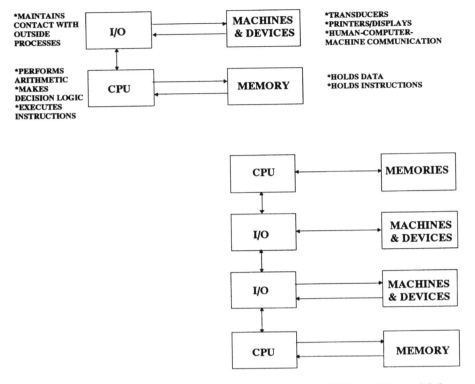

Figure 10.5 Cooperative processes with: a) single CPU and b) multiple CPU.

Figure 10.6 Types of local storage.

- Communication among the processes across busses or I/O channels. Parallel transmission of data is almost always used for the CPU, and serial or parallel is used for I/O depending on desired transfer speed.
- Control to allow the processes to cooperate, sending and receiving of control signals which follow a protocol (formal procedures of conduct). Control signals are utilized to inform neighboring processes about what is to be done and where it is to be done.

MANAGEMENT PROCESSES IN AUTOMATED SYSTEMS 137

- Peripherals with data or instructions coming/going from/to them; data is first collected in buffers, then transmitted to the CPU which sends it to memory unless a Direct Memory Access (DMA) capability is provided. For data/instruction in memory going to a peripheral, the same route is followed. With a common bus, data flow directly from source to destination.
- CPU for operations according to instructions (Figure 10.7); there is an instruction repertoire that it uses. The instruction format contains in bits the operation, e.g., read, write... and operand, e.g., address... . Categories of instructions include:
  - Control instructions.
  - Symbol manipulations instructions.
  - Arithmetic instructions.
  - Register loading instructions.
  - I/O instructions.

Working registers assist in instruction handling, e.g., I/O transfers. Basically, instruction registers and the program location counter are used in processing instructions which are fetched from the memory. The counter identifies the next instruction address to be fetched from memory for execution; it increments the counter by the length of the current instruction. The instruction register holds the current instruction under execution. Execution of an instruction involves a fetch instruction and then an execute instruction. A sampling of instructions within each category follows:

- I/O instructions allow "read"/"write" data and programs from/to peripheral devices:
  - Operations such as read, write.
  - Operands such as peripheral device address, memory address.
- Register loading, memory storing instructions allow:
  - Operations such as load.
  - Operands such as register address, memory address.
- Arithmetic instructions permit us to execute operations on binary numbers contained in registers. For instance, three registers or less are used for the arithmetic operation, e.g., a – b = c. Registers are used in:
  - Operations such as clear, increment, add, subtract, multiply, divide.
  - Operands such as register address.
- Symbol manipulation instructions permit us to move symbols to edit text:
  - Operations such as move.
  - Operands such as move memory address source, memory address destination.

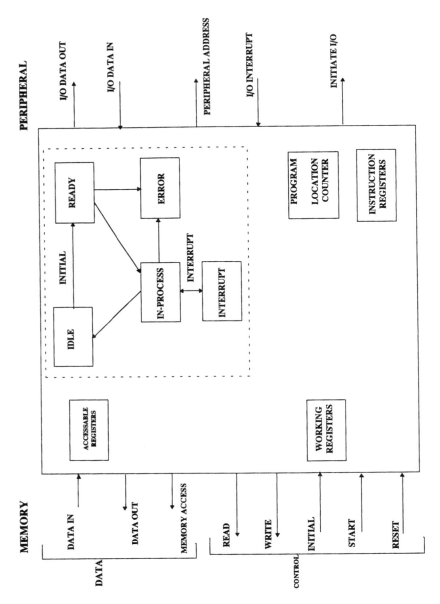

Figure 10.7 CPU ports and state transition.

- Control instructions permit us to change the execution sequence of instructions; there are unconditional control instructions:
  - Operations such as transfer, stop.
  - Operands such as memory address.
- Conditional control instructions transfers occur only under a condition, e.g., testing of values:
  - Operations such as equal, greater, less.
  - Operands such as register address, memory address.

IV. ARCHITECTURE OF A COMPUTER SYSTEM

Architecture is the organization of a structure from building blocks to meet the needs of the user within economic, schedule, and technical constraints. A computer system is built from the blocks described in the preceding sections.

The architecture of computer systems comprise a hierarchy to effect the communication (what is wanted) and operation (what should be done) from the application layer down to the integrated circuit layer; there are:

- Software architecture which comprises
  - Application systems
  - Database systems
  - Language translator (compiler)
  - Operating system
- Hardware architecture which comprises
  - Target system (machine language)
  - Micro systems
  - Logic circuits
  - Integrated circuits

In the hierarchy, one user's processor is another user's processes. At each level of the hierarchy, the user employs the lower level processes to produce higher level processors. Thus, designers of processes must be aware of the use of their processes at the higher level. Figure 10.8 depicts the target system, the hardware circuits, and software instructions that the circuits are to perform. A target system needs a software system to build upon the hardware circuit systems and provide the basic services to the users so as to avoid the need to redesign and to reprogram by each user. Its elements utilize some portion of memory. The instructions down to the circuits are layered:

- A microprogram has microinstructions which tell the hardware logic what operations to perform. It controls communication with an I/O controller; it controls fetching of machine language instruction and data from target system memory. The microprograms utilize the

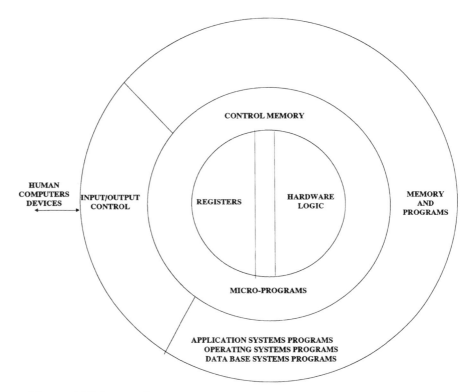

Figure 10.8 Layers in a computer system, (I/O, CPU, memory).

same basic hardware structure to perform processes for different target systems.
- An operating system manages the order of use of the resources of the computer system:
  - It controls access to memory, CPU, I/O.
  - It recognizes, interrupts, and schedules the resources through queues.
  - It satisfies the real-time requirements in responding to critical events.
  - It expedites the operation by scheduling parallel processes.
  - It catalogs the files and keeps a file directory, and manages the creation, modification, and deletion of individual files.
- The database system manages the data in the files and allows the user to make inquiries related to the data. Inquiries may involve data stored in several files; filing may be done by alphabetical order, by date of occurrence, by subject, by project, by author, or by a combination thereof.

MANAGEMENT PROCESSES IN AUTOMATED SYSTEMS 141

- The application system provides end uses of the computer system, be it computation, or real-time control.

The business of computers is to process data. Data represent the varied facets of life. Processing involves manipulation of data to yield information which prompts a decision/action. For manipulation to occur, transport of data (communication) must occur first. There exist organizational alternatives to transport data:

- Memory access; memory - CPU traffic is a limiting element in computer speed. Internal storage takes the form of random access memory (RAM), read only memory (ROM), registers in CPU, buffers in I/O; external discs, tapes, (Figure 10.9a). When all input and output transfers, for instance, are directed through the CPU, the CPU is interrupted from its processing task in order to support the initiation of I/O operations, and the transfers of data/instructions with the memory. When the storage is a disk, the transmission delay is great. To increase the speed, one may architect the system (Figure 10.9b) with a multiport memory and a separate channel to connect the disk directly into memory. In this architecture, the CPU can continue interpreting/acting on instructions from memory while the disk is transferring the data/program to memory.
- Bussing; computer resources may be increased and shared to do larger and larger tasks through a common connection (bus) where each resource has an address to identify it (Figure 10.10). Individual modules can send messages to another module by using its address.
- Networking; connecting computers to allow distributed processing; distribution is prompted by:
  - Physical separation of end users.
  - Overloading of a single computer.
  - Reliability.
- Protocols; among the varied interconnection protocols are:
  - Circuit switching on dedicated physical resources is increasingly in use due to the wider bandwidth enabled by technology.
  - Packet switching on shared physical resources and requiring a layered handshake to avoid conflicts.

As a summary to architecting a system, there are basically jobs set up to do the needed work. The computer system architecture consists of processing units, storage units, interconnecting units, and coordination units. The work leading to a product/service is composed of units integrable into the product service and of a size executable by a basic constituent unit. In automated systems, the basic constituent is the processor. A basic processor includes:

142 SYSTEMS MANAGEMENT: People, Computers, Machines, Materials

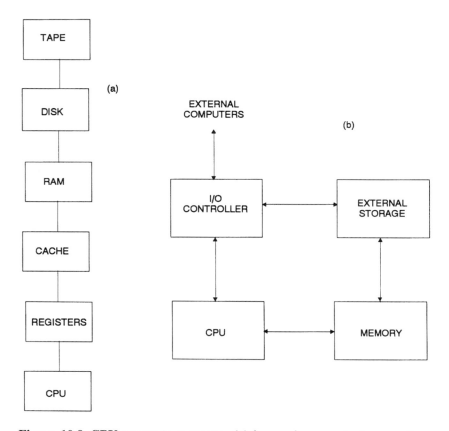

Figure 10.9 CPU access to memory: (a) increasing speed of access from tape to register; (b) direct memory access.

Figure 10.10 Linking of CPU, memory, I/O, and storage on a common bus.

- Central processing unit, where logical and arithmetic operations are performed to execute work at hand.
- Memory, where storage of work in process or final results are shared; memory locality determines how fast work is done; the ratio

of local to global memory accesses for data/instructions determines the speed of operations and extent of required synchronization.
- Synchronization at the local and global levels of activities where process alerts and coordination are facilitated at the lowest level through system hardware implementation and at the processors' interconnects level through system software features, e.g., pipes, sockets, metaphores, messages.

Integration of work units is done through some organizational interconnect. Interconnect topologies for data/instruction transfers are:

- Point-to-point, where a dedicated pathway is used between each processor.
- Bus, where a common linear pathway is used among the processors; this configuration is not scalable and becomes congested as more processors are added.
- Network, where a common ring pathway is used among the processors.
- Combination, where trees, two-dimensional meshes, and three-dimensional meshes are used.

The organization in an architecture aims for maximum parallel processing to compress the timeline in total work execution. Parallelism in the work occurs at many levels:

- Job-level parallelism, where all instructions and data pertaining to a job are executed within a single processor; each processor connects to the job through some links.
- Thread-level parallelism, where a subset of a job's instructions and data pertaining to a thread are executed within a single processor.
- Data-level parallelism, where each data subset is assigned to a given processor, all executing a single instruction at a time as in image processing with a CPU-pixel assignment.
- Instruction-level parallelism, where each instruction subset is assigned to a given processor, all operating on the same data at the same time.

V. DISCUSSIONS

Concepts and definitions are independent of the means used to implement them. In implementation, however, each processor, be it human, computer, or machine, has distinct advantages in the execution of certain processes. Invariably, complex applications such as precise machining, heavy manufacturing, aerospace, chemical industry, commercial market... require a combination of processor types to operate on the physical environment, data, and information environment.

To interface the different processor types, transducers are used as the means of converting one form of signal or representation into another. For instance, human–computer transduction can be accomplished through keyboard and display. Conversion of non-electrical signals into electrical ones leads to ease of automated processing and control. Often, the coordination or control process involves large or complex operating schedules; these schedules can be decomposed into multiple arithmetic and logical operations. The capability of the computer is adept at executing millions of those operations at orders of magnitude beyond human/machine processors and enables it to react in real time to processes needs; nowadays a wide variety of physical processes are controlled by computers.

The computer system processes are made up of hardware as processors and programs as instructions and data contained in storage. Physical processes relay their state through transduction or conversion devices. For example, the state of the process is one of two switch states, e.g., the temperature is below (above) a set value; a part under machining is (not) in place; turning on (off) motors. The computer easily handles interrupts, that is inputs from the transducer indicating a change of state which prompts the computer to hold execution of one part of the program and respond to the new state by executing another part of the program. Another computer capability is the keeping of time and consequently prompting itself to schedule process execution such as sampling of system states periodically or at periodic times. Large and complex processes require large and complex programs usually decomposed into tasks; the tasks are scheduled for execution by an external/internal interrupt triggered by events such as operator request, out of limits variables and time lapse. When simultaneous interrupts occur, a priority mechanism is used.

Processors' types supplement, refine, extend their respective facilities or abilities to process, service, execute, communicate, remember, compute, and conceive. A uniprocessor can execute a single process one at a time. A higher priority process can be attended to through an interrupt. Parallel processing can be executed through temporary suspension of activities; only when multiprocessors are available can parallel processing be executed without process interruption. Processors operate on virtual (data) or physical objects. Objects (virtual/physical) go through cycles of input, processing, and output; objects reach processes through transmission over channels; objects wait for processing in storage. The activities of input, processing, output, transmission, and storage are coordinated through a control process. Control is done using varied strategies, e.g., value limits, timing. Timing signals are generated by a clock, and the control strategy is either synchronous or asynchronous; synchronous control is demanding and clock signal timing cannot be shorter than the slowest cooperating process. The control process prompts processes' states such as initialization, interrupt, error conditions… . The number of system states is finite and the state descriptors are the system state variables. The system states

MANAGEMENT PROCESSES IN AUTOMATED SYSTEMS

transition according to the plan, enacted by the computer program; the logic in developing the plan involves the basic elements of sequencing, selection, and refinement. Program instructions are interpreted into instructions that can be recognized by the processor. In the computer, the general categories of instructions include control instructions, input/output instructions, data conversion instructions, arithmetic instructions, and symbol manipulation instructions. Those instructions allow the user to:

- Summarize data, e.g., median, mean.
- Make decision among alternatives.
- Find relationships such as a) *associations or common agreements*; b) *predictions where relations hold for existing data and new data as well*; and c) *making new patterns to accommodate new experiences*.

Part III:
Computer Systems and Automation

11 UNIPROCESSOR HARDWARE SYSTEM

I. INTRODUCTION

Basically there are three levels of abstractions in the architecture of a computer system (Figure 11.1):

- System functional architecture; it defines which functions are done by computer and which are done by end users; the interfaces are languages and application programs to input and retrieve data; High Order Languages (HOL) provide ease of use but require more storage/time of execution. Varied applications avail themselves of the computer and are expanded upon in Part IV.
- System software architecture; it allows sharing of hardware resources and provides automatic control over the resources; the services include process management, memory management, interrupt handling, file management, maintenance. The services are made available to the user through an instruction set; an instruction set summarizes the hardware characteristics that are visible to the software designer, i.e., register organizations, trap and interrupt sequences, addressing modes, data operators. Extended system software management processes are developed in Chapters 12 and 13, and the database management processes in Chapter 14.
- System hardware architecture; it contains lower level software (firmware) firmly protected in Read Only Memory (ROM) and hardware organization (controls and interconnections of the units). Chapter 11 explains the system hardware architecture for a uniprocessor or single computer thus forming, along with Chapters 12-14, the building blocks for multiprocessing systems explained in Chapter 15.

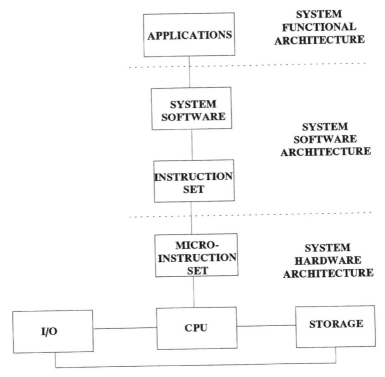

Figure 11.1 Level of abstraction in computer systems.

II. BASIC ELEMENTS OF A UNIPROCESSOR

What machines need to make decisions (Figure 11.2) is logic, storage, and communication (I/O linkage); the outcome of logical operations needs to be storable and transportable; in computers, all are coded in terms of switching circuits (on–off).

The computer's central uniprocessor unit (CPU) performs a few primitive logical functions at high speed (millions/second). Those primitive functions, if repeated enough, perform more complex logic, such as equal to, if, greater than, and branch to. The CPU tasks come from application programs and input devices that tell the machine what to do. Other system programs tell the machine how to do it. System programs are written to enable efficient use of the computing system's resources. To communicate, the central processor is linked through some Input/Output busses. Over the past 30 years, machine performance has improved by a factor of 10 to the 16th, while programming productivity has improved by a factor of 15.

UNIPROCESSOR HARDWARE SYSTEM

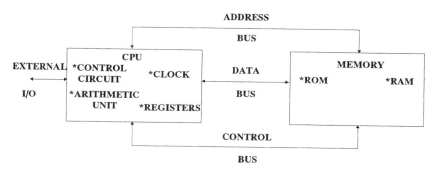

Figure 11.2 Basic uniprocessor elements where instructions on what the user wants done in RAM and how the machine is to do it in ROM.

A. Digital Logic

What elevates the computer above machines is its ability to perform logical operation using hardware. Implementation of digital control relies on three basic operators: OR, AND, NOT. Simple circuits called gates implement them; most digital circuit outputs are on or off, provided by a semiconductor; another is high impedance state provided by a transistor. All parts/components in a gate or on a chip are made by layering conductors, semiconductors, and insulators. The insulator is a glassy silicon dioxide; the conductor is a piece of metal and the semiconductor is silicon doped with the impurity, arsenic. Connectivity of parts is done through metal traces to integrate the parts into gates, gates into chips, chips into uniprocessors, uniprocessors into multiprocessors, and multiprocessors into systems. From chips to systems, the hardware configuration is set with chips on boards, and boards in drawers, and drawers in cabinets thus constituting systems. Program management of such a development is explained in Chapter 18.

Digital progates are the basic elements for constructing hardware processors that carry out the cooperating processes of the system. The resulting integrated circuiting may be application specific or general purpose. The basic logic elements are (Figure 11.3):

- NOT, output is made the reverse of input.
- AND, output is no if all inputs are no; output is no if any input is no.
- OR, output is yes if any input is yes.

High and low voltages are used to signify no and yes, respectively. From these three, higher and higher levels of processes may be organized (Figure 11.4). These gates are the building blocks of logic design. The combination

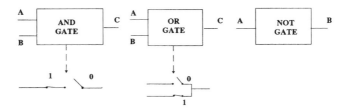

Figure 11.3 Basic gating processes.

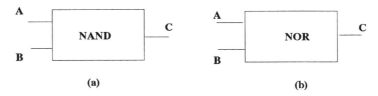

Figure 11.4 Additional gating processes: (a) AND gate followed by NOT; (b) OR gate followed by NOT.

of cooperating gates form gating networks, also called digital circuits/combinatorial circuits. When signals are applied at the inputs, further signals are generated as gate outputs; the total time the circuit takes to produce its outputs is called the propagation time (in microseconds).

B. Basic Units of a Computer

As a uniprocessor, the computer basic units are; a) Central Processing Unit (CPU), b) communication, and c) storage. They interact to jointly perform the processing function. The CPU has:

- An Arithmetic Logic Unit (ALU); it does arithmetic and logic operations and contains high speed registers for temporary storage. It performs binary addition, shift operation, rotation, and Boolean functions of two variables. More complex functions are performed using microprogram sequences. The shift right (left) divides (multiplies) the quantity by 2 to the power n. The shift and rotate are useful in editing, packing, and unpacking items of different bit lengths.
- A control unit receives directives in a logical sequence from the memory unit and coordinates the operations of all other units; it has a system clock closely synchronized to the memory cycle. It implements the instruction set. It avoids using hard-wired logic. The control signals (sequences) are placed in RAM or ROM, and they execute when the

UNIPROCESSOR HARDWARE SYSTEM

appropriate memory locations are accessed. The basic control operations are:
- Instruction fetch from storage to register (and vice versa) and register to register.
- Instruction decode to find the control sequence for the instruction.
- Operand (data) fetch from memory.
- Instruction execute where control signals coordinate the activities occurring at different times; each microinstruction corresponds to a set of signals.

Communication is realized through the CPU. For instance, macroinstruction fetched from main memory is interpreted by a linked list of microinstructions fetched from control memory; next microinstruction has one address contained at the end of microinstruction; fault detection logic for individual errors is detected at the microprogram level. The bus control circuitry provides communications paths (lines) for addresses, data, and controls. In a time multiplexed bus structure, this circuitry has an added feature to recognize when a signal is an address, data, or control. The communication timing sequence in digital circuits is provided by a clock signal; it is the driver of digital processes; it controls when things will happen; and it determines the starting and ending points of logic execution. The rate of clock oscillation establishes the performance of hardware logic. The logic must be designed so that all necessary propagation delays within the circuits triggered at the leading edge of a signal are completed before the next leading edge arrives.

In storage, a hierarchy exists to support different response times:

- Registers can be memory for very high speed internal storage; the CPU has logic gates with response times measured in nanoseconds.
- Main memory has cycle times of 800-900 nanoseconds.
- On line mass storage has cycle times of 10-20 milliseconds.
- Off line mass storage has manual intervention, usually reserved for activities used infrequently.

Examples of memory include: a) *hardware register set*; b) *hardware stack* used in last-in–first-out to service an interrupt and return to the point from which it was diverted; c) *cache* used when all references to memory tend to cluster in localized areas of storage; if the hardware finds the addressed word in cache (hit), execution is quick; otherwise, it is fetched from main memory using Direct Memory Access (DMA); d) *main memory* which is either magnetically polarized to positive or negative positions, thus making it possible to represent the two states of binary 0 and binary 1, or semiconductor memory (flip-flop) realized with transistors; when using semiconductor memory, care must be exercised that critical data/instructions are saved elsewhere in case of power failure; and e) *mass storage peripherals* made of tapes, disks, disk packs; with disks, the drive has either a moving head with a disk arm containing read/write per disk or a fixed head with a disk arm containing read/write per track for rapid access; disk surfaces are broken up into tracks and sectors.

The computer processing functions are described through data and programs: they are structured on the storage media in some so-called files; records are a subdivision of files; a file is like a folder, and a record is like a paper in the folder. A record may contain related or unrelated scalar values. While storage in main memory is rapid, mass storage peripherals are comparatively slow. Recording on tapes is sequential and slow. A more rapid storage mechanism is random access file on storage devices that can process data at separate physical locations on a single volume without passing over the intervening data, i.e., magnetic disk. Such files are organized so that a record key is utilized to denote the address of each record. When the program wants to process a record, the desired key contained in memory cells is sent via CPU and I/O controller to the peripheral device. When found, it is given to read/write the record. With disks, for instance, a logical record may be: a) *on a single track in a sector*, b) *on a single track in contiguous sectors*, or c) *on different tracks/sectors*; one may have a logical record allocated to separate physical recordings. Access time depends on the physical position of data and the programs that process the data; if two blocks of data are physical neighbors on the same track, retrieval of the second block takes 1 millisecond (ms); if they are not and the disk requires motion, then it is of the order of 20-100 ms.

The I/O links the computer parts together and the computer to the outside environment (operator, another device, another computer). Input is a message, and it may be data to be processed, data to be put in storage, or an inquiry for information. Information may be communicated in parallel (all bits in a data word at the same time) or in series (slower). Three basic ways exist for I/O data transfer.

- Programmed data transfer through a program in the CPU that does device rolling and directs the I/O transfer.
- Hardware interrupt from a device after which the CPU directs I/O transfer.
- Hardware directed I/O transfer between external devices and main storage, a direct memory access.

The hardware implementation is more efficient; it time-shares the memory with the CPU on a cycle stealing basis and transfers data in a burst mode. When a peripheral is attached to an I/O, it is governed by a device controller which interacts with the CPU in a well defined manner.

III. UNIPROCESSOR ESSENTIAL ORGANIZATION AND ACTIVATION

The uniprocessor elements are the CPU made up of an arithmetic unit and a control unit, memory, an I/O and peripheral units. The constituents of the five units are shown in Figure 11.5:

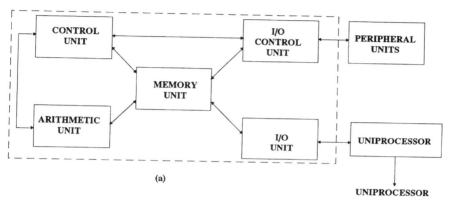

(a)

Figure 11.5 (Parts a, b, c, d, e, f). Basic uniprocessor: (a) elements; (b) basic arithmetic operation; (c) basic control unit operation; (d) basic memory unit operation; (e) basic peripherals; (f) basic disk I/O operation.

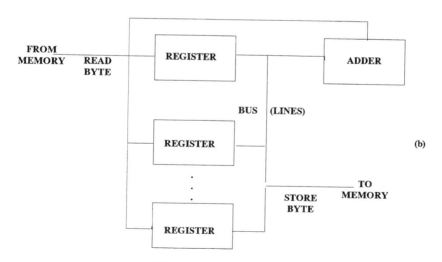

Figure 11.5 (continued)

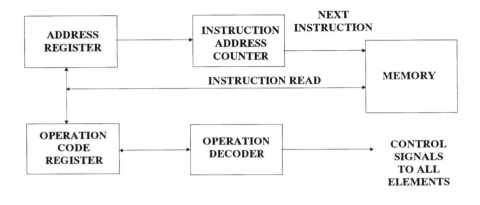

(c)

Figure 11.5 (continued)

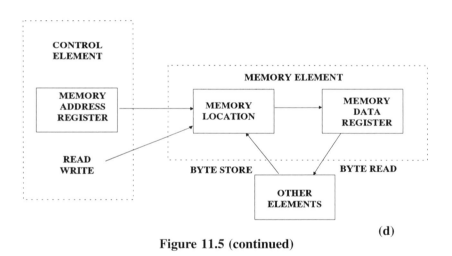

(d)

Figure 11.5 (continued)

- The arithmetic unit is composed of a bank of bi-stable devices (registers) to hold temporary results and a binary adder to perform basic addition, subtraction, multiplication, and division.
- The control unit is composed of a counter to choose instructions, one at a time, from the memory register; a decoder to break instructions into individual commands; and a clock to produce marks at regular intervals. All units respond to command from the control unit.
- The memory unit is composed of magnetic or semiconductor material organized to store programs and data.

UNIPROCESSOR HARDWARE SYSTEM

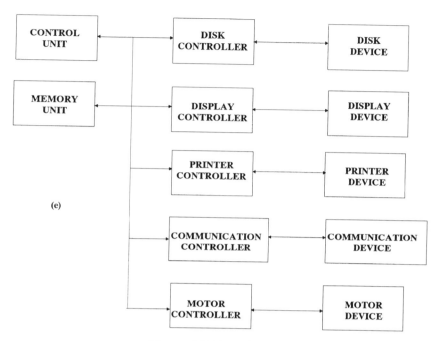

Figure 11.5 (continued)

- The I/O unit is a down-sized CPU memory to coordinate the exchange of data with the peripheral units; unless multiplexed, each I/O control is usually dedicated to a specific peripheral device; the CPU control unit determines whether it is an input or output, chooses the area of memory involved, and then signals the I/O controller (by address) to operate. In serial I/O, the connection is on one line; in parallel I/O, the connection has as many lines as bits in the basic information unit.
- The peripheral unit augments one and/or multiple units in the uniprocessor; it augments the function of the uniprocessor by linking it to other uniprocessors of the same type and/or provides a transduction mechanism to other types of processors.

The units' essential operations and their interactions need to be understood if we are to apply them to specific use. This includes:

- Arithmetic unit; it is where all logical operations and addition are done; subtraction, multiplication and division operations are reduced to addition; registers hold temporary data; the accumulator is the register in which the result of an operation appears; to handle multiplication and division, a double length register is used. Index

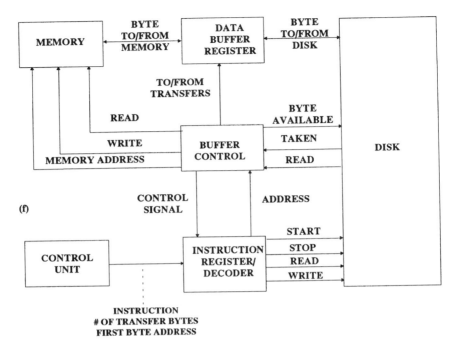

Figure 11.5 (continued)

registers are one part of arithmetic/control elements; they act as counters for repetitive operations that can be incremented for each successive one; index registers can hold memory location addresses for the base, then increment the base whenever one needs the next location using the instruction set in the computer.

- Control unit; the instructions stored in memory tell the computer what operations to perform and in what order; the instructions are read from memory one at a time and moved in the control unit where they are decoded. Then the control unit produces the needed signals and distributes them where necessary to carry out the instruction. Basically there is an instruction stream which operates on a data stream; based on this view, four architectural possibilities exist — single instruction operating on a single data stream, single instruction operating on a multiple data stream, multiple instructions operating on a single data stream, multiple instructions operating on a multiple data stream; the last possibility is implemented in a multiple processing system explained in Chapter 15. A typical instruction consists of: a) an operation code that indicates what must be done, b) a memory address that indicates the location from which data must be read and brought to the arithmetic unit for operation on it. Both instructions and data are stored in memory; they are read

UNIPROCESSOR HARDWARE SYSTEM

from memory the same way, i.e., a read instruction plus address; the control unit recognizes whether the output is an address or instruction, e.g., different busses. A typical flow for an add operation in a single instruction and single data stream follows: a) *a read of instructions from memory and the address (assume 101) of the instruction in memory are sent to memory*; b) *the memory content in the 101 location is sent*; c) *the control unit knowing that it requested an instruction rather than data puts the return in its operation code register rather than the arithmetic unit*; d) *next the instruction is decoded*, suppose that 110 represents add and the operation code for add and the address portion of the instruction is 501; e) *sensing the add code of 110, the control unit knows it must get data from memory*; it sends the address 501 and a read command to memory and takes note that it is now in an execution cycle rather than in a read cycle; f) *the memory sends the content in location 501, and the control unit transfers data to the arithmetic unit*; g) *having the control signals ready for add, the control unit gives the signal to move the contents of the register and accumulator to the adder which adds them and returns the results to the accumulator*; h) *the control unit gives a certain time lapse before it steps the instruction counter by 1 to 102 and reads the next instruction from memory*.

- Memory unit; most random access memory (RAM) is made of chips with bi-stable circuits. These chips are arranged in arrays so one unit of information (word) can be selected at a time. For the byte, 8 bi-stable circuits are arranged in a memory location with an assigned number as its address for reading/writing into/from that location. A register is a group of bi-stable circuits that hold data temporarily for immediate use or for elements to proceed with their operations. The number of memory locations is a power of 2; for 65536 memory locations, a 16-bit word is required to indicate addresses; with 32-bit, direct addressing links the CPU to 4 billion bytes, e.g., a few minutes of high definition TV images; with a 64-bit word, one can address 86 trillion two-hour movies; in indirect addressing, one uses the address of the address. Memory speed is one of 3 factors that determine computer speed. When an instruction is to be read from memory by the control unit, it is available in less than 10 to the −12th seconds after it is requested. Dynamic RAM needs refreshing every few milliseconds and is thus unavailable during this cycle for real-time applications (1 to 5 per cent of the time). Static RAM contains information as long as the power is on. Read only memory (ROM) contains permanent information; there are electrically programmable ROM and erasable ROM. Dynamic RAM stores each bit on a capacitor, the charge leaks out in 2 milliseconds, and they need refreshing. Static RAM stores each bit in a flip flop and conserves information as long as the power is on. A DRAM is four times as dense as an SRAM and is less costly, uses less power, but needs a controller to refresh. For instance, a 4,000 DRAM is structured as a (64 × 64) matrix and is refreshed 64 times; refresh control may be in a) burst mode to refresh all columns at once in sequence; the RAM is unavailable to the processors during 64 cycles, i.e., 64 × clock cycle, say 500 nanoseconds equals 32 microseconds every 2 milliseconds or 1.6 percent; b) distributed or single cycle mode where refreshing a row is done every n microseconds, and thus the RAM is not locked out for a long time from the CPU. Usually, refreshing is done when memory is not busy. Now, for every CPU there are cycles times in the instruction cycle when the memory is not required; a synchronization access is provided for refresh without overhead.

- I/O unit; among other things, this element provides for a) difference in speed between peripheral device and computer; b) difference in data form; c) mechanical units of the peripheral device; and d) connectivity among multiprocessors. Some devices provide only inputs to the computer (keyboard), sometimes receive only output from the computer (display), receive/provide I/O (disk). The data buffer registers compensate for differences in speed by providing temporary storage; it may also provide for changes in data form (data sent in parallel are accepted serially by the device). Buffer control regulates the transfer of data between memory and disk; for instance, the instruction register and decoder provide for decoding of I/O related instructions to disk. When the computer control element reads an instruction pertaining to a disk, it refers that instruction to the disk register/decoder.
- Peripheral unit; type and utility are delineated in section IV C.

IV. INTERFACING UNIPROCESSOR SYSTEMS

Interfacing of uniprocessor system units is accomplished through the CPU. The CPU package links to the other units through its pins (Figure 11.6). With forty pins, for example, at least two pins are for power, two for the external oscillator, ten to twelve control pins for coordination of data transfers (control bus), sixteen pins for the address bus (2 to the power of 16 bits = 64K bits of memory); multiplexing is used on some pins, address/data bus, resulting in slower operation. The busses are made of copper traces. Application specific integrated circuits (ASICS) are designed to rapidly control the bus traffic.

UNIPROCESSOR HARDWARE SYSTEM

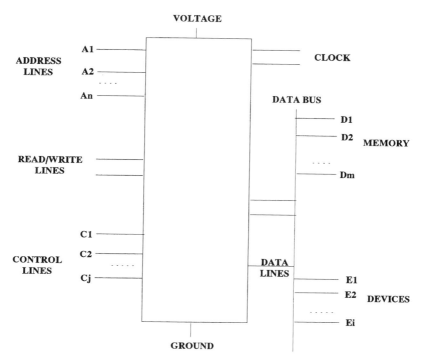

Figure 11.6 Linking the internals of a CPU to external units.

A. Bussing Techniques

Bussing forms an elementary implementation of the communication principles in Chapter 6. The bussing control has four functions: I/O synchronization, memory synchronization, CPU scheduling (interrupt and DMA), and utilities such as clock, reset, and halt. The functionality is realized through an application specific integrated circuit (ASIC) or a programmable one. The basic operations are "read" and "write". The handshake for "read" is a "ready" signal; an acknowledge confirms receipt in a nonsynchronous transfer; the asynchronous transfer allows usage of devices of different speeds. A typical linkage of uniprocessor units is connected through three busses (Figure 11.7):

- Data bus moves instructions and data among the elements and devices.
- Address bus selects the location where the instruction/data are going to or coming from. The address bus has a memory map; for sixteen bits, the total ROM/RAM is less than 64K bits when linear selection is used. With a decoding and memory management unit, this can be extended.

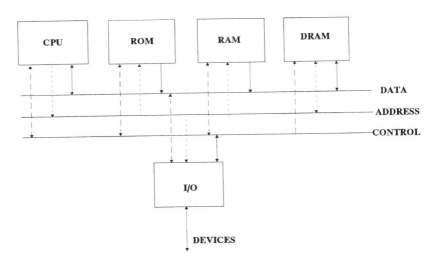

Figure 11.7 A typical uniprocessor connectivity.

- Control bus sequences the operations being performed, i.e., "read" from memory, then "write" to memory or output device.

When compared to point-to-point wiring, a bus connects more than one module/unit together so they can talk and listen; the same is applicable to a system to system connection. There are two basic bus types, parallel and serial (asynchronous and synchronous):

- Parallel busses provide for high speed communication with more wires; typically, one needs a 16, 32, or 64 data bus, 16, 24, 32 address lines and/or control lines. The data lines provide for data transfer in/out of processors; the address lines determine to whom (memory location or I/O port) the transfer is being sent; the control lines are read or write cycle lines, a valid address present line, a wait line, an interrupt line, a DMA request line; a reset line is used to initiate the system after power is down or re-initialization. Signals are used to represent those activities. The greater the number of control lines, the greater is processor flexibility.
- Serial busses use one or two wires to carry all necessary signals; they use standard voltage for the bit and a certain bit repetition rate for the handshake and information transfer; the frequency shift key is where bit 1 is one frequency and bit 0 another frequency; phase shift keying is also used. In synchronous communication with a clock, character codes are imbedded within the blocks; data is transmitted in blocks of many characters at a time called frames

UNIPROCESSOR HARDWARE SYSTEM

where each has a number of fields (one or more bytes of data); types of frames include information frames which contain data for the receiver, and protocol frames which contain data on supervision and management of bus transmission.

Parallel to serial and serial to parallel transfer can be done through a shift register. In either transfer type, a unit may drive one to twenty other units and thus can be loaded by those units; there are bus transmitters for driving the bus and bus receivers for listening to the bus and driving the processors.

B. Input-Output Scheduling

In communicating with peripherals in serial/parallel mode, three basic I/O scheduling techniques (polling, interrupts, DMA) are used:

- Polling; a flag is a bit that, when set, denotes that the device is ready; polling is checking for that flag; it is objectionable in a real-time system since all I/Os have to be checked and some peripheral may expect servicing within a specified time.
- Interrupts; each I/O device is connected to an interrupt line which sends a pulse to the processor which services the device; otherwise, the processor fetches the next instruction from memory. When servicing an interrupt, the contents of all registers are preserved. When multiple devices are connected to the same interrupt line, polling is used to vector the devices triggering the interrupt or a daisy chain is used where the interrupt acknowledge signal from the CPU is passed from device to device until the interrupting one receives it and sends back its identity. Then processor branches to its interrupt handler for that device. If several interrupts are triggered simultaneously, a priority level is used, e.g., level 0-power failure and restart; level 1-display A; level 2-display B; level 3-disk; level 4-teletype. Interrupt may be masked by the user during certain program execution times. Multiple interrupts may be handled by preserving the registers for each interrupt as servicing the higher priority one is being completed. While a processor is servicing an interrupt, a stack (section in RAM to store the processor's register contents) is used; to resume operation, the processor knows the start beginning address.
- DMA; when the software routine is replaced by hardware, we have a Direct Memory Access (DMA). A DMA may suspend a processor and steal memory cycles from the processor. The DMA takes control of the data bus and transfers at high speed blocks of data from disks.

164   SYSTEMS MANAGEMENT: People, Computers, Machines, Materials

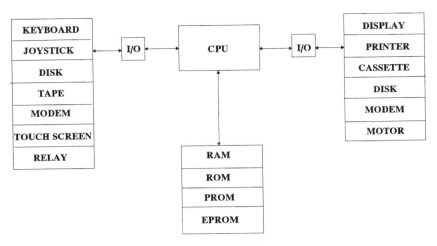

Figure 11.8 Varied I/O peripherals to CPU.

C. Peripherals

Not much utility is accrued from a uniprocessor until it is interfaced to its outside environment through peripherals. Peripherals include (Figure 11.8):

- Keyboards; they are pressure- or touch-activated (20 milliseconds response) switches arranged in a matrix fashion. There are two types: one is encoded and contains the hardware necessary to detect which key was activated and to hold the data until a new key is pressed; the other is non-encoded and contains the software in a processor to analyze the data. Roll-over is the problem when two or more keys are pressed at the same time; one solution is to ignore or take the first or last key pressed.
- Light Emitting Displays (LED); they indicate status to the users and are based on light emitting diodes when current passes through them; a matrix (n × k) LED has n rows k columns of LED.
- Teletypes; they are serial mechanical I/O devices and operate at 150 or 300 baud.
- Mouse; when the roller is displaced, motion is detected and a corresponding pointer movement is executed on the screen; pointing and clicking on an icon activates a software module and allows a graphical user interface to the computer as compared to a command line interface enabled through keyboards; roller motion may also be used to vary parameters' values within a program.
- Stepper Motors; when activated, they move their respective shafts by a step, e.g., 2 degrees.

UNIPROCESSOR HARDWARE SYSTEM 165

- Cassettes; they save and reload programs; the binary information in memory is coded in two frequencies, e.g., 3700 and 2400 Hertz; a phase lock loop is used to differentiate between the tones; recording is done at some 1500 bits/second.
- Cathode Ray Tubes (CRT); they display information for viewing by humans; information is projected through lighting dots on the screen to create the image; the better the display, the more dots there are per given surface, e.g., dots per character, characters per line, lines per viewing surface. For text presentation, characters are generated on screen using a dot matrix, e.g., 5×7 dot matrix/character; each character is represented in the CPU by its 0,1 code; a chip is used for mapping 0 and 1 into the dot matrix using a ROM look up table; the picture (frame) is refreshed 60 times/second using a DMA; a character may take 0.5-1 microsecond; the 5 dots corresponding to the row contents are gated into a shift register and are clocked out in serial form to the video output. To refresh a screen, dual buffering insures continuous system operation; the DMA has to fill one of the buffers before the other buffer empties. The CRT controller chip gives the addresses of characters that need to be refreshed, the row selects (7 or 9 rows) for display in sequence, the video timing to synchronize the signals, the display enabler, the character pointer (cursor) output, and the light pen input. In an intelligent CRT, a processor is dedicated to do graphics, scrolling, paging, editing, moving of blocks, and formatting; with the cursor, light pen, or keyboard commands, manipulation of information on the screen can be done using this dedicated processor which rearranges the characters in memory so they will be displayed in the fashion/order desired by the operator. A CRT screen can hold 24 lines of 80 characters per line; memory is used to hold a number of pages of typewritten texts; in addition to text presentation, graphics are added.
- Disks; they provide medium to large capacity of magnetic storage; the storage is formatted with, for instance, 77 tracks numbered 0 (outermost) to 76 (innermost); one track is used for indexing, 76 are left for data. Each track is divided into sections. In hard sectoring, 32 holes are punched defining 32×128 byte sectors. Soft errors happen and re-readings rectify the situation. With hard error, data are lost. Error detection/correction uses CRC. A smart controller is usually added to the disk and enables symbolic file naming, automatic space allocation, file editing, I/O (dual or memory) buffering, various interfaces, file directory management, space reclamation, various access methods (sequential, random, direct), file copying, error detection and auto-retry on soft errors, and diagnostics; the features include:

- Track 00; it has essentially the table of contents and index reference. Typical information includes file name, type, start address and length, date when created and last updated.
- Space reclamation; when files are deleted, gaps remain; when space is needed, a compaction of gaps takes place.
- File management; it allocates sections to file name; it deletes files, opens/closes file, copies, reads/writes in various formats, updates/modifies, runs diagnostic tests.
- File access method; it may be sequential where continuous sectors are used, random where any byte may be accessed even across sectors, direct where any sector may be accessed bypassing the file management system.

So far, the discussion has presumed a single data stream leaving a single file on a traditional disk drive at rates up to 5 megabytes per second. Newer disk designs allow for parallel streams from an array of disk and drives; parallelism multiples the rates and creates data protection against future through redundancy.

D. Error Detection, Correction, Fail-Soft, and Maintenance

As parts are connected into units and units into a functioning uniprocessor, growth in the likelihood of errors and failures happens with time. Their detection and localization can be performed through observations at the connectivity points, and those selected points vary with applications.

When interfacing a uniprocessor, many devices/components make errors. There are three basic detections of errors: parity, check sums, and cyclic redundancy check (CRC) characters. For correction of error, there are basically two ways: hamming code and cross parity. Detection and correction include:

- Parity where one bit is added to a byte to make the number of bits odd or even; note that two bits changing from 0 to 1 or 1 to 0 will not be detected.
- Check sums where a one-byte check character is added at the end of the block; it can be generated by adding all the bytes in the block together, using an exclusive OR.
- Cyclic redundancy where the 8 bits of a word are treated as coefficients of a polynomial P; $B(X)$ is divided by a generator polynomial $G(X)$ with $B(X) - R(X) = Q(X) \cdot G(X)$ where $Q(X)$ is the quotient and $R(X)$ is the remainder; the remainder, $R(X)$, is appended to $B(X)$ and is called the CRC bits. If an error occurs, then $B(X) - R(X)$ is not exactly divisible by $G(X)$; a chip may be used for CRC.
- Hamming code where for an 8-bit word you add $(\log 8 + 1) = 4$ bits. The 4 extra bits will be parity bits for the different subgroups

UNIPROCESSOR HARDWARE SYSTEM 167

of the 8-bit word. Thus you can detect and pinpoint which bit is in error.
- Cross parity where hamming to each byte and across a block is applied.

Failure occurs when errors are not correctable. Fail-soft is degradation where a component fails while the system continues to operate at lower performance. Failure types occur in sensors, control devices, powers, uniprocessors:

- Sensor failure is detected if unreasonable values are being picked up; if this continues over n times, it is disconnected by weighing the input by 0, i.e., $W_i = 0$.
- Control failure is detected if an unreasonable command (outside the bracket) is issued. A retry is used; if it persists, there is a permanent failure. Then a back-up strategy is used which puts the system in an operation mode and a diagnostic or alarm is generated.
- Power failure halts operation and may occur at any I/O of the three stages (transformer, rectifier and filter, voltage regulator) that take the main line voltage of 120 down to plus or minus 5; failure is noted when instability around a rating is measured beyond a stated percent. With power reset, a bootstrap program is used to restart the system in a table driven scheduling mode where the list of actions to be executed is stored in a table; battery backup can provide immunity for a limited period.
- Uniprocessor failure halts computing operations; a backup one is needed to reinstate operation rapidly until maintenance is done. Failure is noted when reference instructions are not properly executed; instructions include: a) *addressing* such as jump, branch, call, return, halt; b) *data moving* such as move, store, load; and c) *data processing* such as OR, AND NOR, arithmetic, shift, clear, rotate, increment, decrement... .

Eventually maintenance and replacement are needed to reset a uniprocessor system in operation. What goes wrong varies, e.g., wiring fault (short and open circuit, wrong connection), component failure, software bugs, or noise/interference. Testing identifies and treats the failures where:

- Multimeters measure voltage, current, and power, and logic probes verify logic levels; those probes indicate levels rather than time.
- Digital analyzers (logic analyzers), such as multi-channel oscilloscopes, allow observation of multiple interfaces in the system simultaneously before and after a trigger signal; those analyzers emphasize timing information or state information; this presents the

flow of the system's program by monitoring all important circuit components.
- In-circuit emulation allows to get inside the CPU and watch what it is reading and writing; it allows the display of internal registers and checks them against what is expected.
- Signature analysis systems rely on the original behavior of the system and predict what went wrong by relying on a fault tree. Each interface in the system will have its own signature when it is working properly.
- Comparison testing compares a board or device whose performance is in question to a known good one.
- Self diagnostics execute a worst case sequence and observe the results. For instance, the CPU manufacturer supplies some critical sequences of instructions which have been found to stress/fail the system in some cases. This program is run whenever the processor is idle to verify health of operation; check sum and CRC are used to test the memory; feedback check to an input trigger is used; reasonableness checks are also used.
- Stored responses use a computer to emulate or simulate the device or board under test. First, the real system characteristics are stored in tables, then comparison to stored responses is made when the system is subjected to same stimulus.
- Simulation is the functional replacement of a hardware device by a program; for the same inputs, the program gives the same outputs as the device; this is supportive of logic debugging. Such simulation identifies early in the design cycle integration problems prior to product development. Usually, simulation is realized stepwise; first a system-level simulation verifies the system specification and the design concept; next, boards are individually simulated followed by ASICS for functional accuracy; then interoperability of ASICS and boards is verified; finally micro-code and diagnostic routines check for hardware/software mating.
- Emulation is simulation performed in real-time; with emulation, there are break points, halt, re-start and access to register contents at the same time they happen in reality; this tests for any real-time conflict within the system.

V. DISCUSSIONS

The uniprocessor provides the essential building block to the second industrial revolution, information and automation. It began with discrete parts (resistors, capacitors, inductors, transformers, tubes, semiconductors, transistors), and grew into integrated circuits (ICs). From ICs to boards, drawers, cabinets, and systems, the conception of new applications is only limited by the skill and imagination of the human organization. As developed in Part IV, the applications extend from industrial systems, to military systems, commercial systems, consumer devices, and health systems:

- Industrial systems; basically the processor has a) *analog inputs and outputs* with sensors of processes' parameters such as temperature, pressure, fluid motion, and others; and b) *a control loop* to regulate processes and flows through control devices. The processor improves the control strategy based on sensed data, tries out new alternatives, performs reasonableness checks of every input/output parameter and rejects outliers, disconnects errant machines or devices and re-connects as they become fit. The processor makes decisions through rejection and acceptance bands and improves the information quality through filtering and estimation processes.
- Military systems; the processor provides the basis for improved operational capability through the implementation of sophisticated algorithms in radar, sonar, infrared, optical or laser systems; the processor performs a variety of work functions including control mechanisms, as in flight control and similar demanding control requirements.
- Commercial systems; the processor provides for component activities of data processing, telecommunications, and office automation as well as their integration into an overall information resources system; the processor provides for: a) *applications development, technical services, corporate database*; b) *data, voice, text and video communication, local and wide area networks*; and c) *internal and external data services, word processing, intelligent workstations*.
- Consumer devices; the processor replaces electro-mechanical and hard wired logic in a variety of devices such as washing machine cycling, heating control, color TV tuning, word processing, electronic games; it performs reasonableness checks and matches the planned requests with the process at hand, and activates and controls the machine evolution. With this trend, it is a simple matter to link the devices into a system that can be programmed and controlled locally or otherwise through a geographic network.
- Medical systems; the processor performs non-intrusive diagnostics using algorithmic and heuristic techniques and controls precision instruments in surgical actions; the processor monitors human systems for vital signs, such as cardiac rhythm, and detects abnormal functions and sets alerts or automatic compensation. Those systems can be extended in value through geographic networks.

Processor technology has been evolving rapidly enabling the performance of increasingly more sophisticated functions in less space and at less cost. Processing power has become the least expensive resource in a system; it is being introduced in virtually every task and every endeavor by automating most of the processes that require "intelligent" processing. Projections continue to favor dramatic decreases in processor price and space for a given processing power; the chip's yield and area are increasing regularly:

- Price is connected to yield, the percent of good processor chips in a production lot; yield has been on the increase with early problems in production processes, such as contamination, process control, mask defects or alignment being overcome continuously. It is not unlikely that the large-scale computer selling for a million dollars in the past will cost one dollar sometime in the near future. At present, microprocessors are a hundred times cheaper than they were ten years ago for similar performance.
- Higher densities of more parts per single chip of memory logic are achieved by: a) *shrinking the design into smaller chips* or b) *increasing the size of the chip*; the part dimensions are decreasing progressively toward the actual limit (0.1 micron) of the lithographic process which fabricates them; ideally, this allows some 1 billion transistor circuits per chip, a far cry from the 10 thousand transistors per chip in the early seventies. Practical, economic, and manufacturing considerations are leading companies to project 100 to 250 million transistors per logic and memory chip, respectively, by the decade's end. Meeting the challenges of submicron circuit design is the continuous introduction of integrated tools for simulation, preliminary layout, timing analysis, delay calculation, placement, routing, and testing. Meeting the challenges of electron punch through the extremely thin barrier (0.1–0.25 micron) of the electronic switching process, are two experimental approaches, one using multiple barriers into the emitter of a conventional bipolar transistor, and the other using alternating layers of very thin silicon separated by layers of silicon alloys; these approaches inhibit the uncontrolled electron flow and permit the traditional control of switching circuit mechanism.

12 MANAGEMENT PROCESSES IN UNIPROCESSING SYSTEMS

I. INTRODUCTION

Management is effected through communications among the involved parties using an understandable language. The management and communication processes in Part I are now set in the context of a uniprocessor or computer system which implements the synthesis processes in Part II. The managing and managed resources now are the hardware (processors, storage, interconnects) and source code software (instructions, data) units of the computer system which form an automated system.

In uniprocessor or computer systems, a hierarchy of translations occur between human (source code set in high level language) and machine (low level on–off switching, 0–1) language to effect communication and cooperative participation in the execution of the tasks at hand (Figure 12.1). Compilers, built by human specialists, take the source code from a high level language and translate it into object code understandable by the machine.

Compilers produce machine language from source programs; it is done once interpreters produce machine language of source programs piece by piece and is done repeatedly. The ultimate goal is to have the computer understand ordinary language, e.g., English. Some success has been realized with query languages; they allow a request from a database of the data that meet specified criteria; one does not have to tell it how it can be done, just what needs to be done. Ordinary programming language may not have all the required features and various appendages are needed. A data management language handles data storage and retrieval. A data definition language organizes the storage of the data.

Figure 12.2 depicts a top level computer organization and operation. The control directs the activities according to instructions contained in a program expressed by a language form. The total set of programs is the programming environment. Varied generations of hardware and software have been applied to the solution of problems:

172 SYSTEMS MANAGEMENT: People, Computers, Machines, Materials

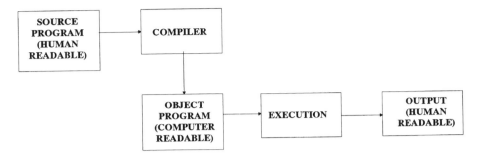

Figure 12.1 Human–computer readable programs.

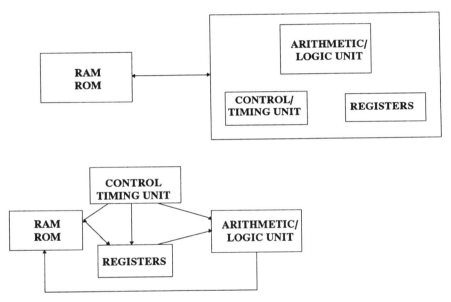

Figure 12.2 Computer organization and operation.

- Hardware; first, the vacuum tube; second, the transistor; third, integrated circuits; fourth, very large-scale integration; fifth, parallel processing.
- Software; first, assembly; second, early Fortran; third, procedural; fourth, non-procedural; fifth, natural language.

Hardware has clearly enjoyed more of the benefits of the maturing hardware and software generations. Hardware designers have long applied advanced computer modeling tools to lay out integrated circuits, circuit topologies and their integration into boards followed seamlessly by automatic manufacturing and verification tools. Software, on the other hand, has suffered

MANAGEMENT PROCESSES IN UNIPROCESSING SYSTEMS   173

flow gaps; principally, design methodologies are still maturing and coding remains people intensive; software reuse and its ensuing productivity increase are gaining momentum helped by object-oriented technology applications.

II. ELEMENTS OF HIGH-ORDER LANGUAGES

Procedural languages (Fortran, C, Pascal, Ada) have two elements: data and procedures (step-by-step instructions for arriving at the desired solution). Non-procedural language is a language that allows the programmer to describe the result without specifying how to obtain it. There are basically two different language structures, object oriented language and descriptive language:

- Object oriented language; objects are set, each having its own data and procedures. The programmer treats each object as an independent entity, which may be residing in its own computer. It receives messages from other objects, manipulates them, and sends messages out. Each object understands messages pertinent to it; the message involves the set of procedures to be executed. A group of objects that share certain properties and behave similarly form a class. A class sets a property of all objects so that they respond to messages by calling procedures to manipulate their variables. For instance, consider a display object; a subclass of an object may be the methods to draw arcs and lines on the screen; for example, a rectangle has variables such as location, side, length; a polygon has location, sides, length, angle. Each object keeps track of its own properties. To get information about an object, a message is sent to it.
- Descriptive language; it hides all procedures from the programmer. It has a database of all the facts and general principles governing the facts. The goal query is given by the programmer and the computer searches to see if the goal is true. This language is used in expert systems. The method of search first tries to match the user's query with one of the rules, it then substitutes specific values from the database into the rule. If the result matches the terms of the query, the search is a success.

There are three basic commercial computer languages that others are a variation of:

- Fortran; it is mathematical formula translating system to the computer with the important "do loop". Fortran is not the answer to handling business payrolls where the need is for sorting items in a data file, merging data files, and generating reports.

- Cobol; it is common business oriented language; it does simple arithmetic, +, −, %, and does sorting, merging, and uses English syntax.
- Algol; it is an algorithmic language. Dissatisfaction with Algol led to Pascal. Pascal emphasizes logic, divides the program into small logically arranged packages and eliminates the GO TO statement that made writing easy but reading difficult; ADA was influenced by Pascal.

Along with application programs, there are programs that make programming easier:

- Operating system (OS); as a traffic cop, it directs the inputting of data from the keyboard/disk, the hardware actions upon the data, and the outputting of data to disk, printer, display, etc. It direct's the CPU activities. There are pieces in OS that deal with special functions such as keyboard or disk.
- Editor; it allows one to input text and modify it.
- Subroutines; these are sets of lines of codes that are used repeatedly by different programs and are put in a library.
- Macros; it is like a subroutine, but performs more than one related function.
- Tool kits; these are several kinds of programming aids.
- Compiler; it translates high order language into an instruction set.
- Assembler; it translates instruction sets into on–off (0-1) voltage levels.
- Linker; it is a program that joins separately or independently developed programs and allows them to run as a single program. Subroutines are placed directly in the program.
- De-bugger; it is a program that helps you find where/when you made a mistake.

A language is a level of abstraction with symbols:

- Machine language, 0 and 1.
- Assembly language; manipulative operations, such as add, subtract, move, compare, jump, branch, on objects such as integers, characters, real numbers.
- High order language (HOL); it masks the internal working of the computer and permits concentration on the problem and its solution. It uses reserved words from the English language to stand for many lower level processes. There are two types of composites: a) action operation composite, including expressions statements, functions,

MANAGEMENT PROCESSES IN UNIPROCESSING SYSTEMS 175

blocks, tasks; and b) data object composite, including arrays, records, files, lists, strings, matrices.

Programmers use operator object composite to develop the algorithms and the data object composite to structure information and form databases. The algorithms simulate the activities of real life objects represented in the databases. Thus, there are two types of objects:

- Data objects; high level languages provide programmers with the tool to describe a data object; we have the instance of an entity. The type of a named entity specifies its attributes, its structure, and the set of operations it can undergo; an entity may be John with height, weight, sex, attributes; the type may be scalar, numeric, character, logical (true-false), or a pointer (address). HOL must be able to process and represent different types of data: a) *integer*, e.g., 1, 2, 3; b) *real, decimal*, e.g., 1.6 or exponent; c) *string*; to hold letter and number, e.g., month 12; d) *boolean*; to hold one of two values, e.g., true or false; e) *array*; to hold items of the same data type, one declares size and type of each array, item position is marked by an index and is selectable through that index; f) *record*; related elements of different types with identifier, e.g., Name: Hassab; Born: 1941; Married: true; g) *array of records*; item position is marked by identifier and index which are recursive trees (linked lists), the basic building blocks are a record object consisting of data components and pointer components which point in turn to other record objects; h) *data typing*; this describes the type of data and kinds of values the variables in a program are allowed to have. To each data type, there is a type of operator that can act on it.
- Operator objects; they are mechanisms for manipulating data objects: a) *numeric*, e.g., addition, subtraction, multiplication, division, absolute value, square root... and functions thereof; b) *logical*, e.g., and, or, and then, or else, if; c) *relational*, e.g., equal, greater than, less than, not equal to, and combinations thereof; and d) *string*, e.g., and, to move in sequence from one operation to the next.

Flow control of operator objects may be:

- Sequential, when the next statement is taken.
- Selective of a set from multiple sets, e.g., branching, if...then.
- Iterative, which repeats a set of statements until a condition is satisfied, e.g., do loop.
- Transfer to anywhere in the program, e.g., go to.

Once data objects are operated on, the results are assigned to data objects through assignment statements. The collection of all statements constitutes a program. Programs are data object declarations and action statements operating on the data objects and controlling the flow of operations. Related action statements and data objects are grouped into units:

- A block is a unit that must be executed in sequence.
- A subprogram is a unit where the flow of execution between units need not be in sequence. It is also separately compilable.
- A module is a group of related data objects and subprograms; modules also have well defined interfaces. There are two types: a) *package*, a mechanism for data abstraction; and b) *task*, a mechanism for functional abstraction and concurrency; with concurrency, tasks are active at the same time; one task may not continue until it obtains some input from another task (rendezvous); then the two tasks require synchronization.
- An exception is a distinction between normal and abnormal events/conditions and the use of exception handlers to handle them when they are raised.

A basic program is depicted in Figure 12.3. Programs are not peculiar to computers. There are many instances of programs; it may be a paper being worked on as it passes through an office system until it reaches the stated objective; it may be cars driven to follow a route until they reach their stated destinations; it may be a large coordinated group of tasks, as in Chapter 18, aimed at reaching a set of goals within stated constraints.

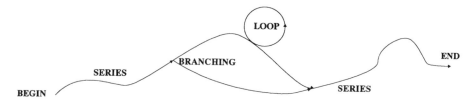

Figure 12.3 Basic program operations.

III. HUMAN–COMPUTER COMMUNICATION: COMPILER ORGANIZATION

A compiler is a program that interprets or relates a high level language program from a human into a machine instruction set. A compiler includes a scanner, a parser, a type checker, a code generator, and an optimizer; its front-end section performs vertical semantic and syntactic analysis; it also generates

MANAGEMENT PROCESSES IN UNIPROCESSING SYSTEMS 177

intermediate representation of the language; its back end allocates storage, assigns registers, and generates machine code. The front end comprises:

- A lexical analyzer which scans the program for operators and key words and forms tables. The scanner reads each character of the source code and determines the meaning following the grammatical rules. Each key word in the program has a fixed meaning to which the scanner attaches a token (begin, end, if, operators, number...); punctuation helps the translator to understand the structure. For words in the program that are not key words, e.g., program name, the scanner attaches a label to them and lists them with their attributes.
- A syntactic analyzer takes the tables and forms trees using a parser. The symbol table describes the objects, their types, their values, their structures, their names, and the environments in which these objects are visible. The parse tree describes the manipulation of these objects and the flow control of the manipulations. The parser takes tokens as they come out from the scanner and arranges them in a tree to allow the computer to decipher the structure (logic) of the program. The tokens may be parent, child or a child of a parent who has its own children. Groups of tokens serve a common purpose under a common block; blocks mark the major pieces of the program. A particular sequence of tokens can result only in one tree. If the parser encounters a sequence of tokens that don't fit the rules, an error is flagged. With the interpreter, the parser sorts enough tokens to make a logical expression and executes it.
- A semantic analyzer performs type checking and associates each operand with its type; now the organized tokens are examined as to violations of rules of language like type checking; this serves also to give information about tokens, i.e., meaning depends on data type.

It is desirable that this intermediate representation be independent of both the high level language and the machine's architecture. This helps when an existing compiler is modified to support another language or is ported to another machine.

The back end comprises:

- Register allocation determines which data objects should be stored and allocates registers to variables with high counts.
- Storage allocation and addressing are done for static data objects and temporary locations for storing intermediate results.
- Code generation decides which instruction to use, their sequential order, and which variables to assign to registers; those are kept in a code template; much of the code generation is machine dependent;

this phase also optimizes the code sequence for a sub-tree, i.e., replacing two jump instructions with a single jump and eliminating identities (x = x1). In optimizing for efficiency, it searches the target code for extraneous commands and, for commands that cancel each other, paying close attention to loops and subroutines due to their repetitive use. The code generator dismantles the tree to generate instructions, it maps the tokens into sequences of machine instructions that will direct the computer. This is done by a process known as traversing a tree. This obeys the rules of the language used to write the source program. One approach is to begin the translation from the top left of the tree, then move down to the right around the tree's perimeter doing the children first before coming back to do the parent.

A model of compiler implementation follows. This is the structure that implements the concept and rules of the language. For instance, Fortran has a fixed storage allocation, i.e., no arrays with dynamic boundary. The model elements include:

- Data objects implementation; to implement them, the compiler needs to determine their routine representation, their access mechanisms, and their storage allocation. Representation is set as bits (values) in memory locations with data objects, names, and their addresses; data objects of different types require different instructions to interpret their representation in linear memory space. Each data object has a lifetime (time the name is attached to memory location) and a scope (part of the program over which a name is usable). Types are the structure of data and the set of operations allowed on data; most machine languages are typeless; the type is determined by operators applied to it; during compilation, type checking is done:
  - Built in types include integer, real (fixed and floating), Boolean, and character; these correspond to built in instruction types in the computer.
  - Scalar types include built in types and enumeration types, i.e., a set of words represented by integers at run time like a type of day (Sunday, 1; Monday, 2; Saturday, 7); sub-types (Ada types with range constraint); or derived types (Ada types with inherited characteristic of parent).
  - Structured types include how data structures are represented in one-dimensional memory and facilitation of access to them. Arrays are components stored in consecutive memory locations and are accessed by an index relative to the beginning of the array; the index is a measure of distance; the high level language index denotes the location in memory where you find for the

address of the first component number of memory bytes per component, number of components in the kth dimension. Records are stored by examining the size of each component type variable of the records and its position in the hierarchy of the record; the computer generates the set for each component relative to the beginning of the record. For access types, data objects representing pointers (integers) are put in a set of bits in a memory word; the value is the address of a memory location where the data object pointed to by the pointer resides. Common array, record, and access types can be used to access other types including matrices or two-dimensional arrays, strings, stacks, queues, linked lists, trees: a) *queue* is a list of items in which items may be inserted only at one end and deleted or selected from the other end; b) *ordered lists* may insert and delete anywhere in the list, not just the ends; to avoid bumping, each element is linked to the next element through a pointer, thus avoiding the necessity for physical storage in adjacent locations; double-linked lists can be scanned forward and backward; c) *trees* are built using records and access constructors; the nodes and leaves of a tree may be linked by pointers; traversal of those trees uses algorithms.

- Operators implementations; high level language operators are implemented onto corresponding machine instructions that are executed on the digital devices discussed in Chapter 11. For instance, a relational operation is executed through a compare and conditional branch. A multioperand expression in high level language is implemented by a sequence of instructions; the order of sequence observes the precedence rules of high level language; what the operator should do is encoded in the operand, i.e., real numbers or matrices.
- Flow control implementation; like high level language, the control of execution flow is sequential. It is implemented in the program counter which contains the address of the next instruction to be implemented; selection, iteration, and go to statements are implemented by loading the result of the instruction in the counter.
- Environment implementation; subprograms are high level language mechanisms to describe the environment (scope) in which data objects are visible. At run times, an environment is represented by a set of locations in storage called the activation record; the context of an environment is all the data objects making it up. Environments may be nested within each other and thus visible (referenceable) within the next. The compiled program needs to have all the addresses of the referenceable activation records. Parameter passing between environments requires the caller subprogram to push the data parameter onto the top of the stack before an environment is

switched. With concurrency, while one user program is waiting for an I/O to complete, another user program may use the CPU; the two programs are running concurrently. Ada calls its concurrent modules, tasks; in synchronization between tasks where at a predetermined point, a task needs to accept/send entry to another task, and the other task must reach that point or the first part must wait, depending on the availability of processing units; tasks may be in a wait/run state and in queues waiting for rendezvous. Once an entry is accepted, the kernel switches the task from wait to run state.

- Exceptions implementation; abnormal conditions or events may occur during the execution of a program; if a handler of the exception does not exist, the program terminates; the handler's catalogs are set in tables.
- Storage allocation implementation; so far we have discussed data structures of various types and how they may be represented in machine storage and how they may be grouped into data areas to represent an environment. For static allocation it is done at compile time, where addresses are given to each object; this requires knowing all possible objects, structure and size, and single occurrence. For dynamic allocation, it is done at run time, by code generated by the compiler; the objects are dynamic; it supports, for instance, recursion. There are two types:
  - Stack handles last- in first-out order for allocation/release of storage good for recursion and procedure calls.
  - Heap does not depend on any specific order, and requires a higher overhead; records and arrays are stored using heap and pointers; the pointers are kept in a stack.

For addressing modes, HOL programmers declare data objects and their names, their lifetimes, and their scopes. The compiler assigns storage by associating a name with address. The allocation may be done from a stack or a heap where addresses of data stored in the heap may be stored in the stack.

Instruction set design and compilation are linked. Compilation maps high level language constructs into a single or a sequence of instructions recognizable by the firmware and hardware of a machine (Figure 12.4); each instruction is executed by a set of microinstructions. The machine architecture matches the instruction set to the machine organization and technology; factors include machine cycle time, spectrum of processed unit, memory, and microinstruction execution. A compiler writer matches the designs of the model of implementation to the design of the instruction set; one is concerned with that translation, data types, and addressing modes; one desires the instruction subsets for each to be orthogonal (independent) and not conditional (special cases). A machine with an instruction set that is a superset of the instruction set of another

MANAGEMENT PROCESSES IN UNIPROCESSING SYSTEMS 181

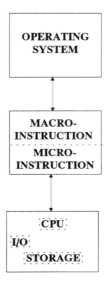

Figure 12.4 Layers in instruction sets.

machine is said to be an upwardly compatible member of the same family, and its software is portable.

Instruction set types include:

- Complex instruction set; each instruction has powerful operation and addressing modes that trigger complex sequences of firmware and hardware activities.
- Reduced instruction set; each instruction is selected to match the cycle time of the machine; it is simpler to design with better cost/performance. If a CPU has to be idle while accessing memory, then a complex instruction set is more efficient unless the memory and CPU activities can proceed in parallel. CPU firmware executes out of a CPU-resident fast control store. Narrowing the semantic gap by means of more complex instruction sets eventually leads to diminishing returns. First, the instruction mix statistics show that only a few instructions are executed frequently. Second, the compiler must do the match to many different instances of the same construct in high level languages.

Categories of instruction executions include:

- General purpose binary integer, e.g., move, clear, increment, decrement, compare.
- Logical and bit operations, e.g., clear, set, test, logical AND, OR.
- Unconditional and conditional branches.

- Jump to and return from subroutines.
- OS instructions, e.g., halt, wait, return from interrupt, trap, breakpoints, top.
- Floating point instruction.
- Procedure call instructions for calling common user created procedures for switching program units.
- Queue instruction for doubly linked queues.
- Index instruction to calculate index of elements within arrays.
- Address passing by a caller to a called subprogram by pushing addresses onto the stack.

Instruction set design methodology proceeds to:

- Construct a design set based on intended use of machine.
- Construct a compiler based on the design.
- Take measurements on code and addressing modes.
- Combine heavily used instruction sequences into one instruction.
- Repeat steps.

The instruction mix is a measure of frequency of usage of each instruction in a set of test programs; the results vary depending on a) *instruction set*, b) *compiler*, c) *application area* (scientific, business), d) *test programs measured*, and e) *test program segments measured*. Measures of overall efficiency include for each instruction: a) *overall mean*, b) *overall standard deviation*, c) *variance due to application area*, d) *variance due to program*, and e) *variance due to segment*. Instruction mix is based on instruction set; by putting instructions performing similar functions into groups, comparison of instruction mixes based on different instruction sets can be made. An order of prevalent computer usage is:

- Instructions concerned with information movement (40%).
- Branching instructions (20%).
- Test comparison (10%).
- Arithmetic increment and decrement (10%).
- Order of prevalent activities of the computer executing program (10%).
- Calls and return (10%).

IV. OPERATING SYSTEM: FUNCTIONS

The local manager of uniprocessor resources is called the operating system. The operating system (OS) is layers of software on top of hardware which lets programmers concern themselves with algorithms and data objects. Standardized calls for the OS services would permit the desired divorce of the software from the hardware and its potential re-use on new improved hardware

MANAGEMENT PROCESSES IN UNIPROCESSING SYSTEMS

platforms. Similar to HOL, OS introduces overhead, triggering the need to move some of its functions into firmware and hardware. OS is a collection of programs that controls the use of system resources. It allocates main and secondary storage and accepts I/O commands from devices and manages files. It forms an interface between applications programs and resources. It insures that all resources are busy simultaneously. The OS multiplexes multiple virtual machines onto the physical machines.

The OS manages memory, processors, devices, and information (program or data). It keeps track of resources, distributes and schedules them, allocates them and reclaims them. Its memory manager operates on virtual memory, main memory, and secondary memory as a single unit; the user views storage as a continuous address space; the memory manager determines which active program should be allocated to memory, when and how much. The processor manager schedules the processes in a job and schedules the processor to execute the processes. Its information manager generates, stores, retrieves data in a file system (open/close a file; keeps track of location, status, size, use and access rights).

The classification of an OS may be done by application or configuration. By application, it relates to having the OS schedule resources and dispatch programs; depending on classification; emphasis on these functions and policy of implementation differ:

- General purpose OS satisfies the needs of batch and time sharing users.
- Real-time OS responds to each event within a prescribed time length.
- Transaction-oriented OS reads and updates databases.

The real-time OS responds in milliseconds, while others respond in seconds. The time-shared OS is a multiprogrammed system that appears to each user as if he/she has the whole system at his/her disposal; memory management provides virtual memory access right. Real-time OS is like time-shared except that the input data is mostly high rate, the response must be quick and must insure that data is not lost due to lack of buffer. To insure quick response:

- Processes reside in memory; otherwise, the system requires a sophisticated paging system.
- Processes are arranged in levels of decreasing priority.
- The system designer is allowed through preemptive priority to set the order.
- High resolution interval timers generate events under OS control in real-time.
- Files are stored in contiguous disk areas to minimize access time.
- The scheduler may be interrupt driven or cycle time driven.

- Too many interrupts can cause time consuming context switches; the context of a process that is going from running to wait state is saved.
- In cycle time, the longest time of consecutive executions must be less than required response time.

A transaction OS is behaviorally similar to a time-sharing OS; both systems require efficient context switches. When classification is done by configuration, classification is then according to the hardware or logical configuration on which the OS runs; initially, we had a uniprocessor OS where a multiprogram OS runs to control and schedule the resources; as technology advanced, processors were connected to multiprocessors and networks:

- A multiprocessor is one in which processors share a single memory; like a multiprogramming OS, a multiprocessing OS switches processes, suspends processes, and responds to interrupts. However, multiprocessors must deal with processes executing in overlapped (not interleaved only) time. A uniprocessor inhibits access to shared resources during execution; a multiprocessor has the problem of being excluded from accessing shared variables at the same time. Thus, synchronization of clocks is required; system initialization, interprocessor interrupts, segregation of processes and data between processors need to be controlled.
- In network operating system (NOS), processors on a network do not have shared memory; they communicate by sending and receiving messages; pipes and mailboxes are useful communication mechanisms. They are message buffers in virtual memory; they act as files and I/O devices and use the same calls; they allow the same interprocess communication as in a uniprocessor. NOS may allow programmers to define explicitly the logical connections among processes and abstract them from their physical connections; NOS allows flow control and message routing and long time delays.

Independent of the differences in application, a common feature of operating systems is their structure into modules that have to be synchronized and protected while they communicate among each other.

V. OPERATING SYSTEM: ORGANIZATION AND ACTIVATION

As the local manager, the OS activates and deactivates processes which are cooperating and competing for resources. The OS has modules that manage files, execute terminal commands, respond to interrupts, execute communication, do timer services, perform scheduling, etc.; there is also a module (kernel)

MANAGEMENT PROCESSES IN UNIPROCESSING SYSTEMS 185

that manages the modules. The OS manages and controls these and their resources (hardware and software):

- Software resources management insures that the software resources are protected, synchronized in their use, and can communicate. There are two types: a) *processes*, a set of grouped activities and b) *packages*, a set of related data structures.
- Hardware resources management optimizes the use of physical resources by system software, thus relieving the application programs of such concerns.

In software resource management, processes are the activation of an executable program unit. Once activated, a process can be in different states either actively executing, ready for execution, or waiting for synchronization signals. The OS defines a context to describe the state (control and status information). Some context information resides in the main memory for quick reference; it is a process table and contains an ID, size, location, events, schedule, and timing for each process. The OS scheduler determines the state and priority of the process and the length of time it has been in the current state; the table is also used to post synchronization and events to/from other processes. The other part of a process context may be kept on a disk with a pointer kept in the main memory table. The access environment is an address translation table that maps the process tasks' virtual address space into physical memory. The disk may store information files that have been opened, accounting and quota information, and the saved hardware context.

In hardware resource management, the hardware context (part of control and status information) is placed in the hardware register for functional and performance reasons during process execution. When the process is forced out (preempted), the OS saves the hardware context in the main memory or on the disk. The hardware context of the next process is obtained from storage and loaded on the machine register. The hardware context usually consists of a processor status word, a set of registers for temporary results, a counter to point to the instruction, and a set of registers defining the access environment (address mapping).

Information structures are made up of data structures (user and computer physical memory space) and control structures (resources and activities of the system). Only certain procedures can manipulate the information structures. The two form a package or monitor and provide a fire wall against intrusion by other software.

In software, there are two types of abstractions: functional and data. In functional, the execution on stream is from process to process, carrying out a function. In data, the execution stream is from package to package, accessing the needed data. The functional has overlapped access environments because the processes share a set of resources, and a protection structure is required. With data abstraction, the package manipulates its resources on behalf of another requesting package, and there is no overlap. Protection in functional abstraction is achieved through privileged instructions, access rights control, and privileged layers. This layering may break due to an error in procedure, and the privileged layer may be accessed. Data abstraction is granular, but demands many more switches from one module and its access environment to another module and its access environment; this requires firmware/hardware for efficient environment switches.

In functional abstraction systems, processes are synchronized with each other by means of events; they communicate with each other through messages. A message is a body of information copied from the access environment of one process into the access environment of another process. Functional abstraction systems provide facilities for sending messages and primitive operations to wait for any message or particular class of messages. Packages communicate with each other by passing capabilities through a communication port. This method is more efficient since there is no need to copy the message.

The way protection is structured to create access environments for the modules impacts addressing, access environment context switch, privileges, rights and type checking, entering and exiting from another access environment, data movement across access environments, multiprogramming and concurrent process addressing, communication and support:

- Switching access environment; when switching context, the data in the internal registers of the processor must be saved. Also, the switching requires a new access environment (segment or capability list of the new context). With hardware support, associative caches are flushed and reloaded or have them already dedicated to the new context. The hardware can save the current processor status word and program counter and loads the new; thus, interrupt and trap vectors are stored on one location area.
- Privileges, rights, and type checking; bits designating the access rights to the segment or object are associated with the segment descriptor or capability. In a functional abstraction system, instructions may be grouped into privilege and non-privilege instructions.
- Entering and exiting from access environment; it occurs: a) *when a routine in one environment requests the service from another environment*; b) *with interrupts*.
- Data movement across access environment; in functional abstraction, the two access environments may interact by having two sets of mapping registers, each belonging to one environment. Special instructions move data across registers. In data abstraction, the capability is transferred from one package to the next.
- Intermodule synchronization; it is done by means of events. The intermodule communication is through messages (copying of information from an access environment to another access environment). Functional abstraction systems pass messages and events; data

MANAGEMENT PROCESSES IN UNIPROCESSING SYSTEMS 187

  abstraction systems pass capabilities; since there is no sharing, the many objects demand a large address space, and switches from one module access environment to another become much more frequent; the advantage is a uniform protection system.
- Multiprogramming and concurrent processes; prevention from accessing the same shared variables concurrently is done through inhibit interrupt up to a certain time limit in a uniprocessor. For shared memory, a lock is set when the shared variable is being used and an unlock when not. A problem arises when a low priority process waits indefinitely and then continues to test the lock, thus wasting processor and memory cycle. A semaphore, signal, and wait concept is used; when a process needs a resource, it is put in a queue and a signal to wake up the process is sent when the value of the semaphore is less than or equal to zero. Another situation requires awakening of processes when an event occurs; when the event occurs, the processes in the event queue are put back into the semaphore queue and allowed to use the shared resource. Implementation may be in firmware/hardware or software. Different OSs support many varieties of synchronous services, but all are variants of a response to an event that requires the system to change its status (run to wait, wait to ready). The systems rely on lower level mechanisms to implement these functions. The first level mechanisms test flags in control data structures, handle system calls and software interrupts, and engage new processes. These mechanisms rely on still lower level mechanisms (context switches, linked list manipulations, stack manipulations).
- Communication and support; communication is sharing of information; it involves the transfer of data asynchronously from one process to another. Events (flag one bit, interrupt, time out) are particular data and form wake-up mechanisms. There are varied ways for communication:
  - Messages; they are interprocess communication where buffers are used; each message is moved at least twice, once from the sender to the buffer and once from the buffer to the receiver; a mailbox is a buffer in virtual memory; with a mailbox, a process does not wait for response. Messages provide for asynchronous communication; remote process calls, on the other hand, require synchronized communication.
  - Shared memory; it avoids buffers; however, it requires mutual exclusion synchronization and tends to be more susceptible to corruption than message passing when more than two processes are involved.
  - Pipes; they are one form of shared memory communication; it is a one way interprocess communication channel that acts like an open file.

- Shared files; files can be shared among users; that is, many users may share information.
- Ports; they are message broadcast facilities to which processes attach and receive their own message.
- Links; they specify the allowable paths over which messages may be sent to processes; links control the interconnectivity processes in a network. A process may own many links and may group a subset into a channel.

In the management of hardware resources, the OS devises multiple virtual machines that are time multiplexed on a single physical machine. As a manager, OS keeps track of status, schedules, processes on the processor, and allocates/deallocates processes from the processor, memory, I/O:

- Processor management and support; management here means assignment of a processor to processes; the basis may be periodic intervals, on demand due to interrupts, or when processes stop while waiting for synchronization signals. The support is in the form of maybe queues for waiting/running processes, scheduling priority levels, interrupt mechanisms, timer facilities, switching mechanism, system calls by which processes can request for synchronization.
- Memory management and support; in multiprocessing, the need exists to use the memory efficiently (no fragmentation of information and no empty spaces). The concept of virtual addresses (dynamic relocation) meets the requirement. A process views its access environment as a software resource and accesses objects within the environment. It shifts addresses. These addresses are referenced to a base address so that mapping between virtual and physical can occur. The hardware support is a set of registers storing the mapping; pages (physical partition of a fixed size) are used for storage (512 bytes, be it RAM or disk). Locality of reference implies that pages of a working set need to be in memory; the others are on demand when needed.

VI. OPERATING SYSTEM: PORTABILITY AND EFFICIENCY

With portability, software on existing processors can be transported to new processors of the same family. The main features of a family of processors are common:

- Instruction set.
- Addressing.
- Instruction privileges.

MANAGEMENT PROCESSES IN UNIPROCESSING SYSTEMS

- Time dependent interactions.
- I/O reads and writes.

Software is portable at three levels: source, assembly, and object. At all levels the data format must be the same in a family of processors (floating point number precisions, addressing, ordering and alignment of bytes, words, long words, structures):

- Source level portability does not require the instruction set of two processors to be the same. Source code level portability requires each processor to have its own compiler, assembly, and loader.
- Assembly level portability requires that the code be run through the second processor's assembler, linker, and loader before being executed.
- Object level portability requires the same instruction set, address layout, and data formats; then disk program from one processor can be run on a second processor.

Portability requires that I/O drivers, the file manager, and the supervisor remain functionally the same, that is the I/O calls should invoke the same type of services:

- The interface between the OS modules and the ported process should remain the same.
- Consistent addressing among processors requires that the virtual address layout and address space organization be the same.
- The timing relationships between various processes and parts of the systems such as I/O devices must remain the same; portable code must not depend on a particular timing relationship.
- I/O read/write must be chosen from programmed I/O commands or memory mapping; in I/O, the same instructions for read/write memory are used for registers.

For efficiency, well used activities are set in hardware/firmware. For instance, a processor spends about 40% of its time in the operating system kernel. The OS optimizes the use of hardware resources, sets up virtual resources for each user and assigns full control to the user, one at a time. Hardware enhancements include:

- Queue administration; rather than using software for search and sort, hardware implements the queue using content addressable memory. This enables efficient search of the memory location containing the desired memory content. The OS posts a request for

service in one machine cycle and returns the highest priority request for servicing in another cycle.
- Context switching; switching the control of the processor from one process to another by swapping the contents of processor registers. This occurs when a process being executed is unable to proceed because it is waiting for response from another process (I/O), because of an interrupt. Some processors provide dedicated hardware to maintain their contexts; they use multiple copies of hardware register sets.
- Interrupt facilities; when an interrupt arises, a time latency in responding occurs (time to change the context of the processor plus determination of the source of the interrupt plus branching into the selected interrupt handling routine). Most systems have a stack facility to help save and restore the hardware context; the stack has a first-in–last-out mechanism which may be extended into main memory. It is possible to have a very fast time response for individual priority interrupts by associating a separate hardware register set with each interrupt level; thus, switching occurs from one set to another. For I/O interrupts, the interrupting device may be identified by software or hardware polling techniques. In software, the routine interrogates sequentially, which is time consuming; in hardware, the polling is done simultaneously. Each interrupting device has a unique index into an interrupt vector table; the address found in the table points to the device service routine. The same approach is used to handle any interrupts, traps, or faults.
- Memory management hardware; virtual memory systems have the complete virtual to physical address translation tables in the main memory. The translation time is reduced by storing the physical address translations most likely to be used in a high speed cache called an address translation buffer. The buffer is used for only one process and flushed for use by another process. The OS has its own buffer. The memory management unit (MMU) examines all memory accesses; if it does not reside in main memory, the MMU sends an abort signal to the CPU which stops execution, returns the registers to their original state, initiates a search of the needed page from secondary memory and copies it in main memory. Those operations are transparent to the user.
- Real-time clock; a need exists to maintain and update a timing list to activate and deactivate processes after a certain amount of time elapses. It is time consuming if done by the OS. Content addressable memory or a microcomputer is dedicated to this function.
- Communication; a microprogram is used to manage all the internal and external communications for a multiprocessor architecture. It consists of receiver and sender hardware, and a scheduler with its wait queues.

MANAGEMENT PROCESSES IN UNIPROCESSING SYSTEMS

All the preceding hardware enhancements may be done in firmware. An OS implemented in firmware may improve its time from 3 to 6 times that of software implementation. Such efficiencies are counter to the ease of portability.

VII. FAULT-TOLERANT PROCESSING SYSTEM

This is computing despite faults. Faults have their sources in design mistakes, and they appear as hardware/software faults; component failures appear as hardware faults, and operator interaction mistakes appear as procedural faults (Figures 12.5 and 12.6):

- Software faults (50% of down time) result from design errors by incorrect combinations of instructions; the interactions between instructions develop many internal states. Additionally, every correction may introduce subtle errors in other parts of the program. Software reliability is the measure that a software system will perform its intended functions for a specified number of input cases under stated input conditions.
- Hardware faults (20% of down time) are triggered by early failures due to improper design, manufacture, usage of components; weaklings are detected by burn-in test. Following that stage, failure is low and constant until the hardware reaches the wear out period.
- Procedural faults (30% of down time) are due to inadequate/incorrect documents and inexperienced operators. With the turnover of operators, those faults continue.

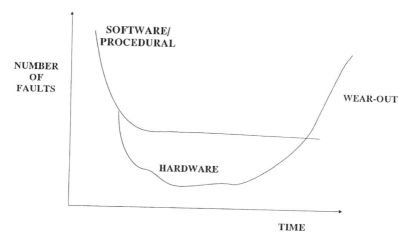

Figure 12.5 Frequency of faults vs. time.

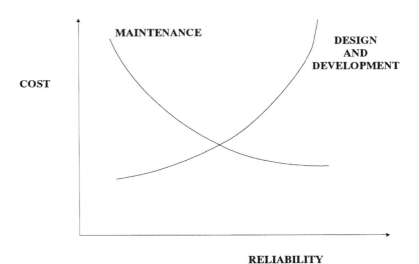

Figure 12.6 Relationship of cost factors to reliability.

The basis of fault tolerance is redundancy. Redundancy allows computers to bypass errors using hardware/software/time mechanisms:

- Hardware redundancy uses an additional circuit to detect and correct errors.
- Software redundancy uses an additional program for fault detection and diagnosis.
- Time redundancy uses a retrial of an erroneous operation.

Hardware redundancy may be :

- Static; it is an over-design by duplicating the components so that if a failure occurs, the other component is good enough to do the job.
- Dynamic; when a fault is detected, diagnosis takes place to a lower replaceable unit followed by replacement of the offending unit. The replacement unit may be active (powered) or passive (unpowered); recovery requires restart from the good data.

Software redundancy may be:

- Static; it is duplicate software on duplicate hardware which use a majority vote to decide what to do.
- Dynamic; it is used with hardware dynamic redundancy; software copies of the state of the system are made during normal system operation and are used for correction; a check point is a scheduled

MANAGEMENT PROCESSES IN UNIPROCESSING SYSTEMS

point in the execution sequence when the system saves its states. When an error is detected, the system rolls back to the latest software copy of the states.

Time redundancy involves a retry to correct a transient error as in I/O transfer.

Fault detection techniques include:

- Hardware fault-detection techniques; they are done by:
    - Replication checks; the critical component is replicated and its output is compared. This technique detects failure, not design fault, in the component.
    - Coding checks; they are parity checks and cyclic redundancy codes to check serial data streams.
    - Timing checks; it relies on a watchdog timer to guard against program faults; it requires that the main program periodically resets it; resetting it happens if nothing unusual occurs to take the main program away from its normal sequence.
    - Exception checks; the program operates within sets of prescribed constraints; if programs have faults, it may throw the system outside constraint; detection of an error raises an exception which calls on exception handling programs, e.g., stack overflow, improper address alignment, division by zero, overflow, illegal operands (software triggered).

For software fault detection techniques, software often relies on hardware exception to detect an error; one must observe the behavior of a system to detect malfunctions:

- Function checking of a process; it is a check on reasonableness of output, given an input.
- Control sequence checking; it is a software fault that causes an incorrect execution sequence. A branch allowed scheme is when a check bit (0 or 1) is reserved for each word in main memory that cannot be referenced by a branch instruction:
    - Relay runner scheme protects against illegal jumps using a password before passing control.
    - A combination of relay runner and watch dog timer is used.
- Data checking; it is done on usual data and programs residing in memory through:
    - Hamming correction code for programs.
    - Check on data structures.
    - Check on parameters passed across programs.

In fault recovery, the speedier the detection, the easier is the recovery, since error propagation is contained. We may have:

- Full recovery where appropriate spare subsystems are available to replace faulty ones. It involves:
  - Error detection in milliseconds to seconds.
  - Switch-in spares to insure continuous operation.
  - System recovery where, following hardware recovery, software recovery is initiated when the states of a system are saved at points during execution; those points form recovery points. When an error is detected, the backward recovery technique resets the system to a previously recorded state and restarts the program execution. Checkpoint is data located in a different processor which has all the information needed to restart the process in the event of a failure in hardware. Another approach is to use multiple versions of the software and use an acceptance criterion of the results. A more involved recovery is to restart/reinitialize the system from a predefined state (cold start, warm start...).
  - Diagnosis and repair of a fault intermingled with normal processes; fault diagnosis localizes the failure to the lowest replaceable unit; it uses test sequences to help isolate the failed unit. Microinstructions are used to test the internal logic operations of a CPU; the result of each microinstruction step is monitored for evaluation. Macroinstructions are used for memory devices and I/Os.
- Degraded recovery; here, essential operations are continued because of lack of spares.
- Safe shutdown; here minimal operations are continued in order to bring the system to a safe state allowing repair.

VIII. DISCUSSIONS

Language is the means through which the communication in Chapter 6 occurs, and the management in Part I are realized. Five interrelated elements make up a language:

- Meaning (semantics); what we want to talk about or represent, real or abstract things. There are real things, e.g., trees; or abstract things, e.g., confusion; characteristics, e.g., red; activities, e.g., moving.
- Linkage (syntax or grammar); sets of rules on how to link objects.
- Medium; vehicle (voice, frown, bark) to say the meaning.
- Expression; what we end up coming out with, what we say.
- Participants; parties to the communication of meaning.

MANAGEMENT PROCESSES IN UNIPROCESSING SYSTEMS 195

In computer languages the elements become:

- Meaning depends on perspective, e.g., user, software, hardware perspective.
- Linkage is the rules.
- Medium is at the basic level, on–off (0–1).
- Expression is the program.
- Participant could be machine–machine, man–machine, software–hardware.

While human instructions are rich in variety, adept at figurative as well as literal expressions and full of diversity, processor instructions are relatively simple and straightforward, and based eventually on multiple combinations of two states, 0–1 or on–off. The gap separating the two instruction types is filled by meticulous decomposition processes from human to processor and machine, and integration processes from machine and processor back to human (Figure 12.7). Lying between the human and the instructions set in programs and the millions of devices that can execute those instructions is a local manager of those hardware resources called the executive or operating system which implements automatically the system synthesis processes in Part II. The functional capabilities of the local manager include:

- Task management; it provides control (suspension, resumption) of processing tasks within the program and provides the task absolute priority based on its priority within the program and the program priority.
- Task scheduling; it selects the next task for execution and sets up that task's execution context.
- Task synchronization; it permits multiple tasks to cooperatively share a set of resources through synchronization and exclusion using semaphores.
- Intertask communication; it provides mailboxes for asynchronous exchange of messages between tasks.
- Memory management; it allocates blocks of memory for use by a program and retrieves those no longer in use.
- Interrupt services; it provides a mechanism to service interrupt vectors from the CPU.
- File management; it provides a filing system, access mechanism, and identification of the physical location of the file.
- Inter-program communication; it provides the capability for users programs within the system to establish communication channels and to exchange messages over those channels.
- Device management; it initializes the interface, permits referencing of devices by logical device name, provides device handlers for

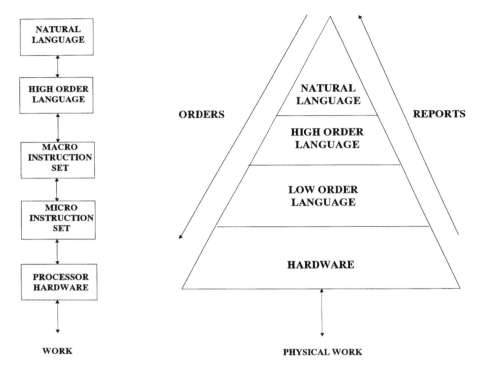

Figure 12.7 Decomposition of human instructions down to microinstructions, understandable and executable by the hardware.

message communication, performs buffer management of I/O data, provides device driver for read and write operations.
- Timer services; it supports timed task scheduling policies and provides for periodic scheduling of tasks and administers the policy for task overruns of their allotted cycle.

13 INPUT-OUTPUT MANAGEMENT IN A UNIPROCESSING SYSTEM

I. INTRODUCTION

Management is focused now at the input/output junctions in a uniprocessing system. Data is the item of transfer across the junctions, data that represent objects or operations on the physical world performed through some transduction mechanisms; in that sense, data are the common currency in flexible automation. The input/output control system is a component of the operating management system and its most important part. The reasons for such an operating manager within the operating system (OS) are to enhance processor productivity in the interaction with its outside environment, in the utilization of its resources and include features to:

- Overlap set-up time of one job with the execution of another.
- Arrange for concurrent processing where the storage requirements of each job are different.
- Separate computer program from devices.
- Substitute high performance auxiliary devices for slower ones.
- Standardize data formats for interchangeability of data among systems.
- Standardize procedures to reduce operator error.

The OS functions are to schedule a job based on FIFO, or priority, while monitoring availability of necessary resources:

- Allocate resources (main storage space, I/O devices, files of data).
- Dispatch programs, i.e., prepare system for execution of a particular program and transfer control to the CPU.

- Assist/replace/communicate with the operator, to cancel improper service requests, to reassign priorities, to stop system in case of control loss, to tell operator of a start/finish of a job, to request mounting new storage, to report errors.
- Diagnose errors/failures and recover device failures (intermittent or permanent) with a small loss of work during operator error.
- Record statistics on usage of resources.
- Store and retrieve data, its main activity.

The operating modes of an OS are:

- Batch processing; it is the execution of one program's transactions before dealing with the next program.
- Multiprogramming; it is execution of two or more programs concurrently using a single CPU through interleaving of the transactions of each program. This is efficient when different resources (CPU, files) are required.
- Multiprocessing; it uses more than one CPU leading to: a) *redundant CPU to improve reliability*, b) *segregation of system functions and dedication of processor to each,* and c) *increasing CPU power*.
- Sharing; this calls for rotation of CPU usage by each user.
- A combination.

II. INPUT–OUTPUT PROCESSES IN A UNIPROCESSING SYSTEM

The processes of input–output in a uniprocessing system involve I/O supervisor, I/O channels, I/O controls, I/O device requirements, opening/closing files, device allocation, scheduling, device inventory, data staging, data storage, and buffering.

Basically, computer systems execute the assigned job requirements (Figure 13.1). They are composed of the same types of components; CPU, storage, I/O, and displays. They have:

- Stored programs where some are interpreted as instructions, others as data.
- A CPU that executes programs from consecutive (unless there is branching) main storage locations; the CPU has a control section to interpret instruction, an arithmetic unit to execute the instructions and registers to hold intermediate data results.
- Main storage to hold programs.
- Certain instructions for control program use only.
- Channels that can do I/O operations with limited CPU attention.

INPUT-OUTPUT MANAGEMENT IN A UNIPROCESSING SYSTEM

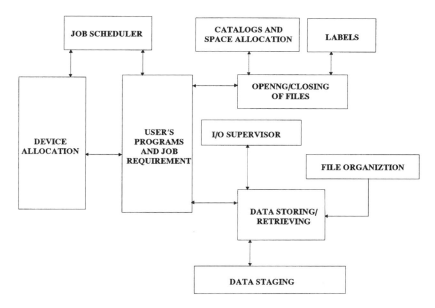

Figure 13.1 Sample I/O activities.

- Control units to attach I/O devices to channels and to control the activities of each type of device.
- Support for the activities of auxiliary storage devices.

The I/O supervisor is the officer that controls traffic at the intersection of the CPU programs and channel programs. He insures efficiency, establishes the sequence of I/O activities, protects data from damage, corrects I/O errors, and keeps statistics. The I/O supervisor has two parts:

- I/O request handler; it validates requests and starts activity if the device is available.
- I/O interrupt handler; it checks for I/O errors and starts activity for waiting requests.

The queue element representing an I/O request identifies the unit table of the required I/O device. The queue may be organized, for rapidity, into linked lists, one list for each channel. The requests within each list can be arranged for processing, first-in first-out, in priority order, or in an order allowing the positioning disk device near data. If an error is detected, an error routine takes over where alternative channels, control units, and I/O devices are used for retries. Error statistics are accumulated so that a deteriorating device condition can be detected.

The I/O channel transmits information between the main storage and I/O devices (Figure 13.2). To control the data flow, a channel executes a channel program. Simultaneous operation of the CPU and multiple channels improves execution time. A channel program includes typically, 1 to 10 commands. They include searching for particular data and transferring data blocks to/from main memory (read, write, transfer, search). CPU programs are more complex since they provide answers to many exception conditions (what if), e.g., what if the device fails to record properly and what if there is not enough memory?

The I/O system controls a) *files to be created or used in decision or control of activities*, and b) *devices to store and retrieve the files*. Those controls link programs to files and devices through I/O device requirements, opening and closing files, device allocation, device scheduling, device inventory, data staging, data storage, and buffering. The maitre d' in a restaurant is the I/O system; you tell him/her of your requirements, and he/she meets them, maybe after some delays. The control statements are used to communicate the requirements for resources by the user to the I/O system. They are kept separate from the user's program. A prestored collection of control statements can be called to use by a simple control statement that calls the collection. A control statement connects program and file. A program is used repeatedly, but the processed data contained in a file is changed for each reuse. A control statement contains the requirement for files and devices. The file requirement contains:

- File identification, name or location.
- Storage utilization, file organization, block size, labeling conventions, space requirements.
- File security and access control through passwords.

The I/O device requirement contains:

- Device type; which one.
- System configuration, number of I/O units required, private storage, shared storage; performance of a hardware/software system can be affected by inappropriate assignments of files to devices.
- Handling source and destination of the file (file routing).

Opening/closing a file connects/disconnects a user's program and its data. A logical I/O control interacts with the user's program; it is the access method. A physical I/O control schedules and initiates activities and responds to computer interrupts. Three mechanisms are used for communicating between user and I/O system:

- System calls where transfer of CPU control to the I/O system occurs from the user's program.

INPUT-OUTPUT MANAGEMENT IN A UNIPROCESSING SYSTEM

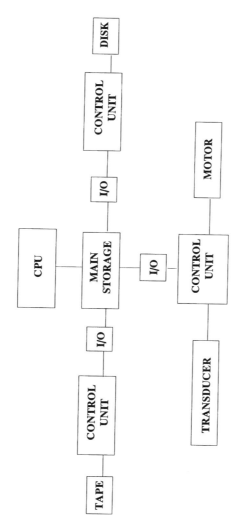

Figure 13.2 Types of devices.

- Tables containing a list of items about each particular I/O unit; when the number of parameters in a system call becomes large or when the same parameters apply to several situations, a table is used to hold that information.
- System exit where I/O control of the CPU is relinquished to user's program.

Device allocation is the process of determining which device will be assigned to fulfill one of a set of device requirements by the user. Two strategies are used:

- Assign a total complement of devices to the user.
- Stage data by system program in a direct access storage device for use by the user; this has a delay in execution.

The scheduler provides for the efficient use of devices while meeting acceptable elapsed time for each service request. It refers to tables that indicate availability of resources and matches resource availability to resource requirements and yields a decision when the system is ready to run a particular job. The methods used to allocate devices are:

- Static allocation which is done in anticipation of need prior to the user's program beginning execution as in the following areas:
  - Demand allocation fills the requirement for devices that can be filled in only one way, i.e., specialized piece of hardware.
  - Decision allocation allocates devices when available devices exceed regular requirements or requirements can be filled in different ways.
  - Public space allocation gives usable space on direct access storage devices.
  - Allocation recovery asks for operator assistance when available resources fall short of requirements.
- Dynamic allocation reduces required number of devices; it gives access to devices during user's program execution (disk communication link); if a request cannot be satisfied, the user's program is delayed; interlocking requests can cause deadlock; deadlock occurs when two programs in execution each want something from the other's device that is already in use. Automatic volume recognition allows operators to make allocation decisions for systems by a volume mounting procedure.

Device inventory is conducted during the system generation process. It tailors the OS for work with a particular computer system; the inventory of I/O devices available to that system is set and represented in tables and bit

INPUT-OUTPUT MANAGEMENT IN A UNIPROCESSING SYSTEM 203

DEVICES #	1	2	3	4	5	6	7	
OPERABLE DEVICES	1	0	1	1	1	1	1	DEVICE #2 UNOPERABLE
AVAILABLE DEVICES	0	0	0	1	1	1	0	DEVICES 1-3,7 BUSY
TYPE A DEVICE	0	0	0	0	0	1	1	DEVICES 6-7 TYPE A
ALLOCATABLE TYPE A DEVICE	0	0	0	0	0	1	0	ONLY DEVICE 6 ALLOCATABLE

Figure 13.3 Inventory process for allocatable devices.

masks. Each unit device is described in a table that contains: device type code, channel and unit addresses, device status (busy, operator action in progress...). Bit masks (0,1) are used to describe inventories of types where availability of a type is indicated. Using logical AND/OR on bit masks, one can establish quickly whether a certain type is available. A sample follows when searching for a given type of device (Figure 13.3). Device status affects the allocation process with unavailability of device to system due to malfunction, special assignment, current usage.

Data staging is the process of moving data from one storage to another in readiness for or after processing. This causes delays and exposure to error. Data entering or leaving the system is staged to provide flexibility in scheduling without having to synchronize all the devices to required speeds. Staging data files provides compact storage of data on one medium and high performance non-sequential processing on another. To make main storage space available for a high priority service request, a roll out may be used that suspends execution of one program while another is executed. Paging is the decomposition of a program into segments (pages); whenever a page is required, it is staged, then rolled into main storage for execution.

Data storage into memory is not the problem; finding it is. The access methods permit the user to say WRITE to dispose of a record and READ to retrieve it (Figure 13.4). The access method program lies between the macro interface and the I/O supervisor and accomplishes buffering, blocking, and device independence using service levels:

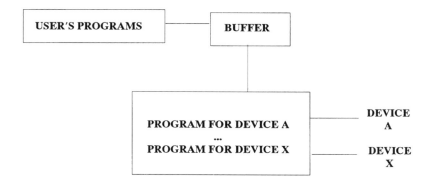

Figure 13.4 User's programs interaction with devices.

- Channel program level where user's program bypasses all access methods furnishing its own program that conforms to the I/O supervisor, I/O device, and file organization rules.
- Block level where the access methods shield the user's program from most I/O device characteristics.
- Record level where the user's call is all done for him.

Buffering provides a cushion against the fluctuation in traffic demand. Simple buffering procedure assigns buffers as required to execute a read/write buffer space. Exchange buffering fragments individual buffers to avoid movement of records within main storage. Circular buffering uses contiguous main storage as an endless buffer.

III. INPUT–OUTPUT DATA ORGANIZATION

In the execution of physical work or transfer of information for decision-making, what flows in/out between a uniprocessor and its peripherals or another uniprocessor is data. Data organization is at the center of I/O activities; this involves its organizational structure, its storage, its labeling, its cataloging, its space allocation, and inventory. The units of data are bits, bytes, records, blocks, files:

- A bit is a one of two pulse voltages, 0 or 1.
- A byte is a series of 8 bits.
- A record in a file can be of three types where records are of uniform length of bytes, a record descriptor declares record length or a user's program declares size.
- A block is a combination of records for handling by a channel.
- A file is a collection of blocks usually related to each other; a file is identified by its location or by a file name. The records within a

INPUT-OUTPUT MANAGEMENT IN A UNIPROCESSING SYSTEM

file may be arranged chronologically as they arrive, indexed, or in a list. A logical file might include a segment of a physical file or may be several physical files. Most operating systems maintain a catalogue about current file locations.

Organization of data depends on the capabilities of the I/O devices used and the anticipated use of the data:

- Sequential organization as on a tape where data are stored in the order it arrives. Sequential organization of data provides magnetic tape-like file processing activities regardless of the characteristics of an I/O device. It allows read and write forward or backward on blocks as the file delimiter. Device independence serves four purposes: a) *it allows the program to interact* with any of several classes of devices; b) *it allows evolutionary changes* in devices/computers; c) *it standardizes object program interface*; and d) *it eases interface definition burden.*
- Indexed organization as on direct access storage where data are stored in the order determined by the relative values in a key field. Indexed sequential organization of data insertion can be done at any time and retrieval is done through a key; sequential retrieval follows sequential key values. For example, Figure 13.5 gives a personal records file where each record is 1000 characters long (name, number, date of hire, birth date, department...); records are alphabetical and index file organizations are made alphabetically. For a new hire, Zebra, one uses the reserve area with an index pointer (address) in the record of the individual preceding the line in the alphabet. There are three major types of processing: a) *file loading* which is storing the initial collection of records and creating indices; the user's program presents records in key sequence using an interface; some files are volatile (changed often) and space storage is provided; files are loaded sequentially; random loading may be done using the reserve areas; b) *sequential retrieval*; and c) *random processing* where the principal benefit of an indexed organization is its ability to carry out individual record transactions in any order; normally, indexes are searched from the highest level down, preceding that search with a search in the area around that of most recent activity.
- Direct organization as when the key and location have a fixed relationship (i.e., same). Direct organization of data is non-sequential storage and retrieval of records; if one knows the key for the record, one knows the location of that record because of a key-location arithmetic or logical relationship, i.e., equal. At times the relationship is not with a record, but with a bucket of records (Figure 13.6). For instance, records 10112 and 20112 belong to bucket 112. Overflow of a bucket may be assigned to the nearest unfilled bucket

206   SYSTEMS MANAGEMENT: People, Computers, Machines, Materials

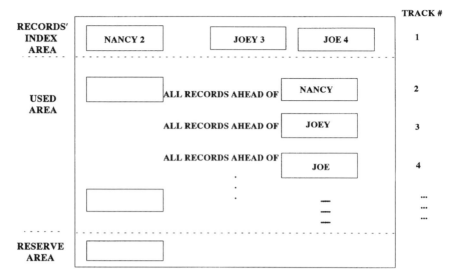

Figure 13.5 An indexed sequential organization of records.

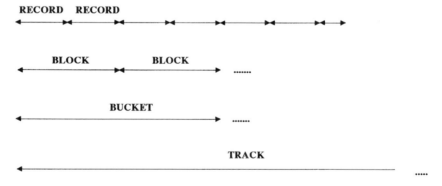

Figure 13.6 Storage formats.

or in separate overflow areas. Chaining overflow buckets reduces search time. The initial set of records can be stored sequentially or randomly. Performance is improved by storing well used records in non-overflow areas. Use of a tickler file (keys for all records that fall into a report) provides a form of logical sequential processing capability. A discussion of key transformation follows: To retrieve a specific record, either one must know the location of the record or search for it. Searching is time consuming, unless information can be extracted from records to form an index; an index can be searched more rapidly because it is smaller, especially if it is cataloged. What is needed is an algebraic or logical transformation of a record into a bucket number. The transformation determines the location. The best known one involves division by a prime number, then discarding the remainder.

- Linked list organization where relationships and order among records are identified through chains of pointers; now we are more concerned with file fabrication and alteration rather than reference or posting; a list organization offers the capability to retrieve records on the basis of two or more key values within the record. Linked list organization identifies relationships and establishes order among records through chains of pointers put in the records. The principal motivation for it is that a group of files which would be maintained separately if other organizations were used can be combined into a single linked list file satisfying a spectrum of user program requirements. Links or chains and indexes are alternative forms of lists; either can be derived from the other. Chains are more easily modified, while indexes are more rapidly searched. Complex chains consist of sequences of pointers. A record at the head of a chain is considered its owner; each record contains identification, data fields, and pointers (Figure 13.7). To retrieve a record, one begins with the owner and follows the pointers. The owner is found because it is a member of another chain, it is indexed, or its location may be calculated from its key. A new record may be added to a chain by following the chain to reach the appropriate point and the adjustment of a few pointers. The tracing of paths follows one of three classes of processing structures, sequential, tree, and network:

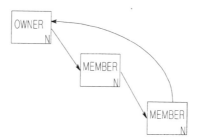

Figure 13.7 Chaining of records.

- Sequential is a relationship that can be represented by a single chain. A college class is an example where the owner is the instructor, the members are student records, and order is alphabetical; if it is indicated that some students belong to a certain organization, then the structure is not sequential.

- Tree structure is a hierarchical structure. A corporation management structure is an example where the president "owns" two vice presidents, each vice president "owns" three departments. Another example is a book which includes chapters, pages, paragraphs, lines, words.
- Network; when it is neither.

Rules for chain structures include:

- Each chain has only one owner record.
- Member records cannot be stored unless an owner record exists.
- A record (owner or member) may be an owner in one chain and a member in another.
- As records are stored, they are linked into their defined chains.
- As records are deleted, re-linking around the deletion is done.

An integrated data storage description follows where:

- Files are made up of pages.
- Pages are made up of records; a page contains a number of records of different lengths; a page header denotes a page number, space available, and line number available.
- A record has a line number on a page; a pointer to any record consists of a page and line number. To retrieve a record, an entire page is moved to main storage where it is searched for the line. So page size is selected according to disk loop size. A record is stored near its owner; it contains identification, data fields, and pointers.
- A chain table is maintained for each type of chain.

Storage consists of a page number and a line number within pages; page and line number are not affected by actual storage address. Page number is mapped to actual storage address. So changing of devices does not impair the linked list organization.

A label is a block at the beginning/end of a volume or file which identifies/delimits the volume or file. The label is machine readable. It usually consists of an 80-byte label and is placed at a fixed location such as disk 0, track 0, block 2 and contains the volume number identifier and its serial number owner identification. There are also file labels; they describe files or parts of files and are created/deleted whenever a new file is created or an old file is extended; it is deleted when a file is deleted.

A catalog is an enumeration of items arranged systematically with descriptive details. Keeping track of files and storage is a big effort; one keeps track of:

- New files created and old files deleted.

INPUT-OUTPUT MANAGEMENT IN A UNIPROCESSING SYSTEM

- Set of volumes containing a file as they change during processing.
- Available space for a file.
- Response to dozens of inquiries/minutes.

The vehicles used to control/identify files and space are the catalog and the volume table of contents. Both are treated also as files since they also:

- Consist of records and blocks.
- Have file labels with name, expiration date, password.
- Have space allocated to them.
- Are processed by other programs.

The user does not need to know where his file resides. A catalog maintains continuous records of the volume used to store individual files. A catalog allows a user to name a file at the time it is created and to simply refer to it by its name whenever it is used. The catalog enumerates files with descriptive details (volume identification of each volume that contains any part of that file).

The catalog structure has requirements which include: retrieval, update, insertion, and deletion in an unpredictable order. Thus it should be indexed in some way, trees and sub-catalogs (indexes). There is a master catalog with sub-catalogs. Each sub-catalog identifies the storage volumes that contain any particular file. Each sub-catalog has a name of its own and contains the names of the sub-catalogs related to it. Such a structure looks like a tree with sub-catalogs fanning out like branches; each level of the structure contains the file name with a qualifier. The physical organization has the highest level catalog in the system residence volume; others are distributed over several devices to improve system performance. A volume containing part of the catalog is called a control volume. Catalog processing includes:

- Building or deleting a (sub)catalog.
- Connecting or disconnecting the collection of sub-catalogs on a control volume.
- Locating a file.
- Accomplishing structural changes by control of program/user's program depending on OS.

The volume table of contents serves to identify and describe all files and pieces of files stored on the volume and to identify all unused storage space. It is used as an extension of the catalog to identify the location of a file and actual extents used by a file. A system user has a user ID code, a password, and accounting information file space.

Storage space allocation is executed by the OS which maintains inventories of available space to fill and requirements for space. Classification of space falls into:

- System residence space; it is assigned during system generation and is not changed during processing; it contains system programs and tables.
- Scratch space for temporary files; it is assigned during system generation and returned to scratch space pool when execution of the user's program ends; those files are never cataloged.
- Other space; it is available for permanent file.

Space inventory is maintained by some means, i.e., an available disk table consisting of disk addresses via available space. This table is updated regularly and is adjusted despite incidents due to malfunction. There, the criterion is to lose space, but not to lose data. Allocation of space is done from the inventory. Allocation routines vary and they:

- Allocate space to the file; it is done before the user's program is executed.
- Extend space of a file.
- Scratch a file and sub-file and return unused space to inventory.

IV. INPUT–OUTPUT BETWEEN HUMAN AND COMPUTER

Interactive graphics link two processors, the computer and the human and provide an effective means for their communication. Here, computers allow for human visualization of objects, abstract or real, rapidly and economically. The data within the computer pertaining to an object are assembled and operated on by the computer for the purpose of displaying its picture for human observations and interactive operations. It is somewhat the reverse process of human/sensor conversion of objects into data for computer operations.

Examples of interactive computer graphics are many:

- Computer data summarization; it uses visual displays for rapid and compact communication of a large amount of data; graphs and charts such as bar or pie charts, line or surface graphs are used to illustrate multiple relationships among the data in varied applications, e.g., management, scientific, engineering, financial. The behavior of physical systems and the status of development systems are often analyzed by constructing graphs and models, especially when large amounts of data are involved; computer-generated models form an effective technique for the study of systems behavior.
- Computer aided design and manufacturing; it allows for design and modeling of individual parts and for fitting together of the parts into the total object and, in turn, the objects into a system. The graphics display is used to view and try out alternate approaches in time/space

INPUT-OUTPUT MANAGEMENT IN A UNIPROCESSING SYSTEM 211

of parts/objects/system supported by a Database Management System (Chapter 14). Since the layout of the parts, objects, and system is known numerically, its manufacturing can be automated by numerically controlled machine tools that follow precisely the surface contours of the part/object/system steering its construction. Such design and manufacturing techniques are used in architectural, electronic, automobile, aircraft, and ship industries where try-out arrangements, virtual (conceptual) prototyping, are made without the need for actual prototyping.

- Computer automation; it displays rapidly pictures of objects where only slight movements in their position from one picture to the next have occurred; the rapid succession (30 pictures per second) yields an animated view of the objects. Animation is used in design optimization of system operation, in training simulators of critical or dangerous scenarios, as air craft landing; in home and movie entertainment; for very slow evolutions, high time compression is applied to feature the dynamics in a scene.
- Computer image processing; it provides for enhancement of picture quality; it uses the sensed data of an object to construct a picture of it through filtering, estimation, retouching, rearranging. Image processing is used in viewing unaccessible objects or systems as in medical tomography, ocean and space profiling, astronomy.

Any general purpose computer can be adapted through graphics software and hardware devices to display desired patterns (Figure 13.8). Typically, however, two or more processors are used, application processors and display processors; the display processor operates on digital data and instructions from the application processor and converts the information into voltages usable by the display device. Display devices range from:

- Cathode ray tubes where vertical and horizontal plates deflect the electron beam emitted from an electron gun to hit the phosphor at the desired point on the screen and emit a small spot of light; random or raster deflection systems are used.
- Plasma panel where the display is constructed by filling the region between two glass plates with neon gas separated into small bulbs; a series of horizontal and vertical electrodes, set respectively at the back and front glass panels, are used to light up the desired point on the screen by applying an appropriate voltage to the appropriate horizontal and vertical wire pair.
- Light emitters where diodes or crystals are used instead of phosphorus or neon gas to display a picture; a firing voltage at the appropriate wire pairs is used to produce light emission.

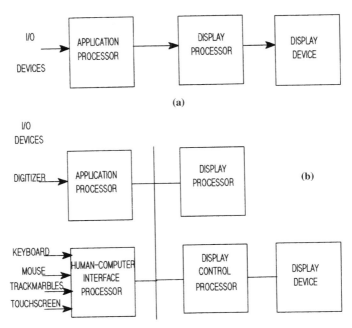

Figure 13.8 Human–computer interactive display: (a) simple diagram; (b) enhanced display system where powerful graphics are done in a special purpose processor.

To display a picture, the application processor provides picture data to the display processor in terms of lighting levels for the coordinate points on the screen; the reference is taken at the lower end of the screen. When application programs define picture points in another coordinate reference, a transformation is done from user coordinates to screen coordinates. Basically the display processor task converts points or line coordinate positions supplied by the application program into voltage levels to apply to the display device; points are the basic constituents of any complex picture; points make up lines and line segments make up any curved line on a picture. To support the application processing load, another processor is added to execute the human–computer interface. To support the display processing load, another display control processor is dedicated to refresh the screen and avoid possibility of flicker.

The basic flow of operations in a display system begins with the graphics commands embedded in the application program being translated into a display file program. Graphics commands for displaying and manipulating objects are designed as extensions to existing languages. During every refresh cycle, the display processor executes the display file program and converts the information into voltages usable by the display device. When a display refresh processor is assigned, the display processor copies commands into the refresh

INPUT-OUTPUT MANAGEMENT IN A UNIPROCESSING SYSTEM 213

display file for access by the display refresh processor. When a display file update is made in response to an input, the display processor insures that the update is inserted between refresh cycles so as not to distort the picture. In a raster scan system, the refresh buffer is called a frame buffer or bit map, and each picture element or pixel is organized as a two-dimensional array of intensity values corresponding to screen coordinates; good quality displays require a screen resolution of 1024 by 1024 pixels with 3 bytes per color pixel which require 3 megabytes of frame buffer.

General purpose graphics routines are available to facilitate pictorial manipulations; these routines provide:

- Input operations; they include the processing and control of data from interactive devices.
- Control operations; they include the initialization, re-set, clearing of the screen.
- Primitives or the basic building blocks to construct a picture; they include generation of natural language and mathematical characters and geometric entities, such as point, line, circle, ellipse.
- Attributes of the primitives; they include text style, shading, color.
- Segments; they include a collection of primitives that constitute a logical unit in an object and permit its processing, i.e., creation, deletion, transformation.
- Views; they include transformation of the object or collection of objects forming the scene for observation from a specified angle.

Many input devices interface with the graphics routines to enable human–computer communication; those include:

- Keyboards used primarily for inputting text and numerics; others are programmable function keys, cursor control keys, customized graphic keys.
- Touch panels used for selection of screen positions with a finger touch based on optical, infared, electrical, or acoustical transduction.
- Touch pens used for selection of screen points based on detecting light emitted from the screen or based on mechanical or capacitor switches in the pen.
- Graphic tablets used for selecting screen positions by setting a hand cursor at corresponding voltage points enabled through a rectangular grid of wires embedded on the flat tablet.
- Joysticks, track balls, mouses used for selecting screen positions through positioning of a cursor.

The varied input devices perform logical operations where: a) a *locator* is a device for determining coordinate positions; b) a *stroke* is a device for determining multiple coordinate position; c) a *string* is a device for determining text input position; d) a *selector* is a device for choosing menu options; and e) a *scissor* is a device for picking picture segments. These preceding types of input devices are all based on manual actuation; oral actuation has not yet found widespread applications, though the human voice is a very efficient avenue to convey instructions and data. Pursuit of computers that can process speech is ongoing, yet state of the art systems cannot process normal conversational speech where they can rapidly figure out the meaning of a string of words based on the context so that the computer can construct the appropriate response. In voice processing, interactive voice response systems, where voice is used to route or guide the user, are most prevalent; voice recognition and voice synthesis are projected to increase in application; voice recognition systems fall into a gradation of difficulty where:

- Continuous speech or natural conversation is most difficult for the computer to understand because of variations in the enunciation of words among people.
- Discrete speech with its pause after each word is less difficult since it gives the computer time to process it.
- Speaker independent, but with a limited vocabulary, permits understanding of the words of an expanded set of speakers.
- Speaker dependent has the computer exercized to understand the voice, pronunciation, and speech pattern of each user.

V. INPUT–OUTPUT OPERATING CHARACTERISTICS

The operating characteristics concern the abilities of I/O systems as to:

- Reliability; tolerance of "goofs" (errors) whether by program, machine, or humans. Reliability rules include a) the program should depend less on relative speeds of devices and CPU for operation, and b) the program should be tolerant of device idiosyncrancies.
- Availability; the features are expressed in metrics, e.g., hours of service or gross availability, e.g., 98%; gross unavailability, e.g., less than one minute; restricted service or graceful degradation in preference to complete failure.
- Serviceability; trained technicians can diagnose and fix the problem.
- Recoverability; three features of an OS insulate a user's programs from adverse effects of change to its environment:
  - Device independence where the user's program performs correctly despite changes in the I/O device serving that program. There, the user's program does not contain commands, instructions, or constants that make the program applicable only to a specific type of I/O device. The design is for a device-independent set of calls,

INPUT-OUTPUT MANAGEMENT IN A UNIPROCESSING SYSTEM

  tables with an I/O interpreter linking them to device-dependent actions.
- Data independence where the user's program should have some ability to cope with changes in the definition of records they process; if not, a change in a field, say from 4 bytes to 3 bytes, would require modification, recomputation, and retest of every program that uses this field. Increasingly difficult levels follow to: a) *tolerate changes in fields* that are not processed by the user's program; b) *tolerate extension to a record*; c) *tolerate changes in the positions of fields within a record*; d) *tolerate changes in the sizes of fields*; e) *tolerate division of a record* into two or more records; f) *tolerate elimination of a field* from a record provided the I/O system computes such data (virtual data since it is not stored). This impacts responsiveness of the system since you need reformulation of data as in emulation of an old processing system.
- Integrity where it involves avoidance of a simultaneous update when one program is updating a file when another program comes in to update the same file. Errors can be introduced into one file when allowing two or more programs to update a record into the file in a way that results in an error in the record. Integrity also involves avoidance of deadlock through staging.

VI. DISCUSSIONS

As sensing devices generate increasingly larger volumes of data, it becomes necessary to develop computer automation to handle the database and manage it as explained in Chapter 14. Most organizations require data for its activation. Data items have value if they are useful to the performance of organizational activities. Data are raw material that is processed by the computer to produce information. That information is of value only if it can be located, retrieved, processed, and communicated in time to allow a decision or action by a human, machine, or device. Storage and retrieval of information are thus of primary importance especially as we move from an industrial society to an information society; though the principles for materials and data handling in storage/retrieval/transport/processing are similar, data as enabled by the computer provide the desired speed, simplicity, and flexibility needed for a more productive society. Increasingly, much of our activities are on data which represent, describe, or record physical reality; more and more physical activities are avoided, delayed, or relegated to operations by devices and machines, controlled by computers operating on data that describe objects and/or events and their characteristics.

In designing computer systems, the human operator must be considered as an extended part of the operating management system (Figure 13.9). Automation, interaction, and manual operations are not independent modes, but are

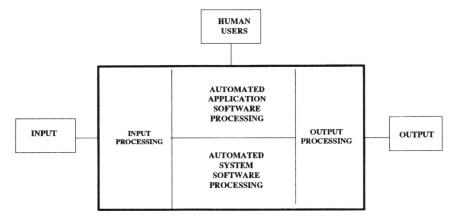

Figure 13.9 Integrated view of human–computer processing.

integral modes of system activities. System design is based on a dual concept involving operators and computers, integrated and adapted to the activities at hand. Regardless of the degree of system automation, the human will always be there to perform critical functions requiring his abilities to conceive, design, develop, and maintain the automation system itself, his ability to process qualitative information, his ability to deal with missing information, uncertainty, and ambiguity. At times, the operator makes errors in the course of interaction with the computer and the computer must be able to recognize the procedural errors and to provide appropriate feedback.

14 DATABASE MANAGEMENT SYSTEMS

I. INTRODUCTION

A database management system provides for the decomposition and integration of the symbolic representation of a system. The managed resources here are data. Data can be text, instructions, or formulas including numbers and functions. Heterogeneous data are encountered in varied applications with a range of multimedia types such as images, text, graphics, voice; data are most challenging to manage on-time and in an integrated way when they are changing in time/space. Management of data involves the processes of planning, organizing, activating, and controlling, described in Part I and synthesized with systematic patterns in Part II. The database manager executes on the hardware described in Chapter 11 and gets to the data through the local operating manager in Chapters 12 and 13. The database is intelligent filing that allows sharing of data rapidly. Data lead to information which leads to decision and decision leads to action. Data are facts about things in our surroundings that we are interested in and, many times, are important enough to have their own manager. The management processes, approaches, and issues encountered in its synthesis are analogous to those encountered with system hardware, software, or material resources.

Data are organized in a suitable form within a related control structure where:

- A bit is the smallest recognizable piece into which data can be decomposed for computer utilization, e.g., a 0 or 1.
- A character is made up by grouping a combination of one or more bits, e.g., an alphabet letter, a decimal number.
- A word is made up by grouping one or more characters, e.g., family name.
- A field is made up by grouping one or more words, e.g., name, address, age, sex, children, income.

- A record is made up by grouping one or more fields, e.g., student transcript; data are clustered about one identifier.
- A block is made up by grouping one or more records, e.g., students in a class, customers in a certain town.
- A file is made up by grouping one or more blocks, e.g., classes in a department; it is the outer boundary of a data grouping.
- A database is made by grouping all the files in a system.

Data grouping is guided by the types of transactions entertained; similar entities are grouped together to speed up the transaction record process. A field in a record is used as the key that allows access to finding that record. Selection of which field in the key depends on the transaction. The key is the order in which data are sorted. Those keys may be organized:

- Sequentially where the key fields for the records are arranged in some logical order, i.e., alphabetical order, numerical order.
- Indexed sequentially where the key fields are sequentially arranged then grouped into a block; the first and last record of a block are marked as such; the block location is given in an index table.
- Randomly where the key field is used to compute the record location.

In processing a transaction, its impact on all the files must be executed; with duplicate files, duplicate updating of those files and thus an increase in processing time ensues; file replication calls for an increased communication bandwidth. The relationships among separate files have to be understood in order to keep an integrated picture of system data.

Initially the data with all its operating details is captured, each type in its own file. Those facts across the files have to be processed to produce information for management. The more the database management provides the needed information for decision and action, the more it becomes an information management system. Human management organizations, however delayered, are unfailingly hierarchical. Each hierarchy distills its operating details and translates them into information. The informational needs up the hierarchy vary:

- Top level managers are concerned with strategic planning where the distilled data from the whole organization need not only reflect the outcomes of past actions but also reflect on the expected results of future actions.
- Middle-level managers are concerned with tactical planning where the distilled data from a component also reflect on how it should be done in the future.

DATABASE MANAGEMENT SYSTEMS

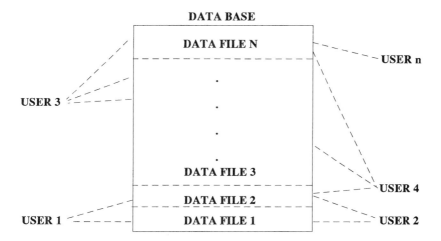

Figure 14.1 Database and users.

- Lower-level managers are concerned with operational planning where the distilled data from a specific project or activity not only reflect on actual results against specified standards, but also reflect on tasks and techniques of performing those activities in the future.

Figure 14.1 depicts a database application and its users. The question is: can a single database be created to service the differing information needs of the three managerial levels? A yes answer yields the solution of the triangles (Figure 14.2). A simple example of a database is the telephone book; you may query it alphabetically only; to sort by address would take a long time due to the way it is organized. A database management system (DBMS) would take the same telephone book data and answer with ease a query, for instance, about a listing of all the people that reside between Tenth and Eleventh Streets on Ellsworth; it organizes and manipulates data; it categorizes related facts; it stores facts in suitable form; it reorganizes facts to satisfy requests; it allows retrieval of data in the order desired, maybe differently from the way it was put in; and it allows alteration to the form of data entry by changing the field.

II. DATABASE STORAGE: FILE ORGANIZATION AND ACCESS METHODS

While a database is a set of records, some management structure is needed to avoid time consuming access to the records. Data management imparts ease of association, retrieval of update and maintenance while insuring integrity.

220   SYSTEMS MANAGEMENT: People, Computers, Machines, Materials

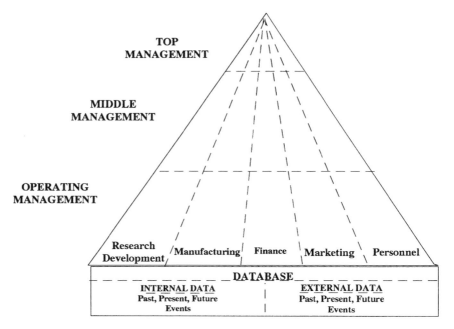

Figure 14.2 The overlapping of the triangles.

Data may be stored on varied media; it may be etched on stone, printed on paper, written on paper and/or a blackboard, or kept in human, electrical, optical, or magnetic media. In automated systems, there are:

- Primary memory; it is storage where a portion of the application programs and the system coordinating/control programs (operating system (OS), database management system (DBMS), and network control program) reside. This storage is large enough to store a portion of the programs and data files needed.
- Secondary memory; it is storage for the rest of the programs and data that is not required at a certain time. A primary memory is like the human brain, and secondary storage is like a book in a library. Data access is sequential (e.g., tape) or direct (e.g., disk). Disk surfaces are numbered (0, 1,..., s,); tracks are numbered (0, 1, ...t,); cylinders are the same track across surfaces (Figure 14.3). Cylinder data addresses are given by disk, track, or cylinder. For fast retrieval, related data are stored on cylinders; timing is dependent on:
    - Seek time; time needed for access arm movement to get to the right cylinder (same number of tracks across surfaces).
    - Head switching; time between read or write.
    - Rotational delay; time for disk pack rotation (2400 rotations per minute).

DATABASE MANAGEMENT SYSTEMS

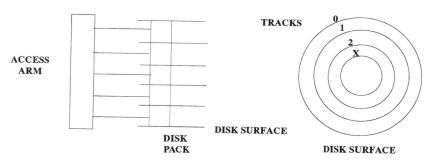

Figure 14.3 Disk make-up.

- Transfer time; time for data movement between disk and primary memory.

Data are collected in record formats; they have usually a fixed length for all records in a file, between 128 to 1024 bytes each, while fields in records have variable lengths. A record in a file is called a logical record. A set of physical records that are brought together to the primary memory is called a block; a block is sometimes fixed, i.e., 2048 bytes. The local operating manager gets those records off the disk in the right order and puts them together to present them as a whole; the local operating manager looks up in the directory where each record is physically on the disks. A track format has:

- An index point to mark beginning of track.
- A home address to denote cylinder number, track number, condition, e.g., defective.
- A track descriptor to indicate number of bytes available, other tracks numbers with this record's data.
- Data records with a) an address marker of record for the drive to know where the record is on subsequent retrievals, b) a count area to repeat home address and record number and length of area to follow, and c) a data area to contain actual data.

Thus, a disk stores a string of bytes. The local operating manager chops the string into files. The database consists of one or more files. The principal function of DBMS is to find the record the user needs from the files; it does this using a key and an index. A key is the order in which data are sorted, e.g., alphabetical. We find the index serially, by balanced-tree or hashing.

File organization is the way we store records for subsequent retrieval. They include simple, linear, indexed, indexed sequential, and hashed. There are four actions that we can do with a record in a file: look at it; modify it; add to it; or delete from it. For any action on a record to occur, the record needs to have a key field, a field with a unique value that identifies it.

A simple linear file is a collection of linear records listed one after the other, e.g., employee name, number, age, birth date, height, weight. A linear sequential file has records stored by order of arrival in physical sequence.

An indexed file resembles a book where:

- The table of contents summarizes what is in the book by major topics and is set in the same order as the material in the book.
- Page numbers are addresses of a portion of information in this storage medium.
- The index gives the location of an important specific piece of information listed alphabetically for search and with a pointer to the page of the book where each resides.

It is of interest to read files in sequences based on values of one or more fields, even though they are not maintained in any physical sequence. The mechanism is the indexed file (Figure 14.4). A file is indexed sequentially when the indexed file organizes the search field (a specific value of which will be used to identify the record) as equivalent to a sequence field (a field used to physically sequence a file). An indexed sequential organization is executed by one of three ways: simple index, multilevel hardware oriented index, B-tree oriented index.

- Simple index; records in file can be accessed sequentially since they are stored one after the other on the disk in that order. Primary memory holds the index and secondary memory in the files; search is done on the index thus avoiding transfer into main memory of the whole file.
- Multilevel hardware oriented index; here, we narrow successively the search for a record based upon physical characteristics of the environment, e.g., disk. An analogous example is finding a man's suit; using directions, we find a shopping mall, we find a department store, we find a floor with a men's department, we find an area with suits and then find a rack with our size. For a disk, suppose we have one made up of nine cylinders with four tracks (0-3) in each cylinder; when a record, e.g., number 2431 is requested, we pick an index in the master index greater than the number; we do the same in the cylinder index and index track leading to the record (Figure 14.5).
- B-tree oriented index; it is a multilevel index concept that is not hardware dependent and is capable of accepting new records with less effort. The storage space contains control intervals and control areas. A control interval is a unit of transfer of data between primary

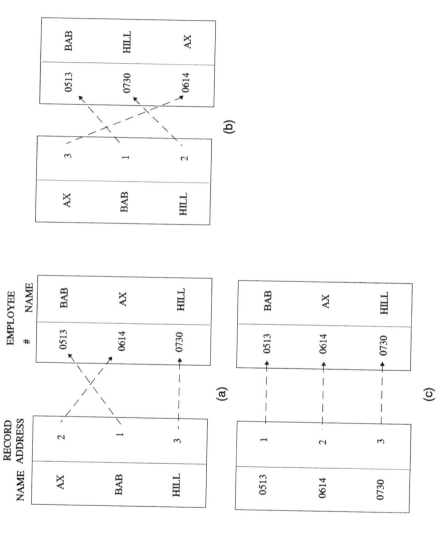

Figure 14.4 Indexed file: (a) file sequenced on employee #; (b) non-sequenced file; (c) indexed sequential file.

224 SYSTEMS MANAGEMENT: People, Computers, Machines, Materials

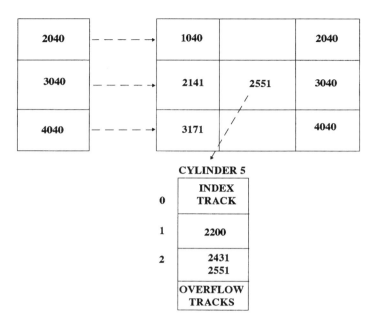

Figure 14.5 Multi-level indexing.

and secondary memory, e.g., a disk track. A control area is made up of control intervals, e.g., a cylinder. The picture of the index and data looks like the hardware dictated index except that now we have control areas rather than cylinders and control intervals rather than tracks. Also, now free space is deliberately left in the file as it is loaded to allow insertion of new records. When no more free space is available then a new free control area would be obtained and the entire control area split into two, and the pointer set between the areas in hierarchical order of index size.

Hashing is introduced to files when speed and direct access are important; it associates a key value with a storage location; the division remainder method is used for that effect; sometimes you may get the same location for a different number; pointers are then used to send you to another location; using hashing, there is no physical sequence of the stored data; the records are scattered. When speed only and not space is of essence, one may take a fixed length of record, e.g., x bytes of maximum record length and reserve it for each of the records; when one wants to retrieve record number y then the physical location is determined by xy.

File organization and linkages are similar to a local bus or network in a system; one connects data together and the other connects processors and devices which process the data to execute a task.

DATABASE MANAGEMENT SYSTEMS

III. BENEFITS OF A DATABASE SYSTEM MANAGER

Initially, time and emphasis in application program development were placed on the programs as compared to data and data structures. Programs got standardized first; data were stored in different formats and in different files and were often not sharable among different programs calling for redundant files, causing program changes as data changed. While access methods allow you to get hold of data, the database manager provides for integration of that data across files. Data describe the state of an enterprise and should be a manageable resource. The question is whether data manager issues are responded to within centralized services, or each application re-invents them; these issues include:

- Data standardization requirement and redundancy within one file and among many files; the storing of a piece of data multiple times calls for multiple storage and multiple updates. DBMS allows, in a single command, retrieval of data residing in multi-files.
- Multiple relationships with one to one, one to many, and/or many to many; simple files leave us with a cumbersome task as we seek multiple associations. DBMS provides the capabilities for multiple associations.
- Security; it enables prevention of people from seeing/changing data they are not supposed to see/change.
- Backup and recovery; corrective action is required following discovery of an error in input data sometime ago which has affected other data; e.g., hardware/software failure. Backup is the process of copying a file, done on a periodic basis. Recovery is the process of using a backup file, and other data, to correct an error.
- Concurrency; it occurs in a multiuser application where two users may try to update the fields in a record (airline seat reservations) and wind up with one being ignored; locking out one of the users is a technique used to make sure that the user knows that his update has not taken place.
- Integrity; it allows update and control of duplicate data.
- Data independence; it separates out storage and access from programs and data. Otherwise, changes in storage structures, e.g., technique/hardware, cause changes in the program; massive time in program maintenance is then spent on working the dependency between programs and data.

IV. DATABASE MANAGEMENT SYSTEM CHARACTERISTICS

DBMS is like the middleman in a transaction. Can you imagine the difficulty without a middleman in a modern society? For instance, in a mail order system, a phone call is made to an office to place an order from a catalog which contains a description of the goods, and the office as middleman tells the warehouse to ship the ordered goods and keeps control and accountability of the goods. A DBMS is a process which specifies how data is structured, controls access to data, provides other essential services, e.g., manipulates data, provides for data integration, reduces redundancy, and provides for relationships in the data. In a DBMS, data definition language specifies the way data may be stored. There are two views of the data; DBMS has at its disposal both views:

- Logical view, the way the programmer perceives data to be.
- Physical view, the way data are actually stored on disk.

A DBMS is based on one of four organizational approaches to structures:

- A hierarchical data structure; each link connecting two record types represents a one-to-many relationship, in the downward direction. Many-to-many relationships can be handled with a combination of hierarchies. Hierarchy and network are navigational systems where data are stored as records and are interconnected by pre-stored address pointers to integrate the files.
- Network data structure; as in hierarchical, each link between two record types represents a one-to-many relationship. The direction however is not only downward and is indicated by an arrowhead.
- Relational database structure; structurally, it consists of independent linear files; one forms useful combinations of data through a functional equivalent of navigation using repeated key fields in the files rather than the kinds of physical connective structures in the navigational systems.
- Pseudo-relational or flat file integrated database; flat file structures are used, but data integration is done through a pre-stored, physical construct as in the navigational systems. More flat files are added to match which records in one file are related to which records of the other files.

Two examples of DBMS structures follow. Figure 14.6 depicts an office with company cars assigned where certain individuals can drive certain cars, and each car has a maintenance record. Figure 14.7a depicts another example of a multifiles system. The files are linked together through unique common items in records using pointers. Figure 14.7b depicts a hierarchical database with one-to many relations with each master having dependents and each dependent having only one master; a dependent has in turn, dependents. Figure 14.7c depicts a relational database with many-to-many type searches;

DATABASE MANAGEMENT SYSTEMS

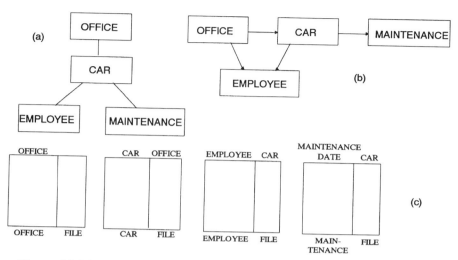

Figure 14.6 Data structures: (a) hierarchy; (b) network; (c) relational.

a fully indexed flat file leads easily to relational database manager (RDBM), but is slow in many-to-many cases. In a relational DBM, relational records point to each other so that essential data are only stored once.

A data dictionary holds the types of items and records, who uses them and with what program. It is used during maintenance and informs users of changes. A schema may be conceptual, which is an ideal definition of data, and internal, which is how the conceptual is implemented on a particular machine.

There are two mechanisms to access and retrieve data; both access methods may be used with any of the DBMS structures; they form the data manipulation language:

- Embedded statements through which an application program issues instructions to the DBMS; in between the go to, if-then-else, $X = Y + 7$ and so on, there are instructions that cause the system to reach out and transfer data in/out of the program to/from the disk drive; the I/O statements go through the middleman, the DBMS.
- Query statements through which a person issues a command to the DBMS and its execution are supported by query language. Query language is an on line construction of commands which, when done according to a syntax, may be interpreted and executed. This replaces the writing/debugging of a program that does the same using a procedural language.

228 SYSTEMS MANAGEMENT: People, Computers, Machines, Materials

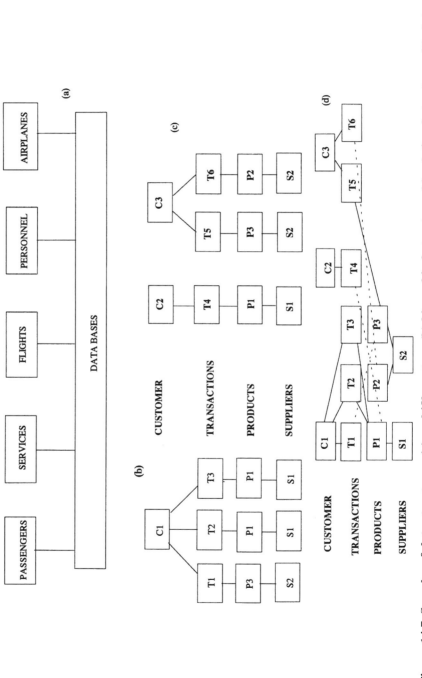

Figure 14.7 Samples of data structures; (a) multifile system; (b) hierarchical structure; (c) relational structure with data stored only once.

V. DATABASE MANAGEMENT SYSTEM ORGANIZATION

A. The Hierarchical Database

A hierarchical structure is a depiction of one-to-one/many relationships; it is represented physically by nodes and lines branching downward. In Figure 14.8, node A goes to B, C, D; C goes to E, F. Storage of a hierarchy is done:

- Sequentially, in the order of top to bottom, left to right, then front to back. Its drawback is lack of space for addition, and one must read through sequentially to find a record.
- Pointer-based, with pointers used for fast access or growth. Usage of logical relationships allows introduction of many-to-many relationships. The hierarchy then becomes a network; programmers write their program based on hierarchies; then they attach the logical relationships. So the pieces of a physical database are hierarchical and simply programmed.

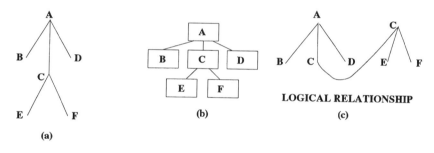

Figure 14.8 Hierarchical representations.

To access the data, one gets to the root node using one of the access method techniques; if sequential access is used, the root node is the top node and the search is lengthy; if the pointer and indices are used, the search is quick. There are agreed call statements which are a set of instructions that programs use to access the data through the DBMS. For instance,

- Find — locates a particular record and sets the current of the run-unit indicator on it.
- Get — brings a record to the user's work area in main memory.
- Store — inserts a record in the database.
- Erase — deletes a record from the database.
- Modify — updates records.
- Connect — links an existing record occurrence.
- Disconnect — cuts link.

B. The Network Database

A network structure has nodes connected by lines where cycles may exist, that is where lines begin and end at the same node and where lines do not always point downward. A tree is a network that has no cycles and where lines always point downward from the root node. Storage of a network structure is depicted using the set which consists of two record types in a one-to-many relationship to each other; sets are stored with pointers. Sets can be combined in different ways to form networks (Figure 14.9). For many-to-many relationships, two sets are created; one set will have one of those record types as its owner, and the other will be a member record type; this is called connection, juncture, link, or intersection.

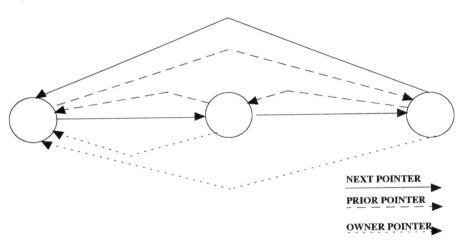

Figure 14.9 Sample network data organization.

To access the data, there are multiple options; for instance, hashing of each set provides fast direct access, but no physical sequencing of the data; so, navigating from record to record can be slow due to the physical scatter of data on disk. The language interface statements are similar to those noted in hierarchical structure.

C. The Relational Database

A DBMS provides for data to be stored non-redundantly while at the same time making it appear to the application program as highly integrated. A relational structure is a table of rows (tables, records). The fields (attributes) of a relation can be arranged in any order. A single or group of fields serves as a unique key; if a relation has more than one key, each is called a candidate key unless one is chosen as the primary key. If an attribute is a key field in one relation and also appears as a field in another relation, it is called a foreign

DATABASE MANAGEMENT SYSTEMS

key in the latter. No redundancy should be across tables among the non-key fields. Storage of relational tables follows one of the file organizations in section II. Across tables, one can perform joining or integration of tables. To access data from a table, one can retrieve it by:

- Taking a horizontal slice.
- Taking a vertical slice.
- Taking both slices.

The language interface is a structured query language (SQL) that involves:

- SQL data structure; the concepts of logical and physical databases are implemented. A physical table is called a "base table". A logical table is called a view and is a subset of one or more base tables; if a query references a view, the information about the view in the catalog will map from it to the physical table where the needed data are actually stored with no physical duplication of data. When data in a table are changed, SQL automatically changes the affected data by that change.
- SQL optimizer; it gives the most efficient way to answer a query; it takes maximum advantage of its indices. The SQL designer includes a set of indices based on the anticipated types of queries. Those indices speed things up, but they are not a requirement to respond to a query. The optimizer checks to see if the query involves columns that have indices and goes directly to only those records that are required.
- The SQL language does commands that can be issued either as ad hoc query commands or data access commands, e.g., ADA.

The varied DBMS include points of comparison as to join/integration, design effort, and update effort and performance:

- Data joins; they are the implementation vehicle for the data integration concept. In hierarchical and network structures, the joins are performed through the pointers when the data is loaded according to those structures. In RDBMS, the joins are performed on-line at execution time by the DBMS.
- Design effort/modification; for hierarchies and networks, decisions must be made as to which of the integrating associations are actually designed into the structure; those accesses are very fast. For modifications, a resetting of the structure may be required. For RDBMS, the design is simpler; the better design, however, insures that the join will find the join fields in the right places when it needs them.

Joins that require full file scan of one or multiple files can be very slow. For modification, the RDBMS is simpler.
- Performance; one can't beat physical pointers in hierarchical/network structures for speed.

D. The Pseudo-Relational Database

A relational system performs on-line joins between files structured as tables. Hierarchical and network structures require some form of a priori linkage between related records of different files before the on-line query. A pseudo-relational has the a priori linkage between files structured as tables; the link files are also a table; the link table may be created at the time of first query and saved (Figure 14.10). That ability to perform joins across files qualifies a structure as a database system; it integrates data between two or more files. An inverted file structure has data storage of simple linear files with indices built over all/many of their fields.

	STUDENT #	STUDENT NAME	STUDENT CLASS	YEAR OF ENROLLMENT
1	101	AX	53	1985
2	195	BENNY	51	1987
3	601	HILL	50	1988
4	1001	HASS	52	1990

(a)

	STUDENT #	PROFESSOR #
	101	102
	601	153
	1001	301
	195	162

(c)

	PROFESSOR #	STUDENT #	CAMPUS
1	102	101	UPTOWN
2	153	601	EASTERN
3	301	1001	WESTERN
4	162	195	CENTRAL

(b)

Figure 14.10 Pseudo-relational data structure: (a) student file; (b) professor file; (c) link file.

A file takes two forms, logical and physical; a logical file is based on one or more physical files:

- A logical file based on one physical file presents the programmer with a view of the data that differs from the physical file and it may contain a subset of the fields in the physical file, may contain a

subset of the records in the physical file, may rearrange the fields as it presents them in the desired order, may rearrange the order of the records based on an index existing over the record, and may contain range bounds checking on numeric fields. Indices are built over the physical files to perform those operations using binary trees; a binary tree structure has two branches at each node except for the terminal nodes; the tree is searched top to bottom, left to right.
- A logical file based on more than one physical file has a join operation to integrate the fields of the records of two different files into records of a new resultant file based on like values in the join fields.

For data retrieval and manipulation, both logical and physical files may be accessed and modified with data retrieval operations like insert, delete, update. Some have higher level operations, such as summation of numeric fields.

VI. DATABASE SYSTEM ARCHITECTURE

Architecture is the structure or highest level organization of a system. It realizes system functions using the preceding design features. Architecture is the interconnection of the external and visible forms of the overall structure, as in a building; in that instance, its design and implementation yield internal structural elements, e.g., beams, pipes...; its realization yields choice of material, e.g., wood, steel, shape, type... . In a database architecture, a database structure describes many populations and their relationships, while a data structure describes individuals in a population and their relationships. In developing a database architecture:

- The system developer sets down the requirements for a whole group of users (programs/people).
- The database administrator sets up the needs for all the groups.
- The DBMS developer chooses the DBMS in consonant with the system developer to marry the DBMS to the computer's local operating manager.
- The application program enters, alters, and deletes data. The software developer designs the programs to interact with the DBMS.
- The operator enters, alters, and deletes data from the database.
- The supervisor extracts information by queries to the DBMS.

In a database, data elements are a representation (description of properties) of a part of the real world (population) which is of interest to us. The population has similar entities, each with characteristics (attributes); each attribute has a field name and a field value, e.g., John – 5'5". A collection of field values for an entity is a record. A collection of records is a file. Data uses include: a) *archival*, where data do not change; b) *retrieval* where data are viewed for information without change; c) *alteration*, where data in an entity are changed; and d) *maintenance file* for the alteration because of a change (addition/deletion) in the population.

Frequency of use and response time are two features of interest. In using the data, a task may activate several application programs; each application program has a set of files that it uses; the files may be used jointly by many application programs. When no DBMS is present, each application uses the files without coordination of who needs what, when, and without assurance that the most recent current attribute is being used. Furthermore, if a record size has grown with a new field, every application program that uses this record must be enlarged. If a placement of fields is changed, every application program that uses this record must be altered. The DBMS provides the solution. As the local operating manager of a computer isolates users from hardware, the DBMS isolates users of data from each other (Figure 14.11). It passes the records through filters to produce the record required by the user in his work area. All internal files are integrated without the help of the application. The subschema integrates all data related to an entity into a single package (segment); a set of related segments is a record.

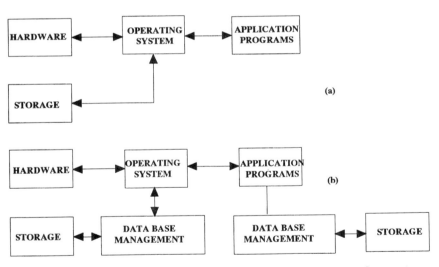

Figure 14.11 Relations among applications, hardware, operating system, and database management.

The schema is the internal data model that describes the organization of all the available data about a given entity and relates it to other entities of importance that reside in other populations; a relation is a quality that binds together two or more entities. The subschema is filtered data from a schema

DATABASE MANAGEMENT SYSTEMS

that an application uses called an external model or data submodel. The database administrator considers functions of record update, file maintenance, retrieval and determines which subschemas are responsible for these functions. The specification of the subschemas is the responsibility of the application manager. The query makes available to the application/user the data it/he/she needs without technical programming knowledge. A planned query is the response to a number of anticipated questions; those may be processed without the need for query processing. In an unplanned query, the DBMS then needs a capability to perform complex activities with the query processor doing logical and arithmetic operations; the answer to a query may be prepared following a prescribed report format.

The DBMS architecture should provide:

- Data independence where data supplied to each application do not depend on the form as it is recorded; all data go through a conversion step while moving from the physical database to the user's work space.
- Limited accessibility where security is insured using a variety of schemes, e.g., passwords, floppy disks with double encrypted authorizations; floppy disk with response to only one terminal.
- Currency where the unwieldy requirement on each application to make all the changes in different files is avoided.
- Consistency where only one schema is the source of all subschemas.
- Single interface where only one subschema is used for each application; it may represent several files which may reside in several populations.
- Retrieval where primary as well as secondary identifiers may be used.
- Query where simple as well as complex ones may be used.
- Reconfiguration where recovery from system crashes or transition from one mode of operation to another are carried out.

The relations within the database architecture involve data as the representation of a small portion of the real world (population), and database as the representation of many populations of relevance to an application family and the relationships among its members. The relations may be one-to-one, one-to-many, or many-to-many. The most important type of relations in a population is the position of individuals/items with respect to each other, the order in the relations. A file is said to be ordered when the position of a record is determined by its key; a record is required to have a field(s) which has a unique value. This field is the key that identifies the individual (item) that is described by this record. A file is sorted when one passes through a file according to the physical location of a record in the file. A linked list provides a pointer in each record to its successor which may not be the one next to it as in an ordered file. This makes addition and deletion simpler. Architecture provides the means to represent the relations in a database. The resulting relations lead to hierarchical, network, or relational databases.

VII. DATABASE SYSTEM DESIGN

Database system design realizes the database system architecture. It includes structured system methodologies with data flows, data normalization, data structuring, entity relationships, accounting of the environment, data dictionaries, and data administration. Database system design realizes the organization of data into a structure to service one or more users. It supports the required relationships among the fields while being within the physical constraints of the devices/system. There are two parts:

- Logical database design, which is: a) *organization in a non-redundant grouping* based on data relationships, and b) *organization of those logical groupings into structures* based on the nature of the DBMS and the users of the data.
- Physical database design, which is a refitting of the logical structures to conform to the performance and operational requirements guided by the user's requirements.

In planning the system, one defines:

- Its functions and the lineup of its subfunctions into modes of operations and submodes, usually done by the user.
- Processes to perform the subfunctions, usually done by the developer.
- Data classes, major types of data (not down to the field) required by processes.
- Information architecture, flow of data between sources and users' processes and how the elements relate to each other.

Processes may be grouped in a table to indicate major groups of data sources and data users. Sources and users lead to information flow within the architecture; for data class 2, in Table 14.1 for instance, process 2 is the source and there are two users, process 1 and process 4. Once this is accomplished, then attention may be directed to division of processes into sub-processes and their implementation on a priority basis.

System analysts serve as an interface between the eventual users or customers of the system and its developers. Their methodologies support a reformulated description of the interfaces detailed enough, but simple enough to be understood by developers. A structured (top-down) system analysis approach is applied which is a marriage of several techniques with a data flow diagram (DFD) to show graphically the movement of data through the system

DATABASE MANAGEMENT SYSTEMS

Table 14.1 A Data Flow Representation

Data Class	Process			
	1	2	3	4
1	Source	User		User
2	User		Source	User
3	User	Source		User
4	Source	User	User	

along with a data structure diagram, decision trees, structured English, data normalization, and data dictionaries. A data flow diagram is a two-dimensional structure depicted with four types of boxes:

- Processes represented by rectangles to transform data.
- Data stores represented by dashed rectangles to hold data.
- Data flows represented by arrows to move data.
- External entities represented by squares to interface with the rest of the world.

DFDs are intended to be non-physical; they do not show CPUs, disks, networks, or people. The presentation of a system data flow diagram is structured in the sense that it is top-down. The system description is given in a top level DFD, with each box blown up in turn into a series of DFDs. All of the DFDs, when taken together, describe the application in a top-down comprehensive and comprehendible manner to users and programmers. Since the DFD will be used by programmers for implementation, this requires a more detailed description of the lower level sub-processes. For that, structured system analysis resorts to other techniques, such as decision trees, decision tables, and structured English for sub-processes.

Data stores grouped into sets with non-redundant data, would be given to the system designer so that it can be implemented directly as physical files. Otherwise, data normalization techniques are applied without regard to the storage medium selected; it forms the transition technique between the system analyst and system designer.

Since all DFDs and associated information must be recorded and maintained, a recording method known as a data dictionary may be used. Systems analysts are the originators of the items in the dictionary. The items are the process boxes, data stores, data flows, and external entities as well as field descriptions. The latter is used and enhanced by the physical design personnel at a later time.

Structured systems analysis provides the interface to structured system design. Each lowest level process box is intended to become a module of code, multiple modules of codes, or combined with others in a single module of code. Those modules reside in a physical implementation. The activities in database design affect the DBMS characteristics of:

- Redundancy; it serves to avoid intra-file/multi-file redundancy; the amount of redundancy (key and non-key fields) varies in the type of DBMS. The relational approach requires certain fields to appear as attributes of different relations while in navigational cases, this field duplication is replaced by pointers.
- Performance; it is a function of a) *CPU processing speed*, b) *disk data transfer rate*, c) *channel speed*, d) *contention of different applications sharing the same hardware resource,* and e) *efficiency of techniques in DBMS.* This depends squarely on quickness of design, the nature of accesses to the data in the main memory, and nature of accesses to the data in the secondary storage (disk I/O); which access to use is under designer control through his choice of hashed, indexed, pointers... .
- Data independence; it is the ability to modify data structures without affecting existing programs. It is a function of a) the particular DBMS used and more so with navigational DBMS, and b) the way the program is written.
- Data security; techniques range from a) *passwords associated with particular users*, b) *passwords associated with particular data,* and c) *various ways*, such as fixed format interfaces, to allow certain users to operate on a certain data; this leads at times to physical separation.
- Ease of programming; a DBMS allows use of the same protocol to obtain data even from physically separate files or distinct records through a call from a high level programming interface. Otherwise, we may have to access data in one file, use it to access data in another file, and so on as necessary to satisfy the query. This last approach is more error prone. Throughout it all, a balance must be struck between a) *combining fields together within one data structure* to simplify the specifications, to retrieve the data and increase data redundancy and b) *adding indices into data* to create flexibility in on line retrieval and in lowering performance during update; a highly volatile file with large connected indices decreases performance.

The methodologies in database design include data normalization, data structuring and entity relationship model. Data normalization and data structuring take as inputs a list of fields and their associations, and arrange fields into files/relations so that redundancy among the non-key fields is eliminated. Putting all the fields into a single file can cause:

DATABASE MANAGEMENT SYSTEMS

- A much higher percentage of fields as key fields; a key value can uniquely identify a record in a file and determines the non-key values as dependent on the key value.
- A great deal of redundancy.
- Problems with the deletion and addition of key fields and with dropping of non-key fields.

With data normalization (Figure 14.12), one:

- Develops a table for each single area of knowledge.
- Eliminates multi-valued fields and leads to one piece of data per field.
- Eliminates one non-key field defining other non-key fields.

With data structuring, one converts from a set of normalized tables to a DBMS:

- For relational data structure, normalization leads directly to RDBMS.
- For hierarchical data structure, the normalized tables are transformed into nodes in the hierarchy using associations among the files (one-to-one, one-to-many, many-to-many) where a) *every branch in a hierarchical structure represents a one-to-many relationship,* b) *a many-to-many relationship is simulated by two one-to-many relationships when loading the data,* where there will be one occurrence of this new special record for every occurrence of the relationship, and c) *access and performance are linked,* and this depends on which fields must be accessible directly by the set of applications, on what number of levels below the key field can be accessed to reach the desired fields. Whatever access method to a key field is used, e.g., sequential, direct, indexed, hashing, the aim is for as much direct access to non-key fields as possible.
- For network data structure, like the hierarchical approach, the starting point can be the set of normalized relations and the list of associations. Here again, chain lengths, access methods types, network complexity, and application requirements affect performance.
- For pseudo-relational, one may merge those tables and create a redundancy. If there is a large number of tables each containing a small number of fields, the next logical linking of files (joining) to each other is done through matched list and/or binary index trees; next the fields of each particular file that must be directly accessed are set.

240 SYSTEMS MANAGEMENT: People, Computers, Machines, Materials

KEYS		SKILL CATEGORY	MECHANIC NAME	MECHANIC AGE	SHOP #	SHOP CITY	SHOP MANAGER	MECHANIC PROFICIENCY
MECHANIC #	SKILLS #							
3	101	ENGINE	SMITH	24	3	MEDFORD	AX	3

DATA NORMALIZATION →

SKILL TABLE

SKILL #	SKILL CATEGORY
101	ENGINE

MECHANIC TABLE

KEY: MECHANIC #	MECHANIC NAME	MECHANIC AGE	SHOP #
3	SMITH	24	3

SHOP TABLE

KEY: SHOP #	SHOP CITY	SHOP MANAGER
3	MEDFORD	AX

PROFICIENCY TABLE

KEY: MECHANIC #	SKILL #	MECHANIC PROFICIENCY
3	101	3

Figure 14.12 Data normalization.

DATABASE MANAGEMENT SYSTEMS

The entity relationship assumes that the system analysis produces the entities (Figure 14.13). The system analysis produces a list of the application's data fields and the relationships among them; normalization proceeds to separate the fields which identify entities from those that further describe the entities. In mapping entity relationship to network organization, the process is straightforward. For a hierarchical organization, one creates one-to-many relationships. For a relational organization, there is design reversal where one dismantles the entity relationship and constructs a set of relations. For pseudo-relational organization, one adds/joins tables or indices to the relational organization.

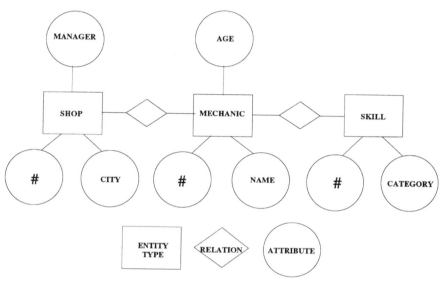

Figure 14.13 Entity relationships.

Accounting of the environment must take place in database design. A DBMS must deal with security, concurrency, backup and recovery, and auditability:

- Security; it involves the protection of data from unauthorized access or physical entry by human and/or computer and includes a) unauthorized viewing; b) unauthorized data modification by human/computer; and c) unauthorized program modification. At times a person may be authorized to view, but not to update; thus, identification tags and passwords to express entitlements are used for humans and programs prior to execution. Interception is also used and involves several levels of algorithmic transformations to effect correct entitlements.

- Concurrency; it involves access to the same data at the same time. There is no problem when concurrent access involves simple retrieval; there is a problem when access involves data modification. For example, concurrent or simultaneous access, as in a multi-programmed system, allows at time, t1, clerk A to read file X to find number 20; at time, t2, clerk B reads file X to find number 20; at time, t3, clerk A deducts 5 from file X to yield 15; at time, t4, clerk B deducts 3 from file X to yield 17; however, at time, t4, the record should show 12, (20–5–3). A solution approach to this problem is locking out any update once one has started; this locking mechanism may cause a deadlock where each of two queries has locked out the other and neither can finish; a solution is to roll back one of the queries to the state before it started execution; one may use the time lapse to detect deadlock. To avoid deadly embrace locks, records nearer to the database, should be locked first.
- Backup and recovery; they are needed when something goes wrong with the data. To execute, two things must take place; first, have a backup copy of the file, and then update the copy keeping track of the old and the change. This special file is called the journal or a log of the changes. For recovery, one goes back to the log and starts recreating the disk that was lost by making the copy and introducing the changes. At times, rather than going through all the changes, only the last change to the database copy is made. Sometimes, other than in a full disk crash, an error is discovered in recently updated data. Here, the error and all the changes based upon that error must be backed out. Some systems have dynamic back out relying on some kind of stable state known as a "checkpoint".
- Auditability; it reviews the activities and results at a later date to verify that the database is doing its functions of protecting the data; it keeps a log of who is using the database and their transactions to verify their entitlements.

Data dictionaries are those files needed to manage a data processing organization. For example, to manage a personnel department, you need personnel files; to manage a parts store, you need parts files. The contents of data dictionaries are entities (something of use to a specific user) or objects, attributes (identifiers), and/or relationships:

- Entities include fields and files, record length and set pointers, a software configuration item, e.g., DBMS, user programs.
- Attributes include key fields; quantitative values, e.g., number; qualitative, e.g., color.
- Relationships include a) *impact of change as to which programs are impacted by changes to the files*, b) *physical residence* and *which files are on which disks*, c) *program data regulations* and *which*

programs use which files, d) *responsibility as which people/programs are responsible to update which files*, and e) *record of constraints as to which fields appear in which records.*

Some data dictionaries are simple files; others are an implementation of the DBMS structure, as another application of the DBMS. In either, they hold the types of items and records, who uses them and what program, and are used during maintenance to inform users of changes.

Data administration is another required element in the design. Data are an important corporate resource along with people, equipment, plant, and money. Data administration is another required feature in the design. Like an orchestra, you need a conductor for overall direction and control; otherwise, chaos results, e.g., redundant data, performance deficiency, maintenance problem due to lack of standards in database design and program construction. Data administration provides the order in the database making sure it exists, but not necessarily performing all the details:

- It reduces data redundancy and insures integrity by having users as much as possible share the same file in the came computer, rather than each having his own copy in his own computer. This is balanced against the resulting performance problems due to the strain on the hardware resources. With sharing, one needs managing.
- It assures the programmer is concentrating on developing the logic of the program and a data specialist is writing code to describe the data structure, where it is, who has access to it, which program has access to which files… . In a shared data environment, there will always be some applications/users that rely on other applications/users to collect/update files for them. Then a data administrator insures that this is happening.
- It insures security, backup, and recovery.
- It is the custodian of the data and arbitrator in case of conflict; data are owned by the user as long as there is a single or primary user and no conflict exists.
- It provides an information resource management.

VIII. DATABASE MANAGEMENT SYSTEM AND SYSTEM OPERATION

System operation with a DBMS proceeds by entering data into the database, defining data for schema and subschema, designing and coding application programs, then running the application programs. For multiprogramming, the memory is occupied by an OS in a fixed area; other regions are allocated to each user, called a dynamic area. All requests about data are made by the application program as system calls. The application program is run with some recurrence; if it is to run once, a query processor provides for the inquiry function, and it is not necessary to code a special information collection request. Data are in the database application library or in a private library supplied by the user; the DBMS is primed when commands in the application program open the subschema referenced by that program. So far, data management has focused on certain types of data; multimedia data, such as graphics in the form of charts and figures, remain somewhat illegible; a picture is worth a thousand words, yet gleaning that information by computer remains a function to be.

Entering data with the expectation of getting it out easily involves:

- Deciding what data to include in the file; this depends on how you want to use it.
- Setting the form for data entry, e.g., table with headings (order number, customer number, item number, color, size).
- Setting field attributes to insure accuracy by machine, e.g., range on a number; cursor not passing through a field on a table such as customer number without it being filled.
- Setting default values to be overridden if a better value is available later.
- Setting file size for actual data in disk.
- Selecting the computer that formats the disk efficiently; the disk maker does not do it.
- Insuring security against loss of data due to power failure, machine failure, and provision for backup and recovery.

Retrieving data involves:

- Selecting fields from each record.
- Selecting records from each file.
- Selecting a subfile containing the part of interest from the original file.
- Selecting to be picked up by the database fields, records, subfiles, and files according to a certain description the user presents. The description is a template against which the search is conducted.

Indexing and sorting are features in the database that determine what and how fast data can be retrieved:

- Indexing is the identification of records without having to search through the whole file; one way is hash coding, which mixes up the characters in a desired field to make it into an almost unique arithmetic number which is held in order within the system.

DATABASE MANAGEMENT SYSTEMS

- Sorting is the arrangement of the file in the database by order of arrival, its rearrangement by name in alphabetical order, its rearrangement by department, etc.

Calculations involve operations in:

- Arithmetic; addition, subtraction, multiplication, division.
- Relational; comparing relative values rather than creating ones, e.g., greater than, equal to, less than.
- Logical; checking on conditions as existing or not, e.g., coexistence, either, absent.
- Functions; sine, cosine... .
- Formulas; combining previous operations.

To facilitate operations, a name gives a list from which one chooses; a hierarchy of menus has an item in one menu that gets a second menu, etc. A user-friendly operation would have a simple guided push-button operation using the menu as a guide. Too many menus can be bothersome to the person who knows what to do and what he/she wants; a balance is in order. The menu offers a list of choices, each backed by a set of instructions telling the program what to do next in the case of an option being chosen. Sometimes, command files are available where stringing a series of primary commands can be called up with a single command. This option is a powerful tool; otherwise, all that is available is a prescribed course of action that the designer has thought of. Both interactive and batch modes are usually set; in batch mode, predetermined instructions are executed by the computer; in interactive mode, the sequence of instructions halts and asks the user for some input.

IX. DISTRIBUTED DATABASE MANAGEMENT SYSTEM

A distributed system performs a user task near the user; it takes a part of the system resource to the job site. A well distributed system should have 95% of the traffic, to be processed by a distributed database system locally. Figure 14.14 gives a mapping of users to process/file; note process 1 and file 3 are only used by user 1 and should be set locally; note, file M is used by all users, thus the need to regionalize it. While distribution leads to decentralization, databases can represent virtual centralization effected through networking and communication.

In defining organizational requirements, one determines the requirements in terms of what information is required, by whom, in what form, and where. This leads to:

- Distributing databases.
- Replicating files so that system performance can be provided.

PROCESS OR FILE \ USER	1	2	3	...	N
PROCESS 1	YES				
PROCESS 2		YES	YES		YES
FILE 3	YES				
FILE M	YES	YES	YES	YES	YES

Figure 14.14 Mapping of users to processes/files.

- Synchronizing records so that they do not become out of phase.
- Determining processing requirements and data storage associated with each user.
- Presenting data in easily comprehensible form to the user.
- Adapting a design philosophy of compatible interchangeable elements and/or complementary differing elements.

A distributed database model follows:

- The database is presumed made up of a collection of relations, R1, Ri...RN.
- The relations are found on n sites, S1, S2...Sn. When a portion of Ri is located at site Sj, we designate that portion as Rij.
- Communication among sites is done by broadcast to all sites or site to site.

A fully distributed DBMS is difficult to build. To connect multiple DBMSs, one needs a control hierarchy to make the local DBMS's characteristics overlay the total system. The DBMS may be implemented as:

- Central, where there is a main computer and disk with users at the same or dispersed sites (Figure 14.15).

DATABASE MANAGEMENT SYSTEMS

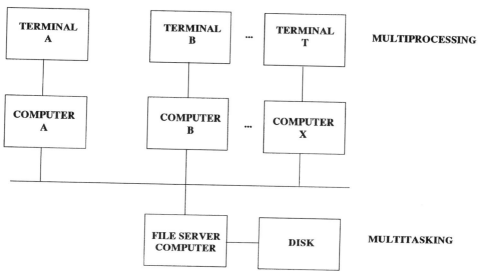

Figure 14.15 Distributed processing.

- Distributed, where there are multiple computers and disks with users at the same and/or dispersed sites linked together by a network; some representative examples include a) *distributed data where a program executing at one node accesses data at another node* by calling another program on that node; and b) *distributed database where files are required to satisfy a query*, as in a relational join, reside at different nodes; either data are brought together or the query is split apart or both to perform the join and send back the response. Most work on distributed databases has been done in the relational context. Recent work stresses an object-oriented approach.

The placement of data assigns which files at which nodes and which files can be replicated:

- Simplest; there is no file duplication and specific files are at specific nodes. This translates into no update problems, but also no backup for recovery and additional communication time delay.
- Most complex; there is full duplication with all files at all nodes. This yields update problems, but full backup; lots of storage is needed, but fast response is provided.
- A variation; all the files are at one node with some of the files at other nodes dictated by a "locality of reference" site that accesses them most frequently. This yields fast response, backup, but some update problems.

- Other considerations; those have been given to break up a file and to allocate one piece to a node; the break up may be: a) *horizontal*, where some records are allocated to a node, others at other nodes, or b) *vertical*, where some fields of the records are allocated at a node, others at other nodes.

In handling distributed data, the techniques are summarized as:

- Central master file and remote slave file; updates are passed to the master for distribution to users.
- A local master file and a slave central copy; updates are passed by local master files to users.
- One logical database where indices to data are stored at each user location so that any data entity can be requested as required; this demands a wide bandwidth communication network.

In updating duplicate data, the system locates the data, e.g., local or on a network. The system updates the local file, then goes around and updates their copies, but problems arise if: a) *the system goes down before updating is completed* or b) *another user updates a local file before the system servicing the first user gets to update the copy*. Then, means to reconstitute the database need to be applied, e.g., locking.

In data retrieval, the user may have a local request or a global request. A local request is satisfied by the local DBMS. A global request results in sub-requests being transmitted to a number of local databases with partial answers assembled to satisfy the global request. The network DBMS contains the directory from which to determine where to send the sub-requests to be satisfied.

Query processing optimization is obtained with respect to a) *total response time*, b) *usage of local resources* (CPU utilization, input/output operations), c) *network traffic*, and d) *parallel computer load distribution*. Each element is not independent of separate optimization. Item b concerns local (central) DBMS; items a, c, and d concern mostly distributed data. An increase in network traffic will improve response time if it leads to more parallel processing. Replicated data improve response time and lower network traffic when the file has a high ratio of read requests to update requests. A query is represented by a graph. It is formulated by a user. It is decomposed into many one-variable sub-queries, then reconverted into a multivariables query. A query decomposition algorithm has as inputs:

- Collection of clauses separated by AND; each clause contains only OR/NOT.
- Location of each fragment.
- Network communication type.

DATABASE MANAGEMENT SYSTEMS

The algorithm executes as follows:

- Do all one-variable queries on all sites.
- Apply reduction algorithm.
- Choose next query piece and select processing sites/variables and transmit.
- Run queries.

Exactly how to choose Ri relations to Sj sites is an open-ended problem until minimum network traffic and minimum processing time are defined.

To control updates, one may:

- Lock all users of that file before update is allowed. This locking mechanism can cause deadlock where two transactions have locked each other out. To solve, one may a) *force one transaction to give up the lock*, or b) *declare one node as the dominant node and send all updates to that node;* in this process, one node may become a bottleneck for a dominant file; in case of failure, the system has to transfer dominance to another file.
- Time stamp the transaction start time and last update and read time of a record and/or a piece of data. Integrity is controlled by making sure that every update stays in sequence; for similar transactions which call for update of the same record, one cancels the transaction with the earlier start and longer time lapse, since the user would be looking at old data by the time this update takes place.
- Use multisite joins which involve a) *sending all files to one site*, b) *sending only records/fields to one site*; or c) *joining the associated files situated at a site and ship the result to one site*.

The problem of data consistency/currency is influenced by the distribution of data and control. Inconsistency is triggered by:

- Delayed updates.
- Messages corrupted or lost in the interconnection medium.
- Interconnection medium variable transit delays.
- Physical processors/interconnection sudden failure.
- Insertion of new physical processors.

Consistency is performed through data locking during transaction processing and time stamping of data. Basically, control of events follows one of two options:

- Centralized at a dedicated control center.

- Decentralized where any one of the interacting elements assumes the role of master for the duration of its transactions with another element. In the decentralized case, control is determined by the events/activities and can run independently and in parallel.

In directory management, system directories or catalogs are needed to describe the physical data, the logical views of the data, and security constraints. In distributed DBMS, one now needs to add to the directory the location of data on node and where the directory should reside; for directory residence, the options are similar to data files:

- Total directory at one site; the update is easy since everybody knows where it is, but communication bottleneck is high, and reliability is low (no backup).
- Duplicated at every site; the update is difficult.
- Subdirectory pertinent to data at one site; then, for the rest of data, one has to search every node. On the other hand, one may add a whole directory at one node, but then the arguments of bottleneck and backup arise; one may add directory at several strategic sites; one improvement for any tradeoff on local and non-local directories is for the node to hold the directory/data found non-locally for a time since it is likely it may need it again.

X. DISCUSSIONS

Database management involves the same principles and issues developed in Parts I and II; the basic difference is in the focus and accent on the managing and managed resources, data. Together, Chapters 11–15 provide the foundation for computer automation of the varied application systems discussed in Part IV. Data cost a lot to collect and deserve special management; anything in lists has a counterpart in databases (catalogs, telephone books, dictionaries…). Distribution is 30% to 50% of the expenditure in an enterprise, i.e., time and capital (inventory management, warehousing, movement of information, goods, or currency – Part IV). Unmanaged distribution is like being transported on a bus with each passenger having a steering wheel and a computer that averages the inputs and controls the steering!

Database and data processing of the future must be able to integrate, share, interrelate a wide range of data, and present a unified interface to the information users; it should:

- Store and manage data in a way to allow rapid updates of displayed pictures (pictorial database).
- Manage digitized images and discern features as in pattern recognition.

DATABASE MANAGEMENT SYSTEMS

- Handle text for word processing.
- Do record keeping.
- Perform usual data processing to extract information in the midst of risk and uncertainty.
- Manage data for applications in computer aided system engineering, design, manufacturing, and office information systems.
- Have a knowledge base about some real world problem and expectation to respond to specific questions requiring inference from the database.
- Store and convert instructions by human, for instance, for processing by machine; this requires storage and rapid retrieval of semantic and syntactic interpretation rules.

An intelligent database manager searches for underlying common characteristics using rules on apparently unrelated data; a human brain has the ability to relate ill-defined data to objects (Figure 14.16).

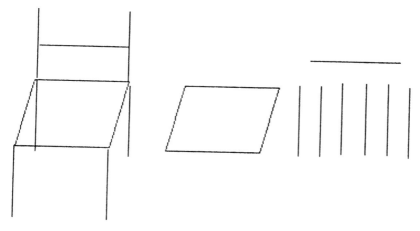

Figure 14.16 Relation of data to objects and their integration.

An object-oriented design methodology in data processing:

- Defines the problem.
- Defines an informal strategy first.
- Formalizes and realizes the strategy.
    - Identifies the objects and their attributes.
    - Identifies operations on the objects.
    - Sets open interfaces for the objects.
    - Establishes objects communication through messages.
    - Assembles objects (new and legacy) into a solution.
    - Implements the operations on objects.

A distributed DBMS is needed for:

- Faster, easier access for data and process through a) *parallelism* and b) *locality* of reference.
- Improved reliability through a) *multiple copies* and b) *multiple sites*.
- Improved modularity through a) *partitioning* and b) *load distribution*.
- Improved control and reporting through a) *organization* and b) *local control* and *global integration*.
- Lower total organizational cost through cost sharing by many sites.
- Faster response to changes of requirements or priorities through local changes.
- Enhanced availability through multiple copies.

Any distributed database, Figure 14.17, must address:

- How to provide an integrated database.
- Where to store the data in the system.
- How to locate data.
- How to control concurrent access.
- How to provide security and integrity.

Problem areas for distributed DBMS design include:

- Resource location as to distributed, centralized, static or dynamic.
- Data as to consistency, security, integrity, reliability, synchronization, structuring, inter-relations, mapping, modeling, accessibility.
- Fault location and recovery.
- Database full or partial redundancy, distributed entities, central copy.

In summary, a partitioned database has no duplication of data; it is made by occurrence or by structure. A replicated database has duplication of data at various sites; it is made by occurrence or by structure. An external schema in a database has the view of the plan as seen by the application. A conceptual schema is the data model defined in terms of entities, attributes, and relationships. An internal schema is the representation of the conceptual schema according to particular DBMS software rules. The directory is a set of tables which drive the internal DBMS programs providing mapping information between one schema level and another. The dictionary is a set of tables which define to the user the schema information in the databases. Database design consists of decisions based on competing requirements:

- How to split files, e.g., geographically, functionally.

DATABASE MANAGEMENT SYSTEMS

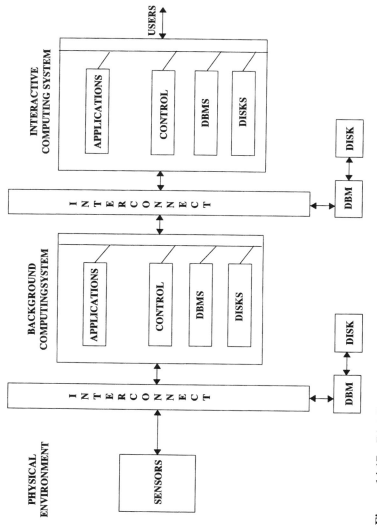

Figure 14.17 Distributed database management system.

- How to split directories, e.g., central, distributed.
- Where to place database management software, i.e., with data, with user.
- What level of communications service is required, i.e., long haul, local area.
- What user access method, i.e., embedded, ad hoc.

Finally, the economic well being of a nation rests on its competitiveness with other industrialized nations as they respectively move to take dominant positions within this formation revolution. The backbone of this revolution is the superhighway of high-capacity networked computers that empower the citizenry with knowledge to emerge in the lead position; this knowledge is embedded in the computer software; heterogeneous data management software is the critical technology needed to provide fast and easy access to the knowledge, to navigate rapidly through mountains of data to find the pertinent detail, to conduct multidimensional analysis, and to execute all that with minimal training. This technology will have a comparable effect on the information revolution perhaps equal to the steam engine of the industrial revolution.

15 MULTIPROCESSING SYSTEMS FOR REAL-TIME APPLICATIONS

I. INTRODUCTION

The managed and managing resources here are people, computers, machines and materials synthesized in a multiprocessing system, by applying the processes in Parts I–III. Multiprocessing systems do more functions in shorter time, with fewer failure times and with self recovery. The three building blocks of a decomposed multiprocessing system are humans, computers, machines/devices (Figure 15.1a). The integration of the building blocks forms the multiprocessing system (Figure 15.1b). The management of the total system leads to an integrated view of Parts I–III with: a) *management of humans*; b) *management of uniprocessors*; c) *management of programmed instructions*; d) *management of data*; and e) *management of machines/devices*.

Multiprocessing systems lend themselves to ease of distribution which in turn allows for on-time access to tasks' execution at low cost, with fault tolerance and stability; a good example of distribution is the telephone system. The on-time ease of access to system resources prompts the need for multiprocessing. Time underlies the desired characteristics of response, reliability, and availability, ease of growth and change:

- Response is time lapse from initiation of an action until the conclusion of its effect. Bottleneck to response time includes CPU time, memory/disk access time, device/manual time, communication time. Tailoring of the architecture to the task at hand is needed to balance the demand for services at each of the bottlenecks. Such bottlenecks are experienced in computer systems, for instance, in: a) *CPU overhead* involved in support of OS fielding interrupts, transferring data, processing messages; b) *memory mapping* needed due to small address space; this causes less processor power (5 to 25%); c) *communication overhead* involved in many activities such

256 SYSTEMS MANAGEMENT: People, Computers, Machines, Materials

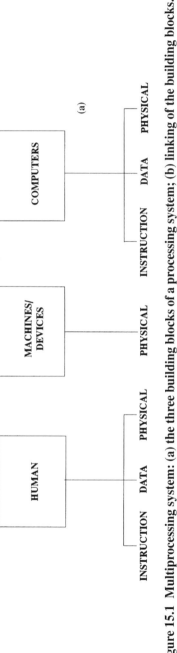

Figure 15.1 Multiprocessing system: (a) the three building blocks of a processing system; (b) linking of the building blocks.

MULTIPROCESSING SYSTEMS FOR REAL-TIME APPLICATIONS

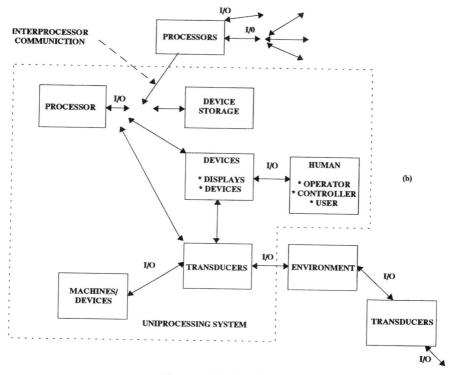

Figure 15.1 (continued)

as mapping the logical address of the receiver into a physical address, formulating a message, decomposing the message into a packet, transferring packets on the bus. While response time is the time it takes a system to respond to an input, throughput is the number of transactions per second that a system can process. Both are statistical in nature. Some specifications, of response time include average response time, e.g., 10 seconds; percent of response time less than a stated time, e.g., 90% of response time less than 12 seconds. Response time must be specified at some level of throughput; the higher the throughput, the higher the response time up to some limiting value.

- Reliability and availability, or fault tolerance denote the longevity of operation; they are statistical by nature and denote measures on the distribution that a system will do its function for a stated period of time. Reliability denotes the mean time between failure (MTBF) or uptime, and the mean time to repair (MTTR) or downtime. Availability is the average uptime divided by average downtime. Downtime gets long when the symptoms of failure appear in a processor other than the failed one; such failures are usually transient and dependent on a unique traffic pattern. Two elements are independent when failure of one occurs and the system can still operate; independence is achieved through parallelism or redundancy. Two elements are dependent if the system fails when either fails; they're in series. Sources of failures with representative metrics include: CPU (MTBF = 10,000 hours, MTTR = 3 hours), memory (MTBF = 20,000 hours, MTTR = 2 hours), disk controller (MTBF = 50,000 hours, MTTR = 5 hours), disk drive (MTBF = 5,000 hours, MTTR = 5 hours), display (MTBF = 10,000 hours, MTTR = 5 hours), power source, power controller, cooling, human operators, connectors, and devices.

- Ease of growth and change or scalability denotes the time it takes to enable the system to support a change/growth from its present characteristics to new desired characteristics. Time costs money and the cost of modernization affects the system price of ownership. Thus, a modular system architecture amenable to growth and change is highly desirable.

II. MULTIPROCESSING SYSTEMS ARCHITECTURE

The system's missions or goals are usually divided into functions, and functions into modes and submodes. This decomposition should tie at the outset the functions, modes, and submodes to the specifications within the system processes; this modular implementation enables the user to tie and untie easily the functions needed in the performance of an explicit service. Thus, progress of workflow is simply tracked through work assignment, routing, scheduling, and process status monitoring. System architecture partitions and allocates the functions/modes/submodes among humans/computers/machines types, selects the processors, sets the interconnects and data flows, defines system control, and predicts system reliability and readiness:

- Partitioning and allocation; the natural split of functions leads to applications and their support infrastructure including communication, database (file handling), operating system (OS), and other support functions. The desired allocation criteria cover a) *stand alone buildable, testable subfunctions,* b) *minimum data flow between processors,* c) *optimum cost of design vs. cost of manufacture,* and d) *minimum project time.* The desired human participation is at the supervisory/executive level to maintain real-time implementation. The implementation of the functions leads to varied types of processes executed by one or multiple processors:
    - Applications processing; an application may be considered as a set of transactions (actions, activities) in response to a given input request; a single processor may contain the application, or multiple processors do, and cooperate to satisfy the transaction.

MULTIPROCESSING SYSTEMS FOR REAL-TIME APPLICATIONS

Transactions, once identified, can serve as a basis for allocation. Parallelism can enhance response time when allocating to multiprocessors. In a client-server architecture, the client requests a service to be performed; the server performs the function and returns the results; clients only know how to request what they need and are not aware how servers execute the request; also, implementation of both client and server is hidden from the rest of the system; client–server computing has three distinct parts a) *the presentation* as to how the application is viewed by the user, b) *the application* of logic as to the way the organization works, c) *the data management* of critical information for the application.

- Data base processing; the file handler converts the logical name into a physical location, extracts the data from the physical file, formats the data and transmits it to the requester. The files may reside in disks or main memory.
- Operating system processing; it executes the task scheduling, memory management, and the interprocessors' communication support.
- Resource management processing; it executes initial program load, error detection and recovery, reconfiguration.
- I/O management processing; it receives and checks messages for errors, edits the message, reformats it into processor format, and queues the message for transmission.
- Traffic processing; it determines the flow path (Figure 15.2); the flow is along a) a preassigned stream, b) a multiplexed stream, c) a mixed stream. Multiplexed streams and mixed streams both require an utilization table in the multiplexers; otherwise, a round robin assignment is used. Systems are usually mixtures of preassigned processors, shared interconnects, and dynamic assignment. In dynamic assignments, the set of resources is assigned as needed by operational or failure requirements in CPU, memory, programs, interconnect, files, peripherals... .

- Processor selection; processors may all be similar or some dissimilar front-end processors, signal processors, data processors, display processors. Backup is more costly for dissimilar processes. Dissimilarity may be due to differences in execution time, instruction set, memory size, word size, etc. For similarity, one may have underutilized processors in the system.
- Processor interconnection; it permits transfer of messages between processes residing in different processors. The physical elements consist of a buffer memory in computer and interface mechanism to move data through an interconnect medium (twisted pair wires, coaxial cable, fiber optic, data busses, shared memory); those media may be dedicated or shared where more than two processes are

260   SYSTEMS MANAGEMENT: People, Computers, Machines, Materials

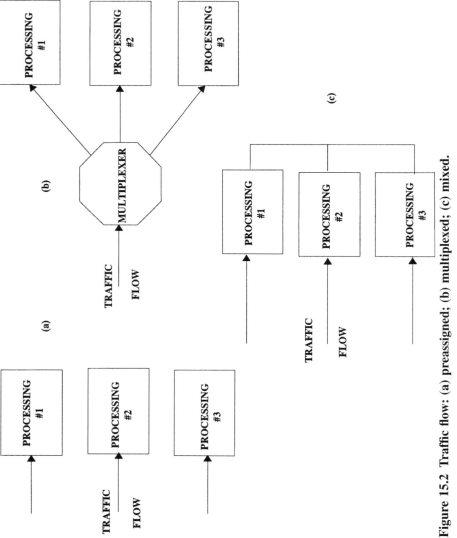

Figure 15.2 Traffic flow: (a) preassigned; (b) multiplexed; (c) mixed.

MULTIPROCESSING SYSTEMS FOR REAL-TIME APPLICATIONS 261

linked. Linking is based on time-division or space-division. The logical elements consist of message assembly, formatting, routing, and error checking. Depending on the intelligence of the interface, much of the logical mechanism is done either in the interface or the processor itself. The design decisions cover routing strategy, distribution of decision-making authority, bandwidth required, distance between processors, reliability, off the shelf elements. Sample networks are illustrated in Figure 15.3.

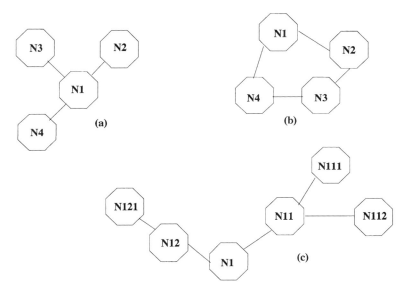

Figure 15.3 Networking structures: (a) star; (b) ring; (c) tree.

- Database organization; it is either central or distributed. A central database is easy to manage (storage, retrieval, update, recovery, integrity). A distributed database has more interprocessor communication. The distribution techniques include: a) *replicated files* where inquiries are alternately serviced depending on utilization; updates are sent to both; replication is good for redundancy and fast access; high replication requires wide communication bandwidth; b) *files divided by function* (purchase file, personnel file) where it may have slow access, less storage, no redundancy, and less update traffic; c) *divided files by related functions* where most transactions are processed by the processor at which they arrive. To find files/records, a directory is used. The location of a file/record is found through a directory; it establishes the disk location. We have a full system directory and local directories. Seldom is a full directory held at each file processor.

- System control; it is the decision-making authority and is organized in a hierarchy (tree), network, or relational way or a combination. The organization is physical or logical where: a) *it provides the system with reports and supports a number of processing modes*; b) *it provides local reports and supports local processing modes*; c) *it provides for data entry.*
- Reliability; this calls for fault/error detection, isolation, correction, system recovery. In a multiprocessing system, interprocessor messages are error encoded and checked at recipient processors to isolate errors. A faulty processor will operate in an errant way or sends incorrect messages; internal self checks are used for detection and isolation. Recovery is complicated by the existence and location of files, and availability of a spare. Single point hardware failures are minimized by putting in redundancy. Software failures are minimized by extensive testing. Fault tolerance depends on hardware redundancy and software checkpoints. Human failures are minimized through extensive training.

In the development of an architecture, there are four principal decomposition and integration stages. On one end is the functional requirements or problem definition stage which basically states the user's needs. On the other end are the system implementation stage and the operation stage. Sandwiched between them is the system architecture requirement-design stage; it begins with a flow down of user needs into identifiable and distinct functions; the functions define system characteristics and are depicted in a block diagram to clarify the textual description and relationship of functions. For each function, its description and relationship to other functions are summarized in a systematic pattern:

- Input data (data type, rates, accuracies, units, limits).
- Processing (mathematical operation, response time for solution, accuracies, text).
- Output data (destination, data types, rates, accuracies...).

The description and relationships are elaborated on in:

- Performance requirement; for each major function, the software/hardware specifications describe the data input characteristics, the required human/computer processing, and the required outputs.
- Interface requirement; it describes the format and content of data transferred between a computer/program and other computer/program/machine/human.
- Design requirement; it describes the items not directly related to performance requirement, such as modularity, expandability, types

of tests, implementation tests, integration tests, quality tests, string tests.
- Human requirement; it describes the requirements for display density, human–device interactions, time for decision-making to allow humans to perform their functions.
- Database requirement; it describes the collection of data that the program operates on, the volume of data, the description of data classes and adaptation from site to site. Depending upon the flow between the functions, the architecture is hierarchical and/or horizontal. In a system, there are many files and cross sections of data useful for the purpose of each file.
- Reliability requirement; it describes availability, on the basis of functional needs; some are more important than others.

To translate the preceding functional requirements into computer requirements, the following steps may be taken:

- A data sheet on a standard uniprocessor is characterized through: a) *benchmarking comparison of processor requirements doing communications*, based on timing of LOAD, ADD, JUMP, STORE instructions; b) *benchmarking of processor requirements doing computations based on timing,* first, arithmetic operations, exponential, sine, cosine, etc.; for a specific algorithm, one may calculate the number of operations of each type, multiply each type by the respective time, and sum all the times.
- Standard processor utilization is set where:
  - Application utilization is computed from (number of times an application executes per second) × (number of instructions per execution)/(number of instructions the computer executes per second).
  - Memory utilization is computed from program size (in RAM, disks), table sizes, buffer sizes (for storage of data in execution, and it depends on message arrival rates and average time a message is resident in the buffer).
  - File utilization is computed from the number of times per second the application makes use of the file. This is equal to (the file accesses per function) × (function calls per second) × (service time per access). A call may include a read or write as one access; an update calls for two accesses (a read and a write); create/delete a record calls for several accesses. A service time includes a disk seek time, a rotational delay, and a data transfer time.
  - Inter-application communication is computed from the amount of data in bits and the number of messages sent per second from one application to all other applications. It assesses the impact

of placing applications in different processes. A lot of messages causes lots of interrupts.
- Overhead utilization is computed from CPU utilization by OS functions, interrupts, task changes, checkpoint functions, and other background functions.
• Mapping of functional requirements to the virtual uniprocessor is now made using the previous approach for uniprocessor utilization:
- The processing requirements are computed in terms of needed CPU time/second.
- Next, memory size is computed; program size plus table size plus buffer requirement for each function are made. Separate estimates for main memory and disk storage are made.
- Next, database files are made to size, organize, and determine access methods. Response time requirements and proposed distribution are evaluated.
- Mapping of virtual uniprocessor requirements onto a needed set of standard computers is done. When the processing requirements call, for instance, for 6.8 seconds of CPU time/second, at least 7 standard computers need to be assigned to the task.

Basically, there are two principal types of multiprocessors, tightly coupled and loosely coupled:

• Tightly coupled processors share the same memory which allows them to operate on the same data or execute different parts of the same program; this tight coupling enables complex analysis of data and solutions of complex mathematical problems.
• Loosely coupled processors do not share the same memory and share data across a back plane/network where access is arbitrated according to a given protocol that guarantees each processor on-time access.

III. SOFTWARE AND HARDWARE VIEWS OF MUTIPROCESSING SYSTEMS

The basic elements of a uniprocessor have a CPU, primary storage, secondary storage (disk, tape), peripherals (printers, terminals, displays), and an I/O. These elements can be organized into a multiprocessor computing system. The two views of the computer system (software programs and hardware) still hold but with increased interaction among them:

• In the software view of the system, the user sees the computer system at the high order language (HOL) level through software; the OS sees the computer system at the software instruction level

and at the I/O software access method which interacts with the device driver appropriate to the device. The OS provides the user with a number of services or system calls and coordinates the system activities; user software portability depends on the compiler, the OS, and the representation of characters and numeric data items.
- In the hardware view of the system, the assembly language programmer sees: a) *main memory* that holds instructions and data; b) *register set* that holds data being manipulated by instructions, used also for stack size indication, context switching parameters, addressing, etc., and c) an I/O where registers hold data for transfer.

The software view of the system includes:

- Multiprogramming; when more than one program executes in the hardware, the OS has to provide:
  - Interprogram protection (memory and file).
  - A scheduler for the program execution.
  - A resource manager to allocate resources (CPU, disks, printers, displays).
  - A deadlock prevention algorithm.
  - The appearance to each program that it has a dedicated computer; multiprocessing is an actual realization of it.
- Multitasking; it provides for processes to cooperate to solve a problem given that hardware is shared. They communicate by way of messages, shared memory, synchronization flags, system calls. Each task acts in response to an event.
- Operating system; the variety of OS provides for: a) I/O to all peripherals, bootstrap, control, communication; b) time sharing. The I/O processing may be: a) *synchronous* or *asynchronous*, or b) buffered to disconnect the pace of slow devices from the computer and to average out irregular traffic.
- Protection; processes must be prevented from corrupting each other; this is done through: a) *assigning memory space to a process* and allowing others read-only permission; b) *assigning dedicated memory* (ROM) to the OS.
- File system; user defines logical structure (records); system software defines physical structure (blocks); contiguous files have the same logical and physical structure. The files may be:
  - Chained with arbitrary physical ordering of blocks with address of (f + 1) denoted at the end of the fth file; it is time consuming since the system must walk through the chain to get to (f + 1). On the other hand, the linkage information may be set into a directory in main memory.

- Indexed where links are provided in sorted order for fast searches; note that a single logical read may be a number of physical reads.
- Scheduling; it is needed with multiprogramming/multitasking. There are three states that a task or process can be in: a) *running/executing in the CPU*; b) *waiting for access to a resource* (disk-printer, messages); or c) *waiting for permission to run*. There are two types of scheduling: a) *priority assignment* (preempts a lower priority out of the run state, and b) *round-robin time slice assignment*. Priorities are assigned using a combination of ways, user assigned, waiting time, time in the run state, program size, time delay in system program itself.
- Process synchronization; resource allocation is designed to avoid having processes deadlocked (blocked) from entering the running state; the OS keeps an inventory status list of all requested accesses to the system resources and blocks all processes from running until their requests are satisfied. A semaphore, a shared data structure, is used to coordinate resource allocation to concurrent processes.

The hardware view of the system includes:

- Memory management; there is a large increase in physical memory. To put an address, a 16-bit computer allows direct access to 65536 bytes. Addressing of equal-size memory ranges (pages) is done and is mapped to a large physical memory. Not all pages are resident in the main memory, thus giving a new feature called virtual memory; the page is brought into main memory for processing when needed, usually 10% of code takes 90% of CPU time. Sometimes user memory and operating system maps are separately located.
- I/O; signals are passed via interrupts and data via programmed data transfer or direct memory access. Interrupts are input signals to the computer denoting that something has taken place (I/O completion, I/O error, terminal request, time run out). The procedure for servicing an interrupt is done in hardware, software, or both. When an interrupt arrives to the control unit in the CPU, the process being executed is suspended and switched out, the specific interrupt is identified and serviced; there is an interrupt latency between interrupt arrival and beginning of servicing. Servicing is done through a programmed I/O or DMA. The programmed I/O has instructions to transfer each byte or blocks of data where the program controls timing and synchronization. This limits the transfer rate; one may get a 500 microseconds overhead for 10 microseconds I/O operation. DMA transfers directly from memory to the I/O at memory speed and is well suited for block transfers. It steals memory time from the CPU. Some systems have I/O channels where multiple

MULTIPROCESSING SYSTEMS FOR REAL-TIME APPLICATIONS

devices are multiplexed to it; others use a bus where devices are attached.
- Peripheral devices; they include disks, tapes, hard copy, devices, and display terminals. Peripherals are usually slow; for instance, disk access time equals head movement time (time to reach the track) plus rotational time (time to reach the sector). To support fast response time, disk systems consist of multiple disks; information can be positioned on disks to minimize head movement with data on the same tracks residing on different disk cylinders.

IV. MULTIPROCESSING COMMUNICATION SYSTEMS

As the functions are partitioned among multiple processors to support the assigned load, a new bottleneck element enters into consideration, interprocessor communication which embodies the convergence of computing and communication. Here communication is the transmission of data whose content is encoded information; the transmission media may be copper wires, cables, wave guides, glass fibers, wireless, or satellites; media selection depends on frequency, bandwidth, distance, damping, distortion, dispersion, interference, noise, security, reliability, and cost. Once the media are selected, the transmission has two principal performance measures:

- Propagation delay; time it takes a bit to reach its destination. It is important for short communications such as control signals, error messages, synchronization signals, and polling signals.
- Throughput; amount of information transmitted per unit time (bytes per second) and can be: a) *point-to-point throughput* between two specific points in the systems; or b) *total system throughput*. N simultaneous point-to-point interconnection throughput is usually less than total system throughput; total system throughput divided by N approximates the desirable point-to-point throughput.

The protocol for an open system communication model has seven layers; not all the communication networks have all the layers. As an auto highway system needs entrance and exit ramps, safety features, and traffic controls, so does a data highway; there is much more to data highway realization than laying cables or fiber optic lines and pushing digitized data through them. A description of the protocol layers follows:

- First layer; it has the physical representations — voltages, currents, impedances, pin and connector definitions, modulation schemes, direction of data paths and physical media, such as twisted pair cable, backplane interconnections, coaxial, fiber optics. The twisted pair cable and the backplane employ direct current signaling (baseband)

where bits are represented by voltage levels and good for short distances; others convert the baseband and employ modems for longer distances.
- Second layer; it has the link level and guards against errors; CRC is added to the message. A typical packet includes a flag, source address, destination address, message header, text, CRC, and flag. If standards are used, off-shelf I/F hardware and software, i.e., protocol, are available.
- Third layer; it has network control with connection setting, call termination, and data routing.
- Fourth layer; it is a host-to-host connection setting, flow control, end-to-end error control; flow control makes sure that the receiver can accept and process data at the rate being sent; this is done through acknowledgment to the sender and buffering.
- Fifth layer; it sets up the virtual circuit to process but does not deal with the location of the process; the host usually supports multiple virtual circuits through multiple sockets (local addresses). Usually, the sender sends to a socket number which, through participation of OS with its directory, finds the requested process; the requested process must have indicated its availability.
- Sixth layer; it deals with data representation to adapt to the host with bit ordering, length of mantissa and its normalization, length of exponent.
- Seventh layer; it deals with application programs and involves system calls.

Communication switching techniques are used in moving data on shared paths from processor node to processor node on its way to its destination. Some switching techniques include:

- Circuit switching where a dedicated path is set between two processors through their nodes; once the circuit is established, data transfer occurs until the connection is terminated.
- Message switching where a dedicated path is not set and the message reaches its destination through an appended destination address; at each node, the entire message is stored, then forwarded until it reaches the addressed destination.
- Packet switching where the message is divided into packets, sent, then re-assembled; routing is done based on addressing. In a datagram, each packet contains source and destination addresses and error checks with no guarantee for unique packet delivery; in virtual circuit, simplex or duplex channels are opened, then packets are sent; no packet source/destinations are used, only the packet sequence number.

MULTIPROCESSING SYSTEMS FOR REAL-TIME APPLICATIONS

System network architecture can have:

- Switch; it connects sender to receiver over a dedicated line as in the telephone system.
- Shared bus; it uses high speed parallel or serial connections over coaxial or fiber optics.
- Ring; it almost always uses serial connection.
- Point-to-point; such connections are impractical as the number of processors increases.

Three basic techniques control access to the network:

- Random access or bus connection technique uses unscheduled time division multiplexing with distributed control at the bus interface node. Each node listens to another node's carrier to see if anybody is using the bus; a transmitting node listens to insure the message is identical; when a node senses a free bus, it starts sending after a set delay pertinent to it to minimize collision; if a collision occurs anyway, an adaptive backoff scheme is set; a buffer at the node is used if a neighboring node sends after it starts sending.
- Token passing (bus or ring), (Figure 15.4), access mechanism awaits its turn for access rather than contending for the bus; a message (token) is passed from host to host; when the host has the token, it talks; a host with a token may poll terminals and allow each to talk; sometimes a priority is set among the hosts in acquiring the token in a bus system, the token is passed to its logical successor; in a ring, each node is in the listening mode and captures the token when it desires to transmit.

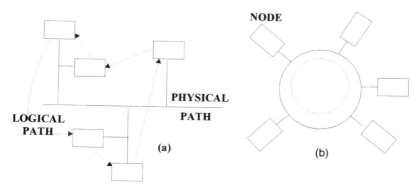

Figure 15.4 Flow path: (a) logical-physical non-overlapping; (b) logical-physical over-lapping.

- Register insertion uses shift registers to read the address field at the start of an incoming frame to determine addressee; the frame is passed onto the host processor or to the next node depending on address destination. When the host is transmitting, the frames to be transmitted are placed in the output buffer and transferred to the shift register when it empties.

Comparative assessment of computer networking is usually based on:

- OSI layers used (physical, data link, network, transport, session presentation, application); protocol between layers (syntax, semantics, timing) and effect on cooperative communication between applications in different computers (network interoperability).
- Network topology (star, bus, tree, ring), transmission media and their effect on reliability, expandability, performance.
- Network interface and effect on ease or complexity in establishing communication between devices.
- Network performance and effect on throughput and response time (Figure 15.5) determined for representative operational scenarios so as to integrate the network characteristics (number of nodes, protocol, number of bits per frame, propagation delay, bandwidth, load) into user measures.
- Network management and effect on initialization, operations, and evolution and their reflection on reliability, availability, survivability.
- Internetworking of homogeneous networks through bridges and inhomogeneous/hybrid through some gateways. Internetworking metrics should provide as a minimum: a) *link between networks*; b) *routing and delivery of data across the networks*; c) *status information on bridges and gateways* and should accommodate differences among techniques; d) *addressing scheme*; e) *packet size*; f) *error*; g) *time out*; and h) *status reporting*.
- Space differentials among the network implementations.
- Cost required to develop and maintain the networks' alternatives.

Connectivity of computers across wide areas is becoming as easy as voice telephone connectivity. Increasingly, computers are becoming impervious to their physical separation; connectivity is carried out initially through backplanes, then local area networks, then wide area networks that use telephone system technology now set on providing multimedia connectivity, using a switching protocol over lines, channels, and satellites.

V. MULTIPROCESSING OPERATING SYSTEMS

The manager for a multiprocessing system has two functions:

MULTIPROCESSING SYSTEMS FOR REAL-TIME APPLICATIONS

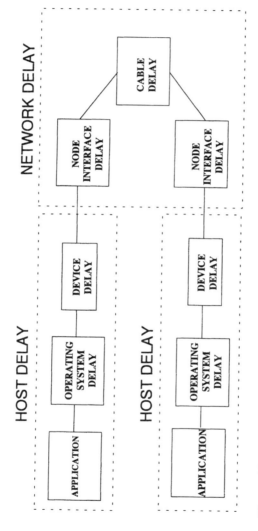

Figure 15.5 Delay model representation.

- Manager of local issues pertaining to the conventional uniprocessor, such as process execution, interrupt handling, and local resources.
- Manager of global issues pertaining to enhancements to the local manager, such as interprocessor communication, error and failure recovery, resource allocation, file systems, multitasking, etc.

A multiprocessing system possesses many specific features:

- Shared memory is used among the processors for high speed multiprocessor communication.
- While conventional OS has a common data area, multiprocessors do not in general have a common data area.
- Processes are executed concurrently in multiprocessor systems and thus quickly, especially when the data needed for each process are held locally.
- The host processor may communicate directly to another host or through switching processors; for the latter, few special commands need to be used.
- Resource control and reallocation are kept for the system (local, global); the global is kept in a directory.
- Multitask communication follows the same procedure as in a uniprocessor except for how the tasks identify each other; one way is through setting of a permanent address (socket); a host may have many sockets; intertask communication is through sockets assigned to tasks; synchronization uses flags or semaphores.
- File access is done either through the transfer of the file to the register or through having a dedicated host to service the file requests.
- Error and failure detection and correction are sent to a central location with automatic and human follow-up rules.

In multiprocessing systems, new issues arise:

- Structure of OS; it denotes how authority is organized and what the governing rules are where:
    - In a master–slave organization, one processor has central authority.
    - In a horizontal organization, the processors have identical functions with little inter-partition communication.
    - In a dedicated function organization, each processor is assigned a specific function with attention to minimum contention for resources; traffic distribution determines the viability of this organization.

MULTIPROCESSING SYSTEMS FOR REAL-TIME APPLICATIONS 273

- Interprocessor communication; the OS involvement depends on the network; it is limited to interaction between the applications programs and the network; with no network, some of the seven layers in the ISO need to be done by the OS in the host. Since most networks operate asynchronously and data are transferred to processes unexpectedly, buffers are used to avoid loss of data and to make sure that the process is in local memory.
- Multitasking; when extended across a network, additions have to be made in a uniprocessor; one must find the process of the program within the file system through its name, transfer it to the appropriate processor, start its execution each time the process is activated in a different processor; a different name is assigned or both the processor number and process name are jointly used.
- Interprocess communication; it is executed through a communication network or shared memory. In a communication network, each processor in the network has sockets; each socket has some tasks attached to it; a process may be attached to different sockets. For interprocess messages, the SEND command contains a socket of process ID. At times, a socket at the host acts as the server which uses a table of processes to inform the sender and receiver of a message; from then on, conversation proceeds independent of the server. In shared memory, the mailbox technique is used; the messages are left in the processor, and the process examines the mailbox at its own pace.
- Process synchronization; the value of semaphore at a single node is transmitted to other semaphores in other processors for synchronization; otherwise, a process count is used.
- Resource control; a single or multiple processors contain the total system resource control with resources named either with unique global names or locally for each processor. An order for a resource is made with a unique value.
- File systems; three techniques are used: a totally distributed file system, a special remote file access, and a file transfer. In a distributed file, there is usually no file across processors. If the file location is hidden from any task, a system with a file directory is needed. A file directory may be centralized or distributed. To avoid delay, it is helpful to have a file attribute, e.g., global, local. The nature of the file is provided in the open command. The name of a file in a category must be unique. File formats must be specified. Privileges of a process accessing a file can be spelled out. In remote file transfer, a consistent version of the file must be captured and would require locking out updates. For large files, this process can be quite time consuming.

VI. MULTIPROCESSING DATABASE MANAGEMENT SYSTEMS

The components of a DBMS in Chapter 14 are highlighted through its primitive operations, multiple access, locking of resources, reliability and distribution. A DBMS provides for organization, storage, retrieval, updates, and protection against loss, e.g., component or system crash of information. Organization is done by file and records. When elaborate relationships are not structured between the data elements in a record, the record is called "flat" with a fixed format. The files may be centrally located or distributed. Distribution can enhance performance by locating data near the user. The components of a DBMS include:

- Access software for creation, recovery, file management; data elements in a record are classified by element type; records are made up of element types belonging to a single entity object (person, place, thing).
- Data base schema, a record that identifies each element type, element restrictions (range...), record structure, access control, date of last updates.
- Indexing structure to access information by content rather than by location; a linked list is a thread through the data elements.

The primitive operations in a DBMS include open/close a database; find a set of records; open/close a record; get/delete a data element/record. With indexing, records specified can be collected in a list for later review. Multiple access to the same database records can occur. There are two types of records, static and dynamic. Static databases are referred to, e.g., telephone directory; dynamic databases are modified during operation, e.g., airline reservation; dynamic databases can suffer from overlapping updates. With an OS, file accesses are coordinated. With a single logical transaction in a database, many file transactions occur, and the OS is unaware of them. An approach is to treat the individual components in a database as allocatable resources by the OS. Otherwise, some synchronization is used:

- One user can access a record at a time.
- No record opened for reading can be opened for writing.
- One writer can access the record; all readers are given the old copy.

To effect the above points, locking of the file record or data elements is needed; a list of locked records is then kept. To avoid deadlocks in locking resources for exclusive access, the common technique is to use the ordered request. Each resource has an identification (ID), and the resource's IDs are set into an order. When requesting a response with an ID lower than the one already held, the higher resource's ID must be released; at times, release is made of held resources and the request for all needed resources is considered. Updates of redundant data call for locking; otherwise the database loses synchronization if one of the nodes cannot be accessed, so a clever transaction flow of lock, update, release must occur across the involved nodes. Note that high availability calls for more transactions which hinder performance on the network. Throughout, the interprocessor communication must be error free. When a single component of a database is distributed across nodes in a hierarchical structure, one node will have authority over the database. For distributed authority, a conflict strategy is needed to resolve disagreements or problems are reported to a human.

In examining the reliability issue, two major sources of failure arise; one is due to system hardware, system software, or power failure, and the other is due to DBMS software errors. A difficult failure is when the damage migrates; index structure damage can lead the DBMS to search wrong areas of the file for data. For fault recovery, one may have to undergo a data dump, then rebuild the operation by inputting the records anew; this is time consuming. A faster way is to make checkpoint dumps. If a fault is found, rolling forward from the checkpoint through the transactions is done since the checkpoint dump was done.

In distributing a database, we have partitioning (non-overlapping data) and redundancy; redundancy occurs for reliability or difficulty with strict partitioning. In clean partitioning, the data are divided to put it near the user without redundancy, at least for the data that can be updated. When partitioning with some redundancy, a record may be replicated in all nodes for high availability; synchronizing the updates is done through broadcast; this can have a low performance if updates to local access is a high ratio. For full redundancy and efficiency, the record is duplicated only once, e.g., node 1 has records 1, 2; node 2 has records 3, 1; node 3 has records 2, 3. A well used guide for distribution is based on system function, e.g., purchasing, marketing, employee relation, finance in a corporation or detection, classification, localization, motion analysis, weapon targeting in a radar or sonar system.

VII. MULTIPROCESSING RELIABILITY AND RECOVERY SYSTEMS

Reliability denotes successful system operation for a specified mean time interval; it indicates the ability of the system to provide performance; it is a function of hardware and software design as well as ease of operation. Hardware is usually stressed to weed out the weaklings prior to full operation. High reliability is obtained through redundancy. A failure may be transient (repaired through re-try) or permanent (repaired through re-load in new hardware). Some failure categories include:

- Hardware with processor memory module or bit failure, halts, incorrect results, incorrect inter-processor routing, open/shorted path, I/O devices failure to read/write.
- Database with inconsistent copies; data can't be found/read.
- Code with bugs, with altered code or an altered process can't be found or read.

In recovery, the categories include:

- Error detection; error is local to the processor itself or due to other processors through faulty messages or time-outs; early detection prevents contamination of data structures in the system.
  - Self detection is done through memory or program parity, ROM memory protect, I/O transfer checking, checksums (sum of elements in a data area is updated every time there is a change to the area). The rate of checking is balanced by the need to keep system overhead down. When an error is detected, a message is transmitted to a central error analysis containing processor ID, time of detection, nature of error, nature of data area... .
  - Cross detection is done through the absence of messages, delay in response, incorrect results, (nonexistent process I/O; process I/O is not in that processor; illegal data fields; incorrect message sequencing; numeric fields too large; garbled message).
- Diagnosis; upon error detection, a routine is initiated with a known result to test the elements of the processor. Impasses occur more in hierarchically structured systems than in horizontal ones. For the first, a horizontal logical structure is needed.
- Recovery; any function critical to system operation must be identified; when failure occurs, that function must be restarted through physical reconfiguration. For processor recovery, many possibilities exist:
  - Removal of low priority functions.
  - Doubling up of functions in working processors; the nodes are informed so that interfunction messages can be rerouted; the failed processors' functions have to be restarted from some initial state; the intervening transactions will be lost.
  - Use of a spare processor either a dedicated spare or roving spares; connection of spare is done through recabling, bus switch, multiplexer, communication line switch, dual I/O or common bus I/O.

For software recovery, programs and system data need to be reloaded. Programs are static and kept in a static file. Data is dynamic; some are transient work information, some are application data files, some pertain to other processors. Some of this data is needed for consistent system recovery. A dual

MULTIPROCESSING SYSTEMS FOR REAL-TIME APPLICATIONS 277

system keeps two separate up-to-date copies of all data; this is costly and a critical part of the system is kept in dual processors. Rather than duplicating a system, checkpoints (copies of critical data areas) are sent to a second processor. Checkpoints may be system wide or by processor; their frequency determines the accuracy of recovery and system overhead. Ready availability with pointers of checkpoints decreases recovery time.

For dedicated function systems, the recovery pieces are application, communication, and file handling. To support that system, there is usually a single spare or multiple spare processors. With a single spare, if one processor fails, error messages are sent by the processor itself or other processors to the area recovery manager or spare processor which sends or orders diagnostic messages be sent to the ailing processor; if there is a response error, the processor is ordered to stop operating; the spare conducts a self diagnostic, loads the needed program, and broadcasts a message to all others that it now acting in the new role. During the failure, communication with the processor may be mangled, some transactions may be lost, the database may be left in an inconsistent state. The central recovery manager/operator should run tests to determine if the problem is minor or if back out of data is necessary.

With a single spare, the faulty processor has to be removed from the rack and repaired. With multiple spares, the uncabling and repair can be delayed and a new spare is switched on. In some designs, at times, power may have to be turned off in the healthy processors, and that is why the multiple spares is a desirable feature. A system of roving spares can be complex, so specialized spares are used where a processor is spared by another processor.

So far, we stressed redundancy in hardware, but there are other subtle single points of failures. For instance, an application processor may create a command to update a file that destroys the integrity of the file records; this may be broadcast to multiple files. A protection for this error is to have the file processor check on the reasonableness of the request, and check on the write request in the table that relates an allowable file to application processes.

In summary, hardware transient failures can be cured by a reload and reinitialization of the software. Hard failures affect the instruction set, address counters, memory transfers; it can be cured by switching to a spare. Irreducible errors are caused by software bugs and hardware transient errors. They can, unchecked, contaminate copies of critical files, such as those of the resource manager, router, index files.

VIII. DISCUSSIONS

Early real-time systems were set around a main frame uniprocessor backed up by a second uniprocessor in case of failure; those systems were simple enough to put together. They were soon bottlenecked in CPU, memory, I/O. Since the eighties, design and development of multiprocessor systems have grown. The focus has been on the analysis and synthesis of large (local,

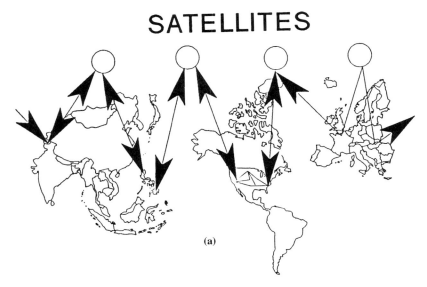

Figure 15.6 Distributed processing: (a) across continents; (b) across regions/localities; (c) across workstations.

regional, global) systems architected to perform multi-functions in real-time applications (Figure 15.6). Facilitating the process, computers are doing many of the management processes including information processing, communication, decision-making, algorithmic controls, and some heuristic controls of machines now relegated to humans. This presents many system design issues to contend with; it is similar to any planning, organizing, actuating, and controlling processes of large multifunctional and distributed operations where the elements are interactive humans, computers, and machines individually/jointly executing the system tasks:

- The multiprocessing systems are called upon to do more functions in shorter times with fewer failures and self recovery. Quality and short time to users, be it information or products, become the essential drivers.
- The functions describing the system once decomposed into processes, have to be partitioned and allocated to an acceptable number of configuration items (human, hardware, software) from the available resources.
- The configuration items have to cooperate closely to execute the processes through an operating plan, a communication medium, an organization, and system controls. Otherwise, the processes' requirements for fast response times, high throughput, high reliability, low cost could not be satisfied.

MULTIPROCESSING SYSTEMS FOR REAL-TIME APPLICATIONS 279

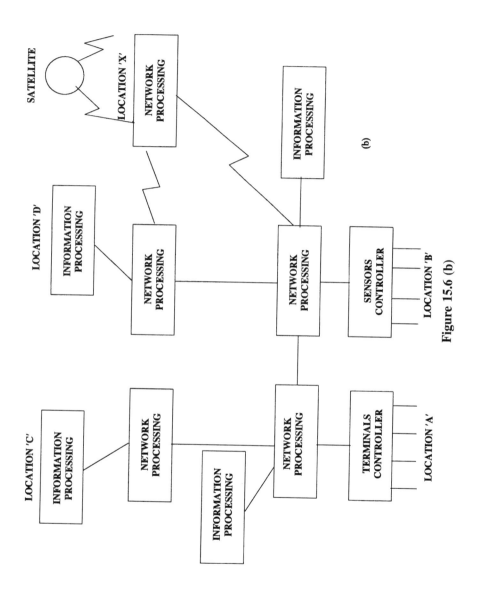

Figure 15.6 (b)

280 SYSTEMS MANAGEMENT: People, Computers, Machines, Materials

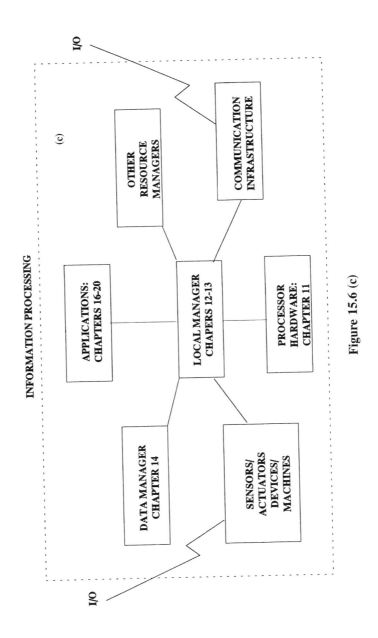

Figure 15.6 (c)

- The interprocesses' communications have to be kept to a minimum, their flow structure lateral rather than vertical; otherwise, the resultant overhead could jeopardize real-time system operation.
- Sharing of resources has to be managed, otherwise contention among the processes could arise.
- The large collection of configuration items has to be testable; otherwise, reliable operation is in jeopardy.
- The systems have to allow for configuration redundancy in recovery from faults, failures, to execute operational reconfigurations, and to be flexible in change and growth.

Multiprocessing results in distributed systems. Distribution lends itself to scalability; scalability is an attractive feature; it saves the cost of re-design and re-tooling and prolongs the utility of legacy system components. However, scaling up a design is linked to the technology it uses and to its failure to predict the upcoming ones. Client–server computing is revitalizing the approaches to distribution; the model is composed of three distinct parts with presentation or interface to the client part and application logic and data management parts acting as the server. Distribution of the parts across networks determines the type of client–server computing, e.g., distributed presentation, remote presentation, distributed application, remote data management, and distributed data management. Success depends on the connectivity software sitting in the middle and called middleware, that acts as the conduit between the parts.

Part IV:
Application Systems Management

16 AUTOMATED MANAGEMENT SYSTEMS AND THEIR ARCHITECTURES

I. INTRODUCTION

The automation of management systems improves, extends, and supplements human abilities in providing products/services effectively. Automation by itself is not a panacea unless the system processes to be automated lead to effective execution of the work. Then automation speeds the products/services processes and relieves the human from the need to have an in-depth understanding of the automated processes themselves. Automation may be a full, associate, and/or advisory type. This chapter capitalizes on the material in Parts I, II, and III, and provides the automation mechanisms in support of varied applications such as those in Part IV. The enabler for automation is the computer; it automates the operations in production systems, in banking systems, in financial systems, in health care services systems, in airline reservation systems, in inventory control systems, in radar-sonar systems, and in food services systems.

Computer systems can execute varied human intellectual functions; they sense, process, communicate, remember, decide, and act. Access to sensing and physical action is enabled through input/output devices which convert physical phenomena or their representations into processable or usable form; physical phenomena are varied and include force, displacement, acceleration, velocity, flow, humidity, temperature, pressure, level, thickness, quantity, rate, electromagnetics, etc. Methods of conversions utilize transduction; a transducer is a device used to interconnect like or unlike systems together. A transducer is actuated by power in one system and supplies power usually in another form to a second system. In electronic systems, the transducer converts a physical phenomenon into an electrical signal and vice versa. Transduction methods include variable resistance, e.g., resistors, photoconductors, thermistors, p-n junctions, magnetoresistances, strain gauges; variable reactance, e.g., capacitors, inductors; optoelectronic, e.g., photoemission, photojunction; photovoltaic, piezoelectric, thermoelectric, electromagnetic, digital... .

Computer processing can copy or surpass human algorithmic processing; it is in non-algorithmic processing where humans are researching ways to enable the computer to emulate and/or outperform them. The basic structure making computer processing automatic is the feedback processing loop with its:

- Standard or criteria of system performance.
- Measuring device and a comparator with weights to register deviation from the standard.
- Feedback to put the activated process back on track.
- Adaptation to a changing environment.

In implementing fully automatic systems, the basic loop closes on itself; it begins with the measurement which is converted, processed, compared to a reference, and fed back for control of the measured parameter. Open loop processing systems will require constant human vigilance to maintain predetermined standards. As the number of measurements increases, more human operators are needed, making the system labor intensive and costly. Computer automation reduces operator need and increases productivity. With computers in the processing loop, the system can have thousands of variables, feedback thousands of loops, and reset all the original criteria any time it is necessary to reach existing or new goals. With computers, the system operates effectively and responds quickly to internal and external events. The following sections present an integrated view of computer automation; underlying processing control structures are given; enabling processes in the computer to carry automation are presented; and automated architectures are presented in a generalized context.

II. REAL-TIME CONTROL IN AUTOMATED SYSTEMS

Technology has enabled the rapid execution of functions without the need for human physical contact or in-depth understanding of the executed processes. This has entailed speed of action, simplicity of operation, and increased confidence in intended output, Chapter 9. These characteristics are found in varied endeavors, be they commercial, industrial, or military. While military systems have all along pressed for the real-time concept of operations (speed), industrial and commercial systems are increasingly accentuating their competitiveness through the application of a just-in-time concept of operations. In all three endeavors, the underlying principle carries through where technology has invariably foregone the need for actual physical contact between humans/ machines and the functions to be performed, and has thus shortened the time of operation in the process. For instance, military operations began on foot

AUTOMATED MANAGEMENT SYSTEMS AND ARCHITECTURES

transport, occurred face to face, hand to hand, using sticks and stones, as weapons and speed of action were limited by human speed; it evolved with:

- Varied transport vehicles to bring humans quickly to combat on land (horses, carriages, elephants, automobiles, tanks...), at sea (ships, submarines...), and in the air (balloons, airplanes...).
- Varied weapons enabling them to conduct warfare at longer and longer separation distances e.g., slings, arrows, fire, powder, rifles, cannons, missiles, torpedos... .

Beyond human view, weapons to target are now carried through automated sensors and processing systems. Materials to products and products to markets are similarly transacted in differing environments. Figure 16.1 depicts a variety of environmental channels a system may operate within. A representation of a generalized automated system is given in Figure 16.2. In its application to a specific system, one modifies the input/output to reflect the mission of the system. Figure 16.3 gives elements found in a generalized system.

A definition of common monitoring and control is required in any automated system. A structured definition provides the system tools to automate platform systems in general. A well-formulated system control and monitoring concept provides:

- Well-defined system states, modes, and transition conditions of the system functions.
- Functional check of the automation concept of operations.
- Integrated system control with feedback processing loops.
- Automation of activities requiring frequent operator actions.
- Redundancy for reliable measurement and control of vital components.
- Local protection, e.g., fuse, for prevention of system damage.

Common control levels and control actions are illustrated in Figure 16.4. In framing a resource management concept, the overall system should have a finite number of states that fulfill the system goals. Each state is defined as a collection of functional capabilities, i.e., subsystem function, mode and submode supported by hardware (dedicated or common), software, and operator. A subsystem or component includes related functions; the mode and submode provide the grouping of subfunctions that fulfill major subgoals and contribute to the definition of the system state; the system modal organization provides to the human organization through workstations/viewports into the system, automatic operation or its interactive (associate/advisory) operation. In its operation, the system will transition from one state to another only if all capabilities required to support that state are available or can be made available. As an example, Figures 16.5–16.7 allocate the total system functions to five

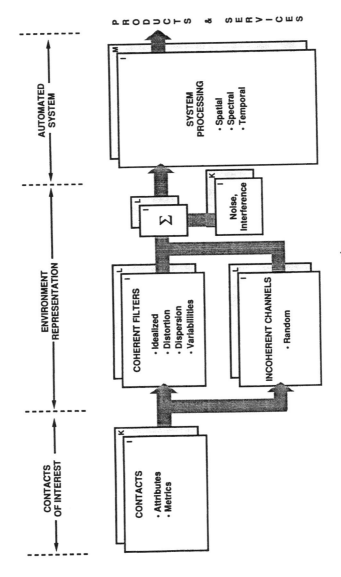

Figure 16.1 Contacts–environment–system representation.

AUTOMATED MANAGEMENT SYSTEMS AND ARCHITECTURES

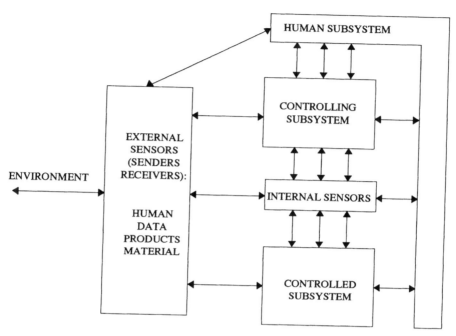

Figure 16.2 A representation of a general automated system.

components and provide a hierarchical flow of resource controls; Figure 16.8 provides: a) *the format for linking the function/mode/submode* in each component needed to fulfill the goals of a state, and b) *the system resources of hardware/software/operators* needed to execute the operations leading to the stated goals.

III. BASIC ELEMENTS IN ELECTRONIC AUTOMATED SYSTEMS

A. Basic Constituents

Multiple constituents are availed of when realizing an automated system, be it real-time or not; real-time adds a restrictive time measure to each of the functional operations. These include:

- Functional specification; any automated system is expected to perform on its own certain specified functions. Functional specifications give the external description of the system with a clear and precise statement of what the system is supposed to do for the users. It has the application areas; the inputs the user and/or the environment bring to the system; the outputs expected from the system; the

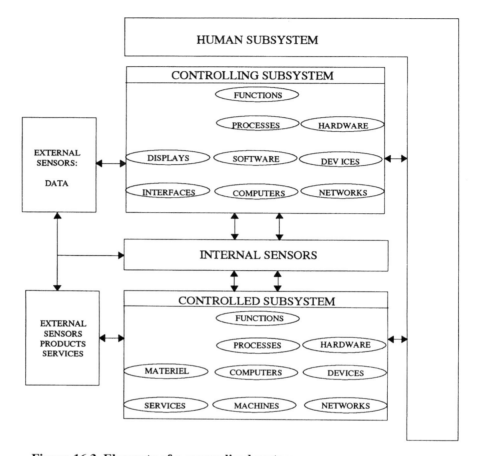

Figure 16.3 Elements of a generalized system.

maximum potential work load; the maximum acceptable time it takes between submitting inputs and receiving relevant outputs; the maximum object/data to be stored; for how long and its accessibility; the mean time between failures; maximum time to repair; and maximum time to reconfigure.
- Functional work breakdown; a top-level breakdown or partitioning of functions is given in Figure 16.9. The breakdowns are linked together through some form of communication, a language. Essentially, a hierarchy of languages, each with its own set of instructions, link the levels together from the functions down to the gates in a computer, for example, where the actual work is performed (Figure 16.10).
- Structuring; the functions are set to be performed by what is called front-end processors and/or back-end processors:

AUTOMATED MANAGEMENT SYSTEMS AND ARCHITECTURES

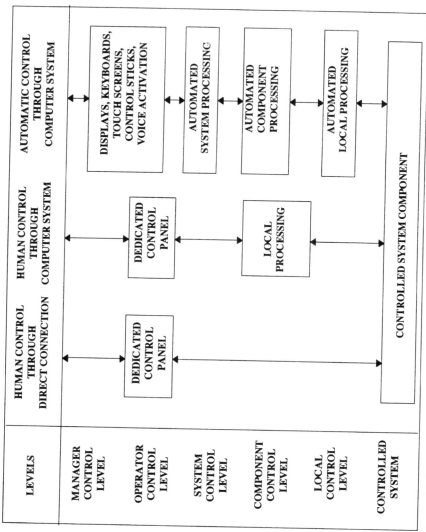

Figure 16.4 Common control levels.

292 SYSTEMS MANAGEMENT: People, Computers, Machines, Materials

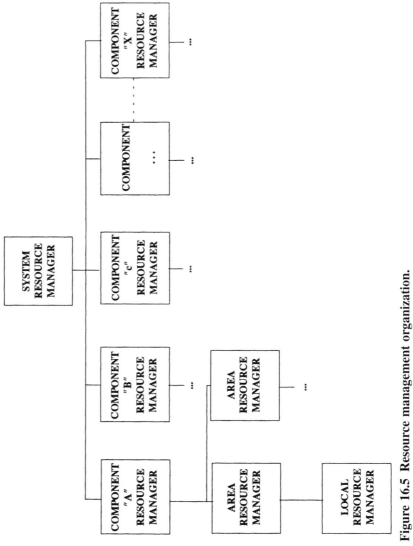

Figure 16.5 Resource management organization.

AUTOMATED MANAGEMENT SYSTEMS AND ARCHITECTURES 293

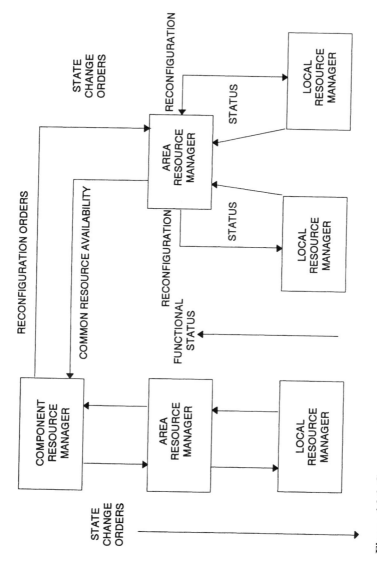

Figure 16.6 Component resource management.

294 SYSTEMS MANAGEMENT: People, Computers, Machines, Materials

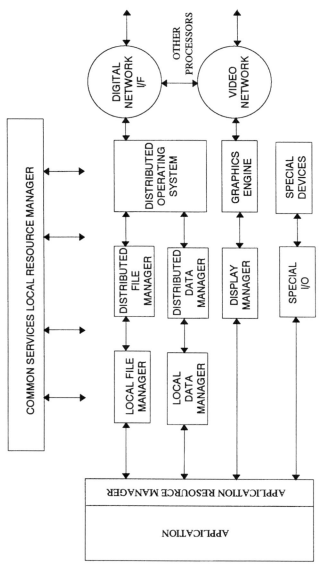

Figure 16.7 Local processor functional organization.

SYSTEM STATES	COMPONENTS								
	COMPONENT A						...	COMPONENT X
	FUNC-TION	MODE	SUB-MODE	HARD-WARE	SOFT-WARE	OPERA-TORS	...	
STATE 1	F1	M1	S1	· · ·	· · ·	· · ·	· · ·	· · ·
	F2	M2						· · ·
	F3							
STATE 2	· · ·	· · ·	· · ·					· · ·
STATE 3	· · ·	· · ·	· · ·					· · ·
STATE 4								

Figure 16.8 System states vs. capabilities template.

296 SYSTEMS MANAGEMENT: People, Computers, Machines, Materials

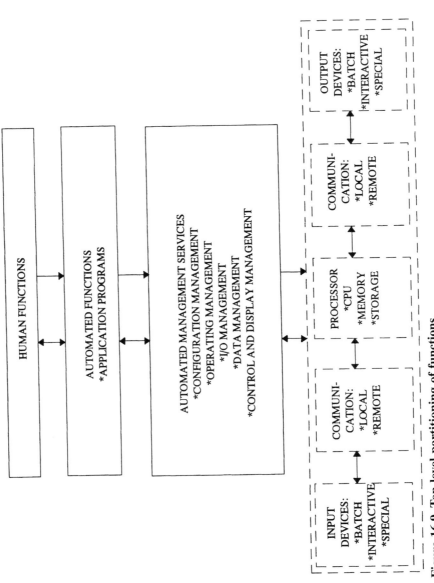

Figure 16.9 Top-level partitioning of functions.

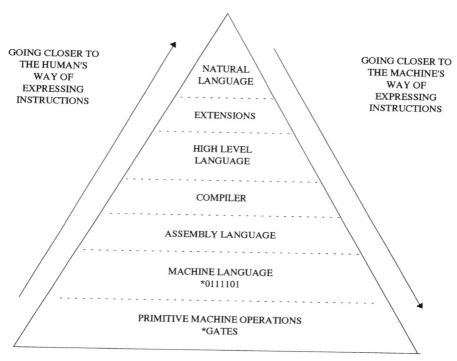

Figure 16.10 Hierarchy of languages to effect communication.

- Front-end processors are sensor driven; they are usually dedicated-function processors where specific functions are preassigned to processors; a particular function is carried out with part of the processing handled by each of several cooperating processors.
- Back-end processors are user driven; data streams from front-end processors are merged, weighted, and routed to satisfy each user's functional responsibilities. With improvement in computing power, front-end processors are becoming more and more general purpose processors of the back-end type.
- Processing; functions are executed through processing. Processing converts data into information. It occurs by human-only (manual-mental), by computer-machine only, or by human-computer-machine in consort. The operations include:
  - Calculation; basic arithmetic operation on which higher mathematical operations are built. They consist of addition, multiplication, delay, and storage to determine a transfer function; the types are: a) *recursive* and utilize previously calculated outputs along with input to determine a transfer function; b) *nonrecursive* and utilize input only to determine the transfer function.

- Detection and location; pinning down the presence of a contact (object) and setting its location in time, space, frequency.
- Classification; coding, identification separation into categories, pinning down the identity of each object in the category and referring to it through abbreviation or code.
- Sorting; putting data in some order, e.g., alphabetical, numerical, by date, by time, by function.
- Summarizing; weighting of data, then adding it.
- Recording; capturing data/information on a medium (paper, disk...).
- Communication; transporting data/information from one point to another. The transport may be done by human, electrically (0-1, on-off), mechanically (production line).

Those operations occur simply, in combination, in some sequence, and/or at varied levels. Each set of operations form one application; an application is a set of transactions; each transaction is a set of actions related to a specific input; processing the input demands a portion of system resources. A symbolic representation of the logical steps in the operations prior to actual code is done in a flow chart; its elements are depicted in Table 16.1. A good implementation of logic should yield stand alone, testable and, therefore, buildable objects of functions within each single processor; testing subjects the implementation to the conditions it must operate under to see if it works.

Table 16.1 Functions and Their Symbols in Flow Charts

Symbol	Function
☐	Input/Output
▱	Process, Calculation
◇	Decision (Yes–No). If–Then
○	Connection of One Point to Another
→	Direction of Flow
0	Beginning/End

At the heart of an automated system are the processors; each has a Central Processing Unit (CPU), its own memory capable of supporting its local operation manager, the application programs, data tables, and a connection to the interprocessor communications system. Basically, the combination of programs provides data and instruction set, timing, and synchronization for the hardware to execute. The central processing unit relies on internal and external memory for storage. The internal memory may be random access with static or dynamic storage or read-only-memory with permanent storage. The external memory may be cassettes, disks, tapes; their drive is all electrical/mechanical

AUTOMATED MANAGEMENT SYSTEMS AND ARCHITECTURES   299

components to read or write data on its medium; it often excludes the controller of flow of data between it and the computer. The trade-off between internal and external storage depends on capacity, cost of drive, cost of media, on-line time, and approximate access time factors. When computer memory is not large enough to accommodate the tasks, two procedures are used, overlay and virtual storage:

- For overlay, the program is split into segments where each can perform a total sub-task and where each segment is brought into memory for processing.
- With virtual storage the operator is relieved from the need to segment programs. A feature known as dynamic address translation will find and take the needed program part from auxiliary storage and place it in main storage.

Devices enable processors to execute physical work. The devices linking processors and human/machines include:

- Input devices; keyboards, buttons, switches, track balls, joy sticks, thumb wheels, analog to digital (A/D) converters, teletype, hard copy, voice input, light pens, optical devices, optical readers, and transducers.
- Output devices; D/A, video, hard copy, printers, teletype, displays, film, binary lights, transducers, motors, and actuators.

Displays play the dominant role in conveying data/information resident within the computer memory to a human and vice versa. How a computer can feed pictures in rapid sequence to a basic graphic display device, when a picture the size of a TV screen has to have 1000 pixels in the horizontal and 1000 pixels in the vertical to begin to be near the quality of the original? A pixel is a picture element of a small square (tile); for a moving picture, the computer must be able to display a new (1000×1000) pixel roughly thirty times a second to give the illusion of a continuously changing scene; this requires the painting of thirty million tiles (pixels)/second. This brute force implementation demands much of the computer power as well as powerful computers. Other approaches can be used and include:

- Predictive techniques where tiles change themselves using their current and recent past states and those of other surrounding tiles. The only input to the tile is a correction signal to adjust those tiles that differ significantly from the prediction.
- Transformation techniques where a display device draws a limited number of basic patterns, such as geometric shapes or fetches them from memory. A controlling computer determines what items to

fetch and how to transform them to approximate any picture. Typical transformations include:
- Moving a figure.
- Rotating a figure.
- Fading.
- Shrinking or enlarging.

Like a cartoonist, drawing a dynamic scene is done on transparency over the fixed background. The computer has enabled the programmed automation of devices' setting, operation, synchronization, and thus the automated multimedia technology and its enrichment through computer graphics, animation, and digital audio. Prior to the computer, multimedia production were stand alone media devices such as television, slide projectors, audio tapes, and video monitors. The integration of the computer's processing power to television has empowered the visualization of information and of complex concepts heretofore unwitnessed.

B. System Configurations

The configuration of a system is composed of the preceding constituents; it is generally arranged according to: space division, time division, and frequency division. The configuration takes the following basic structures (Figure 16.11).

- Hierarchical, where a master processor supervises the operation of "slave" processors on data streams.
- Parallel, where processors operate on data streams in parallel or on the same data for redundancy; parallel lines improve the response time.
- Network, where processors operate on data streams captured by a given processor on a ring according to a preassigned criterion.
- Combination, where parallel and hierarchical coexist or where network and hierarchical coexist with one processor preassigned to be a master processor.

In connecting a processor into a configuration, an interface switch links one onto another. Those communication links may be dedicated and enforce physically the system concept of control; otherwise, the concept of control can be enforced by the logical design rather than the physical. In a general system, the logical and/or physical structures do not necessarily correspond and one may encounter overlays of hierarchical/network/relational structures. A sophisticated switch, for instance, is realized by a processor; the role of this communication process is to interface, e.g., front end, the host processor with the communication channels (Figure 16.12); its functions can be scheduling, protocol handling, e.g.,

AUTOMATED MANAGEMENT SYSTEMS AND ARCHITECTURES 301

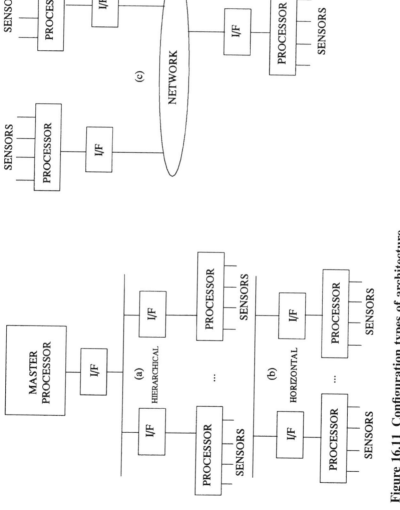

Figure 16.11 Configuration types of architecture.

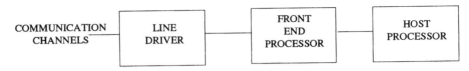

Figure 16.12 Smart interface.

message-to-packet formation, data compression, polling, storage and buffering, data-link control, or formatting.

Communication moves data from place to place by parallel or serial means:

- Parallel; it is used for high speed interfaces; all bits in a word move simultaneously, each on its own wire plus others for control signals. Signal delay, damping, or distortions due to excessive wire length cause machines to malfunction; fiber optic is remedying such deficiencies in copper wires.
- Serial; it sends bits one bit at a time and are reassembled at the receiving end into word. It uses:
  - Asynchronous method; sending/receiving with an arbitrary time between the bits.
  - Synchronous method; sending and receiving at a particular instant of time.

In complex systems, an organized network of data communications pathways is set to link multiple computers/devices together:

- Point-to-point
- Star
- Multipoint
- Ring
- Strapped
- Bus

C. Automated Management Functions

In an automated system configuration, varied services' managers are encountered that participate in facilitating the execution of applications. For instance:

- System hardware management is concerned with the scheduling of hardware system resources including CPU, busses, I/O devices, and memory.
- Network management is concerned with local and wide communication, protocol handling, polling, and interrupt handling.

AUTOMATED MANAGEMENT SYSTEMS AND ARCHITECTURES 303

- Data management is concerned with organization, activation, and control of data structures.
- Program management is concerned with system job flow and program monitoring.

As in any organization, not all managers have exactly the same titles and functional responsibilities; a top-level breakdown falls along application programs and system services programs such as system manager, database manager. System services programs support a variety of applications; application programs are specific to a given application.

System services managers perform a variety of services to support the execution of the applications; they:

- Control resources and access to them.
- Protect users' data from one another.
- Provide data storage and retrieval mechanism.
- Control access to system through passwords.
- Provide resources for backup.
- Provide statistics on resource utilization to allow adjustment of allocation methods.
- Provide device independence, i.e., users need not know the precise characteristics of input/output devices.
- Tell CPU where everything will be stored and how to get it.
- Enable external devices to acquire computer resources whenever they need it.
- Schedule access to system resources; when contention for the same resource happens, allocation is done according to some predefined priority.
- Perform time-shared tasks in the background with a priority which can be preempted by foreground tasks which have higher priority access to resources.
- Control the activities driven by time or event. In a time-driven management system, each task or user gets a time slice of a resource when his turn comes. In event driven or real-time management systems, the system waits for an event to happen and then executes what it is instructed to do. A multitasking or multiprogramming system coordinates requests for resources in response to external events. When contention for the same resource happens, allocation is done by assigned priority. Usually background tasks have lower priorities than foreground tasks. The switching from one task to another takes time, with resulting time loss. When many tasks are called for, the computer may take too long to finish. Interrupts are either:
    - Simple, where a single external device is serviced.

- Vectored, where an interrupt is recognized from any one of several external devices.
- Priority, where an interrupt is recognized from any one of several external devices along with which device has priority.
• Insure that all available resources are maximally utilized in a multitask environment.

The most essential manager in a computer is the so-called operating system manager. It is usually comprised of compilers, control system programs, and utility system programs:

• Compiler; it translates the program used by an application into machine language.
• Control system software; job control statements are processed by control system software; that software performs:
  - Interrupt analysis that determines the cause and directs the computer to take appropriate action.
  - Initiation of I/O.
  - Error recovery software that orders the device to re-read the record that caused the error and/or resorts to more sophisticated methods in its attempt to fix the problem.
• Utility system software that includes:
  - Library maintenance; software that can add, delete, and/or copy programs into or from various libraries located in auxiliary stores.
  - Diagnostics software; software that provides error messages, prints contents of memory, and generates test data.
  - Sorting and merging; reorders data file to suit the purpose, e.g., alphabetically, by department.
  - Job reporting; software for identifying users.

IV. AUTOMATED SYSTEMS ARCHITECTURES

A. Requirements

A generic processing block diagram is given in Figure 16.13 to motivate discussions of specific system architectures. The mix of environments and contacts modulates the processing difficulty and drives the type of technology needed and the adopted architecture. Representative processing requirements are given in terms of data rates and operations per second, (Figure 16.14), whose processing and management need to be satisfied automatically. From a historical perspective, the technology has evolved over a span of five decades, from hard wired analog systems to high order language programmable systems (Figure 16.15). The implied trends in processing requirements are depicted in Figures 16.16, 16.17, and 16.18 in terms of projected operations, physical space, and memory size. Other processing requirements besides FLOPS/MIPS/BYTES are

AUTOMATED MANAGEMENT SYSTEMS AND ARCHITECTURES

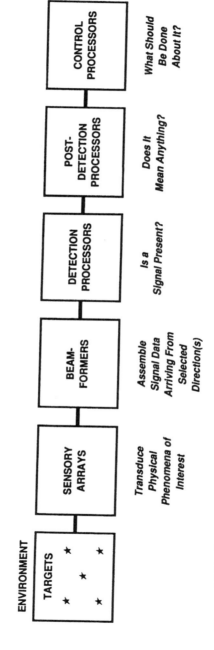

Figure 16.13 Generic processing block diagram.

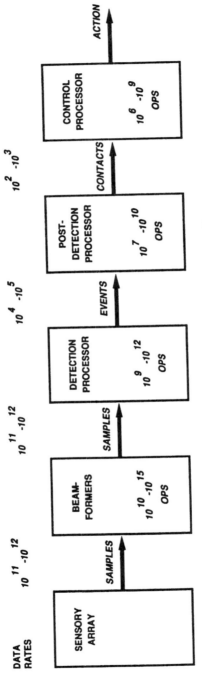

Figure 16.14 Representative flow of data rates per seconds and operations (OPS).

AUTOMATED MANAGEMENT SYSTEMS AND ARCHITECTURES 307

TECHNOLOGY	HISTORICAL PERSPECTIVE (DECADE)			
ANALOG (HARD WIRED)	70	60	50	40
DIGITAL (HARD WIRED)	70	70	60	50
COMPUTER (PROGRAMMABLE)	90	80	70	60

Figure 16.15 Trend toward digital programmable computers with higher order language.

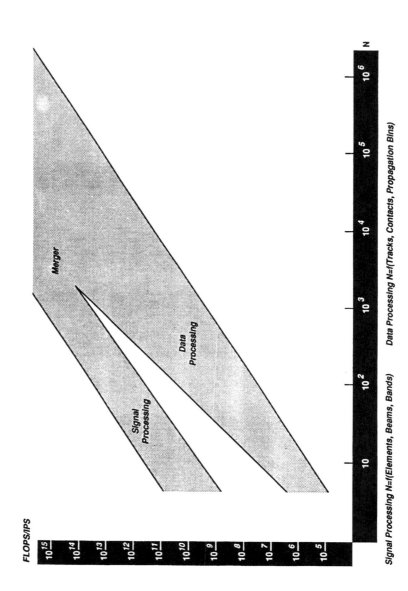

Figure 16.16 Signal and data processing requirements in terms of n-floating point operations per seconds (FLOPS) and millions of instructions per second (MIPS).

AUTOMATED MANAGEMENT SYSTEMS AND ARCHITECTURES

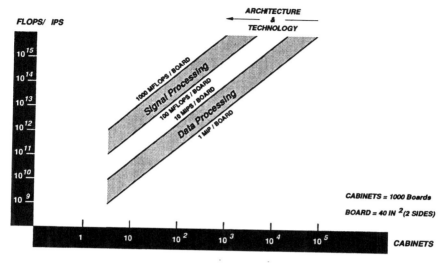

Figure 16.17 Compaction of processing requirements into physical space.

given in Table 16.2. The underlying implication is a requirement for maximizing the physical geography over which system action can be conducted, while minimizing the physical space needed to house the system; a decrease in system space/weight has carried with it a decrease in cost. If system space compaction from the eighties to the nineties is an indicator, where an electronics cabinet was shrunk into a toaster, the path is clear for further progress (Table 16.3) in space and cost reduction.

B. Implementation Approaches

In satisfying the requirements, three technological approaches can be pursued: (a) *representations of the applications*, (b) *improvement in devices*, and (c) *framing of the architectures*.

For the applications, a single processing pursuit solves a problem type only, e.g., a given contact, scenario, or environment. What is needed is:

- An integrated processing approach to respond to the variety of factors. Figure 16.20 gives three functional areas, signal analysis, pattern recognition and expert systems, integrated to estimate the elements of interest in the environment.
- Reduction in computational order, N, for the variety of processes using new concepts, new relations, new conditions (Figure 16.21). The processing form remains batch/iterative/recursive.

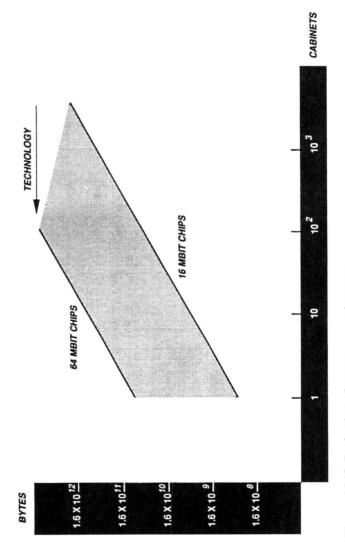

Figure 16.18 Projection of memory requirements.

AUTOMATED MANAGEMENT SYSTEMS AND ARCHITECTURES

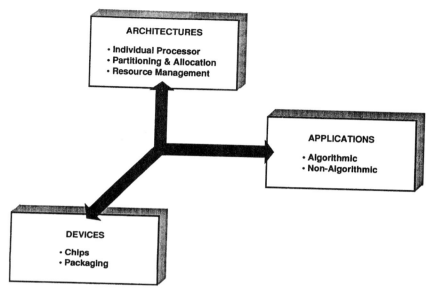

Figure 16.19 Three technology approaches to satisfy the requirements.

Table 16.2 Requirements Beyond FLOPS/MIPS/BYTES

Requirement	Implication
High input/output rate	Superfast internal computation alone is not fast enough
Critical latency	Computational rates cannot be averaged
Wide variety of algorithms must be supported	Architecture cannot be tuned to classes of algorithms
Hostile environments	Mandatory fault tolerance
Low cost product	Not low cost or high quality but both

For the devices, there are two paths:

- In chips (MIPS/ft^3), the technology is pushing toward the physical limit with one more doubling in clock speed and one more quadrupling in memory density.
- In packaging (Wires/ft^3), the technology is pushing toward direct high density interconnect of multiple chips within a package rather than chip packaging first resulting in a) size and weight reduction from 10:1 to 20:1, b) performance improvement with some 75% reduction in interconnect delays, c) reliability improvement with some 75% fewer connections, and d) some plus and minus cooling issues.

Table 16.3 Sample of Compaction from 1980 to 1990

	Compaction (Military Computer Systems)	
	The Eighties	The Nineties
Cabinet Space: ~24ft^3	• 3 MIPS	Processing • 96 MIPS, • 32 MFLOPS I/O • 118 MIPS (Dedicated) • 92 MIPS (Networking)
	• 20 MBYTES	Memory • 164 MBYTES

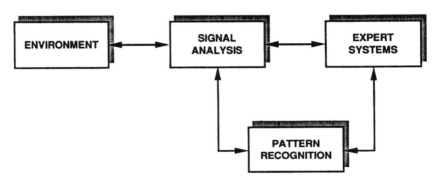

Figure 16.20 Integrated estimation of the environment-processing variables.

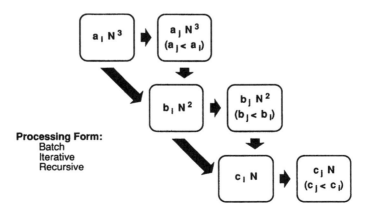

Figure 16.21 Desired trend in computations to reduce necessary requirements.

AUTOMATED MANAGEMENT SYSTEMS AND ARCHITECTURES 313

Interconnect technology plays a major role in heading off performance limitations, repeatedly confronted by system designers as they scale up their architectures. Present focus stresses speed of operation. But, a linear logarithm–logarithm relationship exists between the number of processing elements and the number of pins/wires needed to communicate with the remaining system. When unanticipated, the system can become interconnect-limited; wires connect devices on a chip, chips on a board, boards in a drawer, drawers in a cabinet, and cabinets in a system.

For the architecture, partitioning, and allocation of functions along serial processing would lead to bottlenecks; parallel processing is required; throughput, response time, fault tolerance are achieved by concurrent (parallel) processing (Figure 16.22). Benefits of parallelism can be realized by effective functional decomposition and balancing of architecture:

- Functional decomposition must support parallelism at many levels (by and across sensors), Figure 16.23. The level at which parallelism is effected in functional decomposition affects system complexity, cost, and response time, Figure 16.24.
- Balancing is required between parallelism at the operational level vs. functional level. Unbalanced functional decomposition deteriorates speed up vs. number of processors (Figure 16.25).

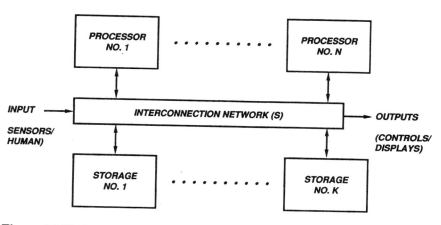

Figure 16.22 A parallel processing architecture.

C. Management of Resources

In architecture, management concerns itself with the optimum uses of resources in response to work load leveling (Figure 16.26). In large systems, a single manager very seldom has charge of all the resources. The need for specialized resource managers is triggered by the decomposition process:

314   SYSTEMS MANAGEMENT: People, Computers, Machines, Materials

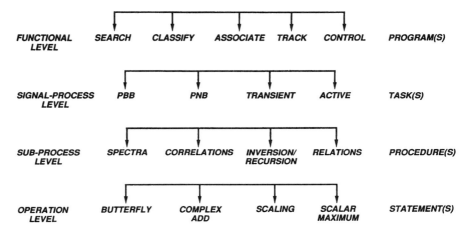

Figure 16.23 A functional program decomposition; (PBB: Passive Broadband; PNB: Passive Narrowband).

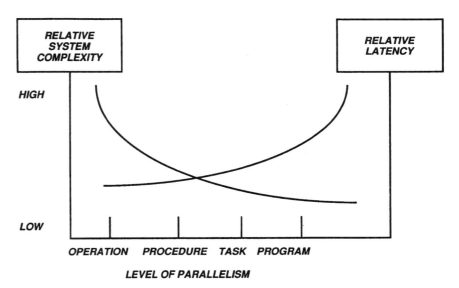

Figure 16.24 The levels in parallelism and their effects on system complexity and latency.

- Decompositions rarely result in subfunctions which exactly fit the resources provided by one processor.
- Decompositions often result in two or more subfunctions sharing a single computer and/or one subfunction split across more than one computer. A special function called resource management is

AUTOMATED MANAGEMENT SYSTEMS AND ARCHITECTURES 315

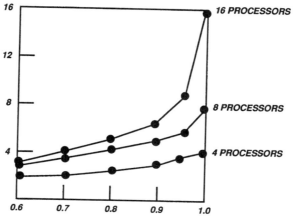

Figure 16.25 Functional decomposition and effect on parallelism.

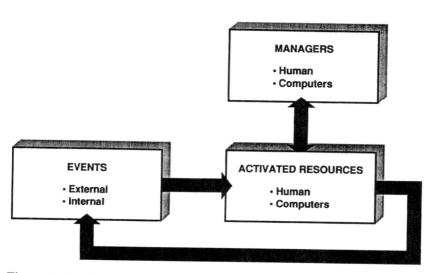

Figure 16.26 Management of resources in response to events.

assigned the requirement of managing the sharing of resources among functions and managing their activities which include:
 - Goals setting.
 - Programs to achieve goals.
 - Organization to carry out programs.
 - Resource manager to activate the organization.
 - Controls of activities to stay on track.

Management of parallel or concurrent use of resources requires certain characteristics to be implemented in the management layer of the shared resources. The layer is set in software intermediaries which harness the primitive power of the hardware in order to execute the essential functions of the applications:

- The greater the insulation of those primary functions from both the intermediaries and the hardware:
  - The greater is the ease of change and growth of the applications.
  - The less is the turnaround time in support of new system roles/goals.
  - The less is the cost of re-doing non-primary functions.
- The greater the insulation of the elements in the intermediaries from each other:
  - The greater is the flexible expansion of the system as needed.
  - The greater is the ease of capture of commercial advances, but...
  - The less is the system response time, which can be a drawback in real-time systems, but increasingly compensated for by increased hardware processing power.

There are three types of resources: hardware, software, and so-called operation resources. Resource management uses the operation resources to activate/create on-line software resources on the hardware resources (Table 16.4). The management of the three resources is given to various types of resource managers. A sample work breakdown among the managers in the system is given in Table 16.5. To share better, the linking of resources needs to handle not only the vertical/downward relationships, but also the lateral/diagonal relationships (Figure 16.27). In activating the resources, the management processes coordinate/synchronize different functions in each system, different levels in each function, and different structures in each level to satisfy the system's goals of speed, reliability, and low cost (Figure 16.28). A sample system organization is given in Figure 16.29. Parallelism allows rapid system change and growth and reduces the need for high risk development with long term payoff. Further parallelism is introduced to process similar data together.

In designing critical architectures, an integrated system engineering process is a necessity. Back-of-the-envelope balancing of the diverse and parallel activities will not make it (Figure 16.30). A system balancing and verification cycle need to occur (Figure 16.31). Some key architecture performance metrics are given in Figure 16.32. Finally, operational engagement scenarios must be used to exercise and validate the architecture (Figure 16.33).

Table 16.4 Computer System Resources

Hardware Resources	Software Resources		Operation Resources
Computers	Applications	Memory	Create
CPU	Program	Partitions	Initialize/Configure
Memory	Task	Pages	Read/Receive
Memory mapper			Write/Send
Timer	Files	Data	Schedule
I/O	Directories	Tables	Start
	Files	Rows	Wait
Displays	Records		Stop
Graphics engine		Displays	Status/Health
Display memory	Communications	Viewports	Lock
Screen spaces	Channel	Windows	Unlock
MMI	Buffer	MMI	
	Messages		
Mass Storage		Timers	
Disks		Time of Day	
Tapes		Alarm Clock	
Special		Elapsed Time	

V. DISCUSSIONS

Mechanization is the first application of physical devices for productive work; varied physical sources power up the mechanization process. Automation is self-directed mechanization through computation. Computers follow the prescribed computation plan, called a program, and execute such computational work as investigation, quantification, understanding, prediction, and decision-making. Thus, effective automation requires first an intelligent re-examination of the system to insure the effectiveness of its processes prior to the application of automation technology.

Mechanization is a first step toward automation. Mechanization is a step up in complexity and in cost over devices, where cost is compensated for through increased productivity. With real automation, mechanization is not only at the machine level, but at the system level; complex sets of machines are managed by other intellectual machines, the computers. Even at the machine level, the computer monitors the behavior of the machine, e.g., an engine, and modifies its multivariables, e.g., firing, gas and air mixture, to achieve greater efficiency. Computers are particularly suited to dealing with multivariables; while machines, through continuous production, provide for "plenty", computers, through job order processing, allow for "variety".

Table 16.5 Managerial Levels and Managed Resources

Sample Work Breakdown Among Managers			Managed Resources						
Management Level	Generic Names	Applications	Files	Communications	Memory Partitions	Data	Display Windows	Hardware Suite
Local (Single Computer) Frequent assignment Temporary use until needed by higher priority	Operating system manager	X	X	X	X			X
	Reconfiguration manager	X					X	X
	Data manager					X		
	Display manager						X	X
Area (Multiple Computers) Modest frequency Grouping by related resources I.E., by sensor stream Resource dedicated until surrendered by function	Network manager			X				
	Reconfiguration manager	X					X	X
	Data manager					X		
Central (Total System) Infrequent assignment System wide Resource dedicated for life of function	Reconfiguration manager	X						X
	Data manager					X		

AUTOMATED MANAGEMENT SYSTEMS AND ARCHITECTURES 319

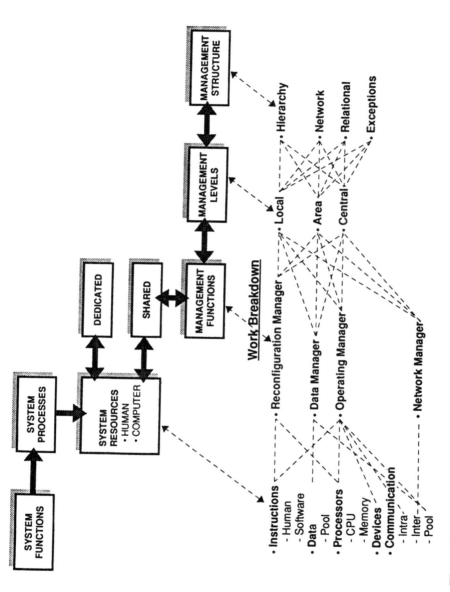

Figure 16.27 Linking of resources and their management.

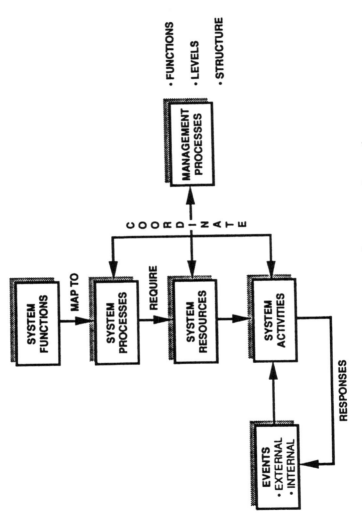

Figure 16.28 Management of system processes in response to events.

AUTOMATED MANAGEMENT SYSTEMS AND ARCHITECTURES

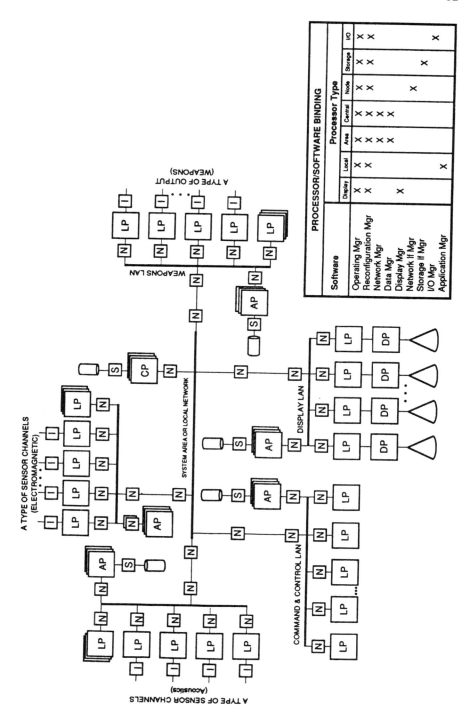

Figure 16.29 A system architecture: N (Node); I (Interface).

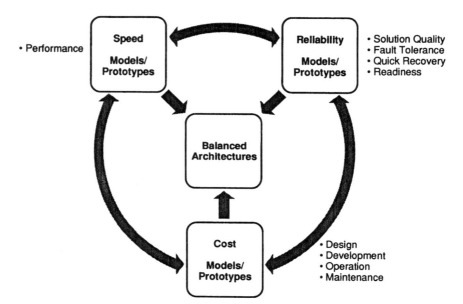

Figure 16.30 Balancing of an architecture.

Computers, in some sense, are a go-between between humans and machines. With complex and multiple machines forming a system, a rigorous process of system definition, analysis, design, development, and verification is followed (Figure 16.34) during realization of such automated systems. With automation, there is greater in-depth and upfront system management needs for planning, organizing, activating, controlling, and communicating.

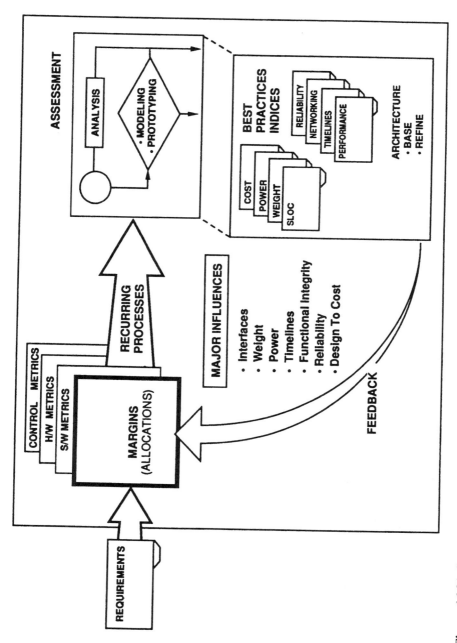

Figure 16.31 System verification approach.

324 SYSTEMS MANAGEMENT: People, Computers, Machines, Materials

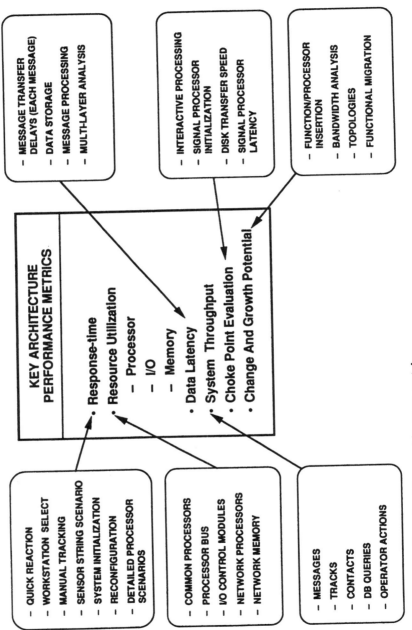

Figure 16.32 Sample of key architecture metrics.

AUTOMATED MANAGEMENT SYSTEMS AND ARCHITECTURES

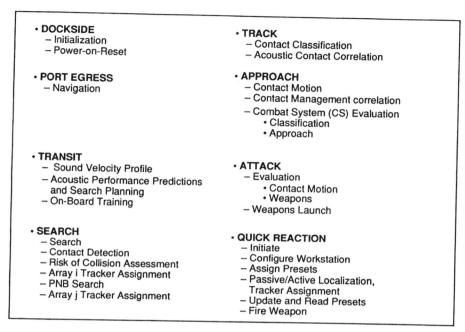

Figure 16.33 A tactical engagement scenario.

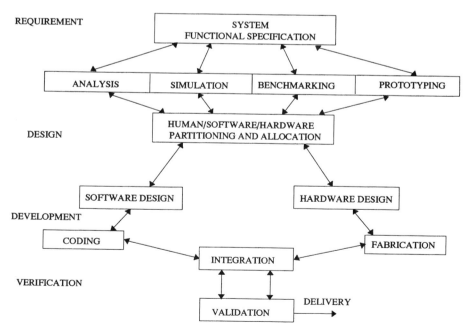

Figure 16.34 Inclusive process for realization of automation.

17 AUTOMATED INFORMATION PROCESSING UNDER RISKY AND UNCERTAIN CONDITIONS

I. INTRODUCTION

Managerial activities involve decision-making between alternative courses of action. Successful selection of the best alternative rests on availability of information usually extracted from data under risky and uncertain conditions. Information processing can be considered as a search for invariants in the midst of transformations, distortion, dispersion, and noise (Figure 17.1). Its synthesis rests on the concepts in Part II, and its automated realization on the computer systems described in Part III.

Information processing breaks down into signal and data processing. Signal and data processing are defined as the filtering of physical data to develop information for decision making; without such processing, the information will remain obscure, and the database management of Chapter 14 can only manipulate obscured information; the computer systems processing in Part III operates on non-obscured data unless a fault/failure arises. Though the principles apply to other applications in Part IV, and to failure modes in Part III, we limit information processing to deal here with the extraction and characterization of two basic but important descriptors of physical phenomena, their location in time and their location in space. Signal processing deals with extraction of information from a single look observation and short time interval. Data processing deals with extraction of information over multiple look observations and extended time intervals. The time length of an observation is limited by stationarity or invariance, present or induced, in the physical data. Three elements join to influence the difficulty in signal and data processing:

- The character of the incoming signal.
- The type of noise and interference.
- The channel through which they propagate and interact.

328 SYSTEMS MANAGEMENT: People, Computers, Machines, Materials

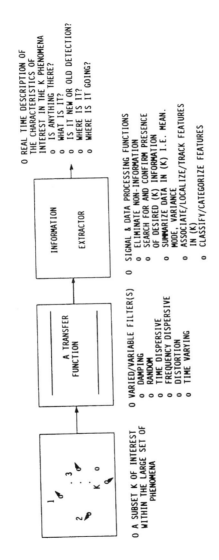

Figure 17.1 General problem and desired solution.

AUTOMATED INFORMATION PROCESSING

To exemplify the concepts of extracting information in difficult channels, the ocean/atmospheric channels are used as the context. Additionally, the physical phenomena of interest are not directly observed, but rather through the signal waves they emit or scatter and which propagate to a distance away, where they are sensed and processed.

Space and time are considered independent in the Newtonian sense here; that is, nothing spatial affects time and vice versa. Space and time are fused in the theory of relativity. Space and time are not an intrinsic property, as mass or force, and have to be specified through a system of reference. In space, direct spatial measurement uses a yardstick where the number of times the yardstick is laid between two points yields the number of yards making up the distance. In temporal measurement, natural time units or their subdivisions are the year and the day, but fluctuate in value to serve as an easy standard; thus, better clocks were increasingly developed based on rhythms of impressive constancy until employment of rhythm in waves as in electromagnetic waves or acoustic waves. In fact, when distances cannot be suitably traveled to lay the yardstick, methods are applied that rely on the wave speed to measure distances, angles, and velocities, thus yielding localization and tracking of objects. Tracking of such objects as stars, planes, and submarines falls into this category.

Radio Detection and Ranging (RADAR) systems and Sound Navigation and Ranging (SONAR) systems are employed to extend the visual and hearing senses beyond human capability. While both are not substitutes for human eyes and ears nor are they capable of recognizing sophisticated attributes of objects, e.g., color, they are able to extract information on range, angular direction, and velocity very accurately as reflected by their respective acronyms. Common to both radar and sonar are the signal and data processing functions which yield resolution in range, angle, frequency, and time. Spatial, spectral, and temporal decompositions followed by weighted integration are applied to the detection localization and tracking functions of objects in either radars or sonars.

II. SIGNAL, NOISE, CHANNELS, PROCESSING, AND INTERRELATIONS

A. Signal Noise and Channel Outputs

A signal is the part of the wave field we are interested in extracting due to its emission or scatter by the physical phenomenon or object under observation (Figure 17.2). The rest constitutes noise and interference to the extraction process. In the ocean, for instance, acoustic waves include (Figure 17.3):

- Human-made noise, including that from ships and submersible machinery. The machinery type is involuntary and has a frequency content dominant below two kilohertz (KHZ), and its character is broadband superimposed with narrowband and transients. There, transient signals have a frequency spread far beyond the two KHZ. The

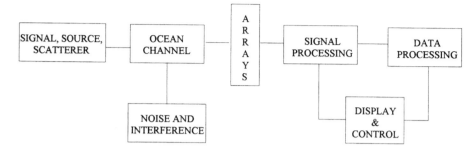

Figure 17.2 Elements of the problem.

active sonar is a voluntary type of emission and has a variable frequency, .1–300 KHZ, dictated by the application at hand. Passive sonar is the processing of others' voluntary and involuntary emissions.
- Thermal, due to molecular fluctuation and is usually detectable above 30 KHZ.
- Sea surface motion with a frequency range from about 1 to 50 KHZ.
- Biologics, due to snapping shrimps and others with approximate frequency range .1–100 KHZ.
- Seismic with very low frequency content of a few HZ.
- Flow caused by objects moving through water, by current flow over uneven bottoms, pressure changes...with normally low to very low frequency content.
- Sea ice formation and its breakage with a frequency content range from a few HZ to a KHZ.

In the atmosphere, similar classes of electromagnetic sources are encountered:

- Human-made, including those of transmitters in support of varied applications such as communications, radio, television, and radar. The transmitted waveforms can be pulse, continuous wave, amplitude modulated wave, frequency modulated wave, phase modulated wave, or other coded pulse-compression. The communications frequency coverage runs the gamut from Hertz (HZ) to thousand gigahertz (GHZ). Conventional radars are operated at frequencies extending from 200 megahertz to 35 GHZ, while over the horizon radar is operated in the range 4–30 MHZ, and ground wave radars as low as 2 MHZ, millimeter radars at some 100 GHZ and laser radars at even higher frequencies.
- Thermal, impulsive and arise from a number of sources such as conduction electrons in the ohmic portions of the receiver (thermal), galactic noise from outer space (thermal), atmospheric noise from

AUTOMATED INFORMATION PROCESSING

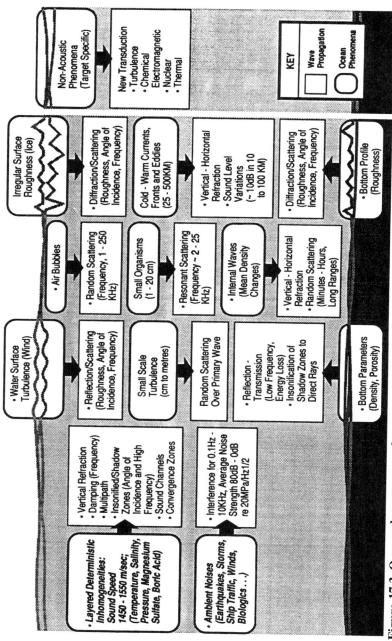

Figure 17.3 Ocean phenomena and wave propagation.

lightning discharges (thermal/impulsive), electrical machinery (impulsive). Noise interval to the receiving system is usually the dominant noise in systems with frequencies above 300 MHZ; in the frequency band (3–30 MHZ), atmospheric and galactic noises usually dominate in operations away from human-made sources.

The channel can be benign or very complex and difficult. There can be many changes in the wave velocities which are a function of both space and time. Those changes map into multipath effects, inhomogeneities in the volume, and roughness at the boundaries. When present, they tend to introduce various types of complications to system operations:

- Damping, where, for a given input, the output is too weak to be detected.
- Random, where for a given input, one of multiple outputs occurs with a certain probability.
- Time dispersive, where the input time duration is spread in time at the output, as in multipath.
- Frequency dispersive, where input frequency content is spread further in frequency at the output.
- Distortional, where the input amplitude spectrum is modified differently as a function of frequency.
- Time-varying, where the character of the output depends on the time the input is applied.

Ultimately, the signal and data processing are influenced by the above factors through the incoming waveforms' time variation, amplitude, phase, and frequency contents. Typically, a starting approach linking signal–noise–channel and processor is the power signal to noise relationship where the linkage is influenced only by power loss (db):

(Source signal level) − (transmission loss) + (target strength (if active)) = (noise level) − (receiver directivity index) + (receiver detection threshold)

This relationship initiates the decision-making process, i.e., detection. To effect detection, the relationship gives us some options, such as looking for stronger signals; operating in less noisy and damping channels, getting closer to the source, building a bigger array, securing a lower thresholding processor. One thing the relationship does not tell us is how to execute these options, i.e., it does not give, for instance, the structure of that lower thresholding processor. Other things the relationship does not provide include contact classification, identification, tracking, localization, and motion. It is obvious that thorough approaches are needed if we are to make more serious decisions, and those approaches lead to increasingly complicated mathematical formulations.

AUTOMATED INFORMATION PROCESSING

B. Information Processing Structures

A general signal and data processing diagram is given in Figure (17.4). Not all elements are necessarily present in a specific application. Sonar applications in the ocean environment include:

- Underground exploration as in echo sounding from the bottom and sub-bottom strata for mineral location and identification.
- Biological applications as in fish detection.
- Undersea navigation and equipment positioning.
- Passive and active sonar from manned or unmanned platforms for object detection, classification, tracking, attack, or avoidance; passive implies observation of a platform's own wave emission. Passive observations mode extend beyond sonar to radar/infrared/visual signals to provide tactical and intelligence information on the source.
- Remote sensing of oceanographic phenomena, i.e., fronts and eddies.

Radar applications include:

- Air traffic control as in airport surveillance radar and enroute radar.
- Sea surface surveillance as in harbors and open sea collision avoidance with other ships, buoys, and land.
- Air surveillance as in airborne, ship-based, ground-based radars.
- Ground surveillance as in artillery and mortar locators, artillery support, terrain following and avoidance.
- Remote sensing of geophysical objects or the environment as the weather, earth resources which include water, sea conditions, ice cover, forestry, agriculture, geological formation, environmental pollution, and ionospheric studies.
- Contact tracking as in air-to-air, air-to-ground, and ground-to-air missile guidance.

The functions in the applications are performed by computer–machine, by human, or jointly by human and computer–machine. In setting performance, the steps include:

- Partitioning of overall system processing requirements into tasks.
- Allocation of those tasks to processors and people to perform.
- Synchronization and control of multiple processors/people to perform the overall system processes.

The simplest human–machine allocation is an array of transduction elements and human ears or sets of eyes; this system still constitutes a baseline for detection and classification of contacts. The addition of basic parameter estimation and displays enables the human to conduct more diverse signal and

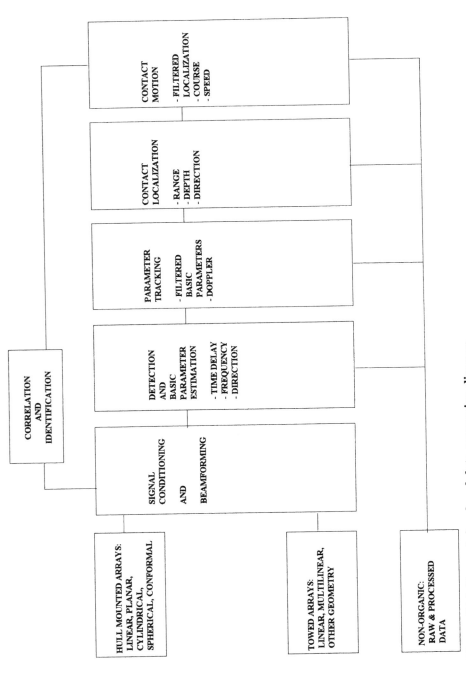

Figure 17.4 General signal and data processing diagram.

AUTOMATED INFORMATION PROCESSING

data processing functions. The computer still has the advantage when handling multiple activities rapidly, simultaneously, and for a long period of time.

Load sharing between human/computer is depicted in Figure 17.5. Figure 17.6 gives a functional block diagram of basic tasks that are usually performed. Typically, the front end of the system, where the data rate is high, i.e., 10^6–10^{12} per sensor per second used to be pre-programmed. With reduced rates at the beamformer output, i.e., 10^3 per sensor per beam band per second, the human interacts sparingly with the computer and selects process parameters. From parameter tracking on, lower rates, as dictated by information content, i.e., (10–.01) per second exist. Now, processing is event driven, rather than data driven; the human inputs commands, selects algorithms, and interacts more fully with the system. The computer processing, organization, and presentation of information to a human varies; it may be done by array, by signal bands, by function, by spatial sectors, or a combination.

In dealing with complex problems, usually spatial, spectral, and temporal decompositions are made to yield elemental problems to solve. Then, the solution to those decompositions are integrated to present a total solution for all space, all bands, and all times. Optimum algorithms dictate joint operations on the data collected by all the available sensors. Often this is not possible. Three principal items prevent this implementation:

- Complexity and cost of implementation.
- Data spatial and time characteristics at variance with the optimum assumptions.
- Geographic dispersal of sensors making it impossible to effect the instrumentation, i.e., mobile and fixed sensing systems; diversity and positioning on a single platform.

In practice, the total problem is done in steps, Figure 17.7. First, the sensors are clustered along geometric shapes, i.e., spherical, cylindrical, planar, linear, and conformal. Each shape forms an array based on a combination of available space on the platform, feasibility, cost of implementation, and required spatial and spectral coverage. A planar array is most effective when the "look" angle is close to normal, i.e., broadside. A non-planar array leads to multiple planar arrays each looking broadside at an incrementally different direction angle.

Typically, an array is divided into two or three subarrays. The beams on each subarray are formed first; then, the required signal and data processing are performed across the beams. For arrays focused on the same space, their resultants are further processed to develop the common picture as seen by all the arrays. Such realizations are achieved at lower cost and are postulated to have performance similar to that of the optimum when implementation of the latter system is feasible. A generalized processing structure is depicted in Figure 17.8. Its implementation may be done in one or multiple computer/display suites. The allocation depends upon the desired physical separation of

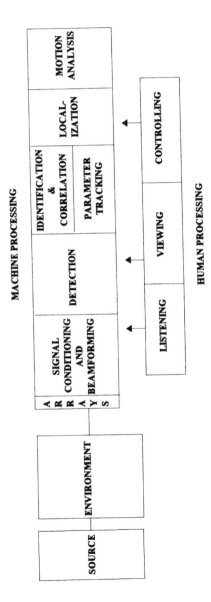

Figure 17.5 Man/machine processing.

AUTOMATED INFORMATION PROCESSING

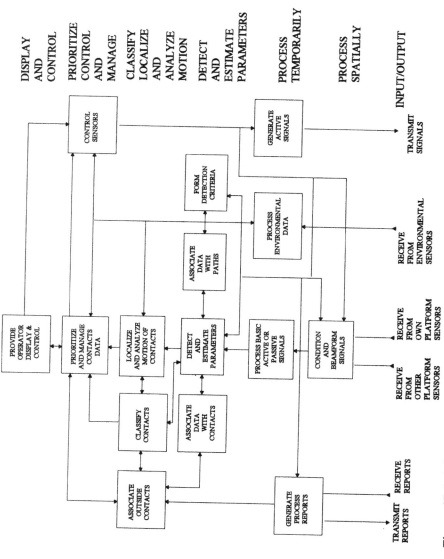

Figure 17.6 A functional block diagram of typical tasks.

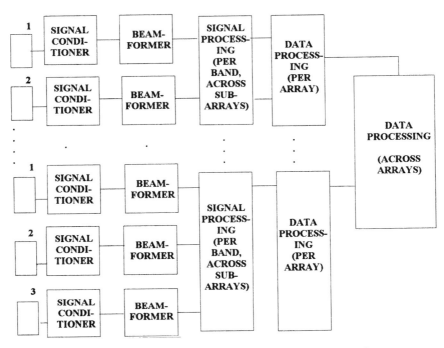

Figure 17.7 Typical signal and data processing implementation.

the processing elements and the reliability, response time, throughput, and memory capacity of the computer and the humans in executing the tasks allocated to them. Depending upon the allocation, the interconnections between the principal blocks may be in software if all functions are in one computer or in actual hardware ranging from a simple wire to a multiplexer or network when functions are in multiple computers. The processors/processing are configured to operate independently or cooperatively on a task (Figure 17.9). We interconnect the required algorithmic operations while insuring their correct sequencing and synchronization of data flow subjected to required control.

C. Information Extraction Through Decomposition and Weighted Integration

1. Spatial Information Extraction

Spatial decomposition is accomplished when the array is set to receive/emit signals from a given direction and to a certain degree reject signals from all other directions; parallel settings lead to simultaneous reception from multiple directions. Figure 17.10 gives a block diagram of a beamformer, a signal level detector, and an estimator of direction/range. For purposes of this discussion,

AUTOMATED INFORMATION PROCESSING

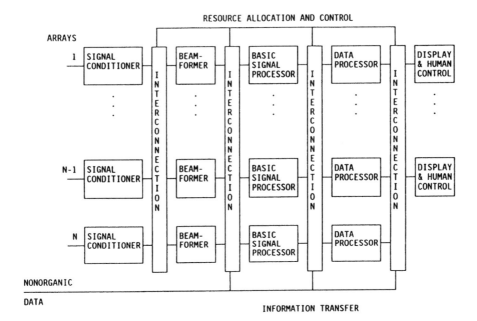

Figure 17.8 Generalized partitioning, allocation, and interconnection of functions.

each array sensor receives a combination of signals and noise assumed to be Gaussian. The delays align positively the various signal components prior to summing. Filter insertion designed to selected criteria, optimizes the performance. Basically, signal processing causes the signal components to add coherently while the uncorrelated noise components do so incoherently so that a larger output is obtained when the array is steered in the direction of the contact and focused on the spatial region where the contact resides. The preceding configuration can be instrumented using generalized correlators between each pair of transducers and weighting of the resulting delays denoting the peak in each. The weighting process constitutes a spatial filtering process that sharpens the estimate of contact location. Formal implementation of such configurations can be cumbersome. But, one can derive simpler implementations with performance levels close to the optimum. In active probing, the replica signal is used in the correlation function.

The preceding discussion centered on a time domain implementation. Structures for equivalent frequency domain beamforming are available. A block of data from each sensor is transformed to the frequency domain. To form a beam in a given direction, the transform from each transducer is multiplied by the appropriate complex phase factor and then summed over the transducers and filtered. When another Gaussian source is present and interferes with the signal detection/estimation process, the structure of the receiver

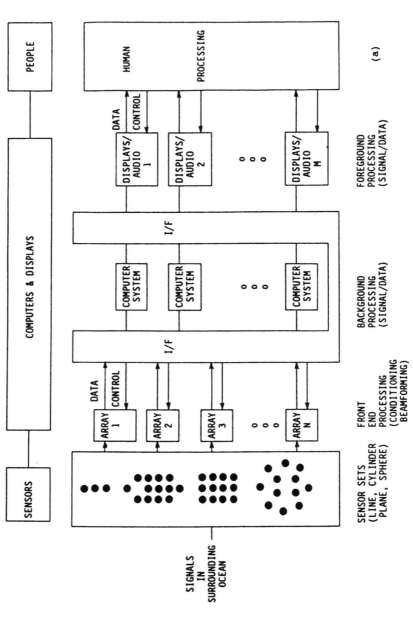

Figure 17.9 General processing to extract desired information: (a) staged structure; (b) data flow as a function of processing stage.

AUTOMATED INFORMATION PROCESSING

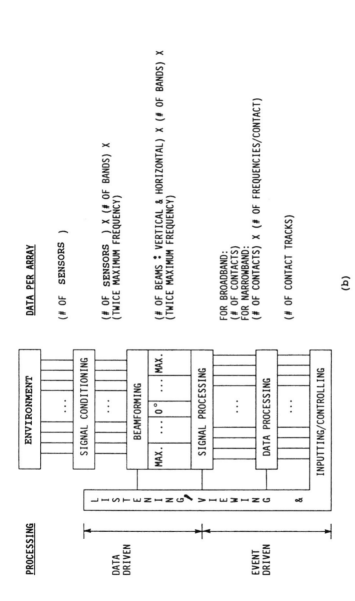

Figure 17.9 (continued)

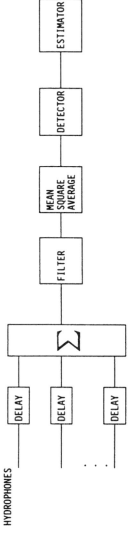

Figure 17.10 Typical block diagram for spatial decomposition.

now contains a null to the interference. It uses the output of a beamformer pointed in the direction of the interference to properly subtract its effect from the received waveform at each sensor.

When non-stationarity enters the problem, new and important issues develop that need treatment. When the source is in motion, for instance, and no motion compensation is made, one limits the size of the mean square average time to create stationarity and avoids smearing of the information. Now the spatial information is viewed in a series of frames that need to be aligned, weighted, and superimposed to develop unequivocal spatial decomposition for decision-making.

2. Spectral Information Extraction

Spectral decomposition is accomplished when the time signal (electromagnetic or acoustic) from a given beam is represented in terms of amplitude and phase as a function of frequency. Selection of bands and/or bins for observation and tracking continues the decomposition into smaller cells. Classical and modern methods have been used in spectral estimation with varied weighting approaches:

- Classical methods (Periodogram, Blackman-Tukey).
- Modern methods (Autoregressive/Moving Average – ARMA, Maximum Likelihood, Prony, Pisarenko).

Integration and application of spectral estimation include:

- Classification of signatures and identification of their source.
- Doppler tracking and estimation of source's speed.
- Varied filtering and estimation.

Just as time and frequency and time and space provide useful tools for system description, so do space and frequency. A variation as a function of space may be expressed as an amplitude function of inverse space, i.e., spatial frequencies.

Those two spectral domain representations are analogous in relevant instances. The split beam tracking in space and frequency line tracking in time are analogous. For a line array of the length L with N equally spaced sensors, the maximum distance l between the sensors is set by $l < \lambda/2$ where λ is the wavelength of the highest frequency. The discrete sensors perform at the spatial sampling of the incoming wave. The Fourier transform for the spatial array is analogous to a Fourier transform of a time series of length T and N equally spaced samples where the maximum sampling time, ts, is set by $ts < \lambda/2$ and where λ is now the period for the highest frequency.

3. Temporal Information Extraction

Temporal decomposition involves the viewing of a dynamic scene, i.e., signal, noise, or channel over a series of observation time gates. So far, spectral and spatial decompositions have been applied to deal with detection and estimation over a temporal observation gate presumed long enough to suppress the blurring effect of noise, i.e., a single look observation. Over that time gate, stationarity is presumed. Non-stationarity causes a blurred or smeared version of the original data when viewed by the signal processor.

A typical non-stationarity originates from receiver or transmitter/reflector motion, and channel or noise statistical fluctuations during data collection. Two options are usually available:

- Limiting the time gate to stationary observables.
- Introduction of a desmearing process.

The first option can be exercised when the noise is low in value and the imposed observation time yields desired resolution. Otherwise, a restoration or desmearing process is needed; it involves modeling of the motion, estimating its parameters, and extracting that effect out.

A variety of practical situations leads to a restoration process which is typically effected over a series of observation gates. Over each time gate, spatial and spectral decompositions are made leading to a snapshot of the dynamic scene. The snapshots form the temporal decomposition of the dynamic scene. When the snapshots are fuzzy due to noise, they are weighted and superimposed along the invariant parameters to produce a clear dynamic picture.

Simple examples involve the detection and estimation of Doppler shifted signals such as sinusoid or time delay between broadband signals. The peak denoting the information is smeared as a function of the motion and the size of the observation time gate. Smearing involves a displacement of the peak, a reduction of its height, and uneven spreading of the peak's curve. Suppressing the Doppler effect involves a reduction in the size of the time gate and/or a feedback loop to null its effect where a balance is struck between dynamic response and needed resolution.

Resolution and stationarity are, in some cases, intimately related. Resolution refers to the ability to separate parameters, variables, and/or functions for measurement. Stationarity refers to constancy in time, deterministic or statistical. Discussions of resolution address:

- Spatial resolution, i.e., separation measure of objects in angle/range.
- Spectral resolution, i.e., separation measure of adjoining frequencies.
- Temporal resolution, i.e., separation measure of events in time.

AUTOMATED INFORMATION PROCESSING

Spatial resolution is inversely proportional to the effective linear dimension of the array performing it; the longer the array and the higher the signal frequency, the sharper is the resolution. However, spectral resolution is inversely proportional to the observation time length; temporal resolution is inversely proportional to the observer response bandwidth. Non-stationarity impacts directly the observation time length, forces its reduction and the related system resolution. Another principal impact of reduced observation time length is the reduced measurement accuracy in the presence of noise.

A simple example tying resolution and non-stationarity involves the resolution of frequencies which are time variant. A simple model is given by the linear relation where $\dot{f}$ is the rate of change of frequency with time. To effect frequency resolution for a given $\dot{f}$ value, the resultant observation time length is less than $1/\dot{f}$. Thus, the higher the frequency drift, the shorter is the observation time.

III. BASIC SIGNAL PROCESSING, DECOMPOSITION, AND WEIGHTED INTEGRATION

Estimates of time intervals and variation of those intervals in time constitute basic measurements in a signal processor for many applications. A time interval refers to any time lapse from a reference, i.e., a period, a delay in signal reception as in active sonar/radar, or delay between arrival paths as in passive sonar/radar. Those measurements are then mapped in the data processor into the desired estimates, i.e., emitting source range, direction, depth, and velocity. Figure 17.11 gives the three basic types of channels with the pertinent time intervals and processing functions. The processor structure has typically a spectral type formulation for its stationary elements and a state space formulation for its nonstationary ones.

Though varied, the signal processing elements and their connectivity to data processing have a common set of features that can be addressed simultaneously (Figure 17.12). Each basic signal processing stage has a set of linear and nonlinear decomposition operations, i.e., Fourier transforms, absolute values, logarithm, tangent, etc. that map the embedded time interval in the input function into a recognizable clue denoting the value of that time interval. The clue may be a dominant peak, a minimum, or a slope. To enhance a deteriorating clue in the presence of various interferences, windowing is incorporated into the basic signal processing stage. Gating is added to limit the search for the clue to the most probable region in the signal processor output. The detected time intervals above a set threshold could then be passed through a short time filter to improve and assess the quality of the estimates. Those estimates are fed back in a loose couple to adapt the parameters of the gate, the level of the threshold and in turn those of the window. The same estimates are also fed forward through an extended time filter to map the time interval estimates into smoothed source position and velocity estimates. The source

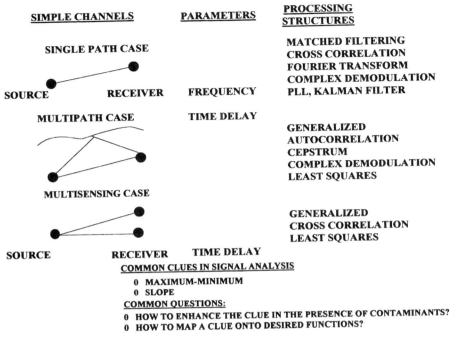

Figure 17.11 A sampling of channels, parameters, processors, issues.

parameter estimates are also fed back for use in adaptive design of gates and windows' thresholds.

A time interval is a basic measurement in many applications. For instance, localization and tracking of a source may be determined directly from the time delay measurements. Bearing, range, and elevation angles are evaluated from measured time delays. In communication, radar or sonar, the time delay estimates provide the possibility for echo removal. In deconvolution, the time delay information is critical to the separation and estimation of the probing signal and channel response. In classification, the periodical content of a source helps in its identification.

Though the applications are diverse, two basic types of time delay estimation problems emerge that depend on spatial diversity in the channel. Diversity refers to the reception of the signal element through different propagating paths, thus giving rise to a delay in signal arrival times for different path lengths. In one basic type, at least two propagation paths become reflected, diffracted or refracted by an inhomogeneity in the channel, thus leading to their intersection at one spatial point. Now a single sensor picks up the composite signal which may be processed by autocorrelation or cepstrum to estimate the time delay parameter. In the other basic type, the signal is emitted in the quasi-homogeneous channel so that it traverses a continuum of nonintersecting paths. Then, at least two spatially separated sensors are used to realize the diversity condition. Generalized cross-correlation techniques are then applied to estimate the time delay parameter. When multipath signals are present, serious problems can arise with the time delay estimation by a cross-correlator. Antimultipath techniques are generally ineffective, especially for sources within a beam width of the reflecting surface. Instead, the strong multipath conditions are capitalized upon for the determination of source range and depth through processing of the time delay estimates obtained from the cepstrum or autocorrelator.

Other applications depend on time diversity where the incoming signal and receiver are linked by a single path, i.e., active sonar/radar, monitoring of a signal frequency content. There, the delay is referenced to a point in time, i.e., emission time, beginning of a period.

A. Source, Channel, Processor

Figure 17.13 describes a basic system and its functions. The source signature may include wideband, narrowband, and transient elements. The scatterer returns a modified emitted signal. In either case, the signal traverses non-intersecting or single paths and/or intersecting or multipaths. The channel may be distortionless or display distortion/dispersion and may contain noise/interference. The receiving sensors, after beamforming, are effectively concentrated into an effective center or spatially distributed centers. Traditionally, signal analysis of radar–sonar systems has separated the signal from the data processing functions and has marginally included channel characteristics in the analysis process.

Two basic channels are usually viewed, that of homogeneous propagation and that of multipath propagation. For each type of channel, signal processing methods have been evaluated analytically and experimentally in the literature. Since both types of channels may occur alternately or simultaneously in practice, a joint view is beneficial; both types share much of the structure and concepts.

In general, the signal emitted by the source or the scattered signal reaches the receiver through different paths as direct, reflected, and/or diffracted waves. Reception of the signal is picked up at a single or at multiple spatial sensing points. The processing techniques depend on the number of receiving points and the number of arrivals at each point. In one basic type, multiple propagation paths become scattered by the obstacles in the channel which lead to their intersection at a sensing point; to dissect the resulting composite function, generalized cepstrum, autocorrelation, or complex demodulation have been applied (Figure 17.14). In the other basic type, the propagation paths do not intersect at the sensing points; then, spectral estimation and/or generalized cross-correlation with a reference signal may be carried out at each sensing point, and/or generalized cross-correlation across sensing points (Figure 17.15). Using recursive or batch filters, the pertinent time intervals from each channel are smoothed and mapped under some weighted least square error criterion into the desired source range, depth, course, and speed.

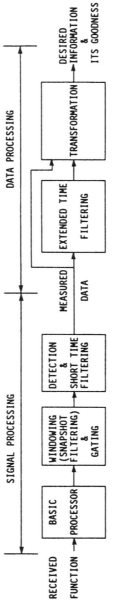

Figure 17.12 Canonical signal and data processing functions.

AUTOMATED INFORMATION PROCESSING

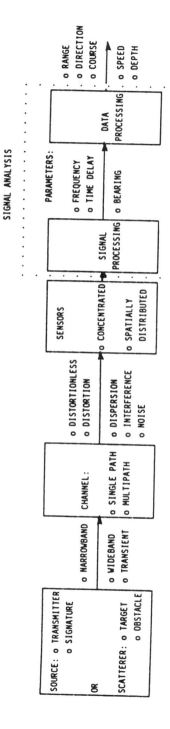

Figure 17.13 Characterization of the problem.

350 SYSTEMS MANAGEMENT: People, Computers, Machines, Materials

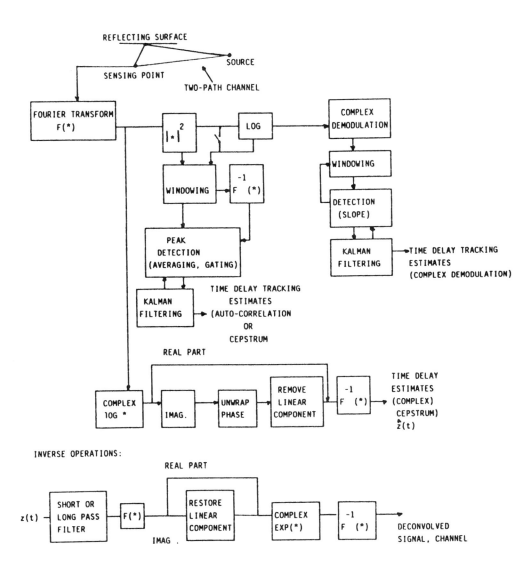

Figure 17.14 Signal processing in a multipath channel.

In general, signal analysis and design are optimized to some average and idealized operating conditions in the source and channel. When variances arise, they cause long interruption in useful data and a wide gap between detection and source tracking. Weak links result in the system chain from such a design. Sensors, channel, signal, and data processors are then disjointed in the parameter estimation and source tracking chain. The realization of potential gains

AUTOMATED INFORMATION PROCESSING

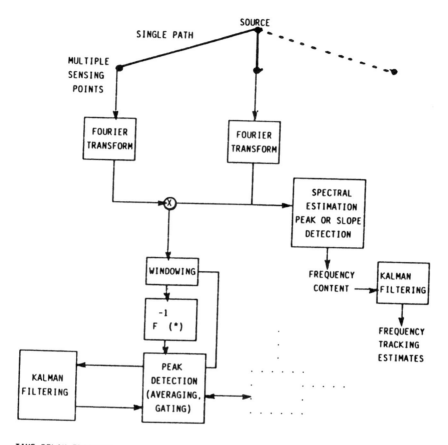

Figure 17.15 Signal propagation for separate sensors. In active mode, one sensing point has an exact replica.

in performance can result from functionally cooperative, mutually supportive subsystems having real-time access to a shared information base. Figures 17.14 and 17.15 give a sampling of such subsystems. At times independent or dedicated designs have been developed for a subset sensor configuration, single source characteristic, i.e., wideband signals, and a fixed and simple weighting characteristic. This prevents the sharing/selection of appropriate subarrays out of the total sensor suites as determined by source signature and range, array speed, noise distribution over arrays, channel conditions, and the adaptation of the weights to more complex situations.

B. Weighting Through Windowing and Gating

Windowing and gating have been the subjects of extensive weighting studies for both types of channels. Windowing attenuates the spectral content of a function being processed. Gating attenuates the processor output where the search for a clue to the desired parameter is undertaken. Techniques for the measurement of time intervals are varied because of their differing limitations and their unacceptable performance under general conditions.

Windowing lowers the operational threshold of the basic processor. The windows are designed to remedy or compensate for physical conditions that affect unfavorably the performance of a given signal processor. They are dependent upon signal spectra, noise spectra, and channel parameters. Unfailingly, addition of a properly designed window has extended the region of satisfactory performance of a given conventional signal processor. It should be stressed that the windows must be designed to suit the processor at hand and the situation under consideration. Analysis is needed to determine the cause of deficiency and its remedy. Otherwise, improper windowing may be applied that would have the reverse effect of deteriorating the performance instead of improving it. Often the input parameters for the design of the window are not known a priori, and adaptive techniques are used to execute the implementation. The ultimate guide to the functioning of the window is in the added enhancement to one of the two basic clues denoting the value of the time interval parameter to be estimated. For correlation and cepstrum, the basic clue is a peak in the output; and for complex demodulation, it is a slope.

The simplistic scheme of independently selecting the dominant clue from the total signal processor output can deliver erratic time interval estimates, whenever adverse but temporary conditions exist at the input. For initialization, an ensemble average over a number of signal processor outputs is taken to enhance the clue against the mean background noise. Where the clue is identified as having sufficient signal power over noise power, a gate is centered at the corresponding output region. Ultimately, gating defines the spatial sector where one thinks the source resides. The characteristics of the gate and the threshold level are provided by recursive and non-recursive least squares filters and log likelihood ratios operating on the raw time interval estimates. Such gating enhances the robustness of the processor against fades and limits the clue search to the most probable region in the processor output. The filtering has other benefits as a detector of source maneuvers through jumps in time delay rates, and passing on this information to the filter that is estimating the source state dynamics. A successful stabilization process of the estimates allows for automatic and quasi-optimal processing of the data to estimate source location and motion.

IV. ACTUAL AND LOWER BOUND LOCALIZATION ACCURACIES

This section gives an introduction to the issues in actual and lower bound localization accuracies. Several factors influence localization accuracy when using passive or active arrays of sensors. The arrays receive delayed versions of the emitted signal. These delays are used to determine the signal source location. The precise values of these delays are a function of:

- Signal and noise spectra.
- Filtering process.
- Number of sensors.
- Observation time.

The observation time is a controlling factor on the delay errors. Just how long it must be to reach a lower bound depends critically on the other three factors. Just how long it is in practice is balanced by the non-stationarity in the physical element and the rapidity with which a solution is required. In the limit of allowed very long observation time, a lower bound on the error can be reached as set by the Cramer-Rao bound. Under realistic conditions, the observation interval is short and the resulting snapshot estimate varies from the lower bound.

The general problem decomposes into two classes, linear and nonlinear. Typically, active probing yields a linear problem while passive observations yield a nonlinear and consequently more difficult one. A common solution approach has been to:

- Estimate the delays according to some criteria, optimal or otherwise.
- Transform the filtered delays to contact location, a nonlinear operation for passive observations.
- Assess the delay errors on localization accuracy.
- Contrast the results against some absolute lower bound on the errors.

Figure 17.16 gives summary cross-sections of the general problem. These cross-sections give a good perspective on the encountered problems. In practice, the total set of problems is decoupled into horizontal and multipath problems. Those include:

- For horizontal passive observation problems, the sensing elements are typically grouped into one, two, or three subarrays and beamformed. Then generalized cross-correlators are used to measure the time delays between the sensing centers. These hypothesized configurations give similar results to the optimum system, but at much lower cost. The optimum processor has to be configured, for a linear array

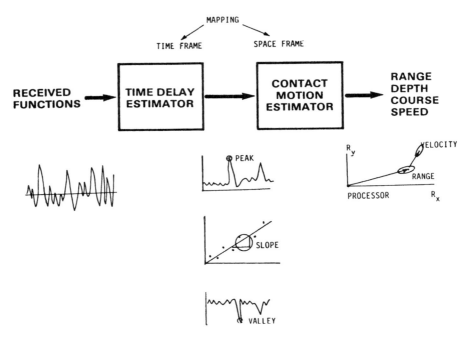

Figure 17.16 Basic elements and transformations.

with M-element sensors, as a large set of M(M–1)/2 generalized cross-correlators computed between each pair of sensors. In both implementations, the time delays are further processed to determine the range and bearing estimates. The grouping of the sensors influences differently the quality of those estimates. The best way to configure a limited number of M linear sensors is: a) to set the sensors into (M/2, M/2) clusters at greatest separation to obtain best bearing estimates; b) to set the sensors into (M/4, M/2, M/4) clusters at equal separation to obtain best range estimates; c) to set the sensors into (M/3, M/3, M/3) clusters at equal separation to obtain best location estimates. In general, the platform's shape, dimensions and available space constrain the way the sensors are eventually clustered.

- For passive multipath problems, generalized auto-correlation, cepstrum, or complex demodulation are used. Now a replica of the signal, though corrupted with noise, is not separately available as in the horizontal channel. This entails a lower quality estimate of the delays when dealing with the additive noise effect. As to data processing of the delays to estimate range and bearing, analogous processes to those in the horizontal channel are encountered.
- For active probing in either horizontal or multipath channels, the preceding problems in signal and data processing are minimized.

To estimate the delay in echo reception, the generalized cross-correlator operates now with a noiseless reference signal. For data processing, a simple linear relation ties the contact range to the measured delay. As expected, an improved performance is realized. In reverberant/distortional/dispersion environments, those advantages are diminished.

The Cramer-Rao inequality is the principal analytical tool used to set the lower bound on the covariance in the delay vector. The Cramer-Rao inequality asserts that any unbiased estimate of the vector parameter has a mean square error lower bound that can be approached under very general conditions, in the limit of very long observation time. When the estimator yields a biased estimate, forms of the bound do exist. This bias occurs when the time delay estimates are mapped through nonlinear operation. For such cases, not only the estimate is not true, but biased; also the variances are increased by twice the square of the bias when calculated up to second order.

The Cramer-Rao lower bound gives the local or small error performance of parameter estimators. Physical instrumentation, i.e., generalized correlator, maximum likelihood estimator, exists to reach this bound for sufficiently high signal to noise ratio and/or long observation time. When the conditions are not obeyed, the errors become global and spread beyond the local vicinity of the true value, and it is not possible to reach the Cramer-Rao lower bound. When the errors are large, constraints or gates on the interval where the parameter value may fall are helpful in controlling the large errors in the estimates and reducing them to local errors. Probabilistic and heuristic approaches have been applied to the design of such gates.

In practice, the time delays are usually measured using a generalized correlator with the averaging time short enough to presume local stationarity, but much longer than the correlation of both signal or noise. When the time delay measurements are slightly noisy, with zero bias, first order Taylor expansion analysis yields, even for the passive observation class of problems, an unbiased estimate of both range and direction. For that case, it does not matter whether the transformation and filtering is applied to the range or to the time delay. However, as the time delay estimates deteriorate further, or the range between source and array increases and/or the physical array orientation points toward the contact, the nonlinearities in the transformation from time delays to source range and bearing become significant in the passive case. Then, the range and direction estimates become biased.

Two types of nonlinearities exist where one is significant when the time delay variances are large and the other when the second partial derivatives in the source–array geometrical relations are large. With the delays negatively correlated, an increase in the range bias and a decrease in the direction bias result. Also, the variances on the range and direction estimates, when calculated up to second order, are increased by twice the square of the bias.

Those biases may appear inherent to wavefront curvature ranging and thus present an irreducible limit to system performance. In actuality, a bias compensation scheme to the signal processing which yields a snapshot estimate of source bearing and range can be developed to alleviate this limitation. For the data processing problem of successive estimates, the reduction of the bias as well as the variance is accomplished either through:

- Appropriate mapping of the time delay estimates onto invariants in source trajectory parameters over which filtering is performed to remove the bias limitations. Effectively, this mapping imparts stationarity to the problem thus allowing an increase in the averaging time to reduce the time delay errors for the system beyond that permitted in the signal processing stage. This approach has entailed a nonlinear formulation of the tracking problem.
- A compensation scheme to remove what has been considered as a fundamental limitation on the performance of systems which employ the snapshot ranging relation. There, the source tracking phase is reformulated as a linear estimation problem, thus avoiding the difficulties pertinent to nonlinear estimation approaches.

To avoid excessive nonlinearities, range gating can be exercised. This amounts to discounting time delays whose differences fall below a high-pass gate; such gating is set as a function of expected system performance, environmental conditions, and operation of interest.

Finally, the discussion has concentrated so far on nonlinearities triggered by delay errors. Other errors, i.e., sound speed fluctuation or stretch in intrasensor distance and sensor position uncertainties, cause similar nonlinearities and similar behavior is observed when estimating the source location.

V. BASIC DATA PROCESSING DECOMPOSITION AND WEIGHTED INTEGRATION

This section discusses the problem of estimating the location and velocity of contacts via observation and processing of data. The received function includes the desired signal and unwanted noise. Embedded in the body of the received signal are differences in signal arrival times data (time delays) as well as variations (Doppler shifts) in the signal. These differences and variations in the data are functionally dependent upon contact–observer geometry and environmental conditions.

Contact localization and motion analysis (CLMA) systems make use of the received signals' time delays and their variation in time to estimate a contact's location (Figure 17.17). These processing systems basically comprise a received signal's time delay estimator and a contact motion estimator (Figure 17.18). A generalized definition of time delay is used to mean either the

AUTOMATED INFORMATION PROCESSING

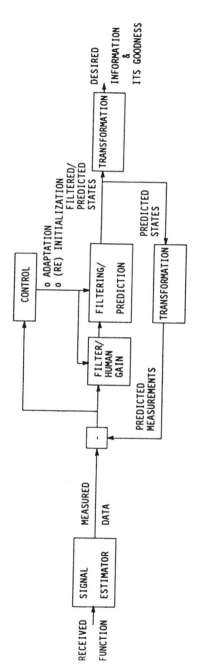

Figure 17.17 A data processing structure for estimation of desired information.

difference in signal arrival times between two raypaths, the difference in time delay between signal emission and reception, or the time lapse before a signal repeats itself. The time delay estimator maps the received signal data into recognizable and measurable clues of the time delay (a dominant peak, valley, or slope on a curve, for example). These clues are further processed by the contact motion estimator so that estimates of time delays are smoothed and mapped into values for contact range, direction, depth, and velocity.

CLMA systems process data spatially and temporally. That is, they process data received simultaneously at spatially separated sensors, as well as data received during sequential observation intervals spread out in time. The total system gain is the sum of both spatial and temporal gains. Spatial gain is influenced by such factors as size, number, placement, and configuration of sensors in the array. Temporal gain is influenced by the quality of the received data and the manner in which it is processed in time. In principle, simultaneous optimization along both spatial and temporal dimensions is desirable for optimum system performance. In practice, considerations such as array stabilization, cost, spatial coherence, platform dimensions, etc., limit the achievable spatial gains, while the requirement to yield an acceptable solution within a given time limits temporal gains.

Elements in a CLMA problem may be stationary (e.g., no relative motion between contact and observer in a homogeneous environment) or nonstationary. When the elements are stationary, processing is straightforward and is accomplished by a continual integration over the observed contact clues until desired accuracy in the solution is obtained. When the elements of the problem are nonstationary (e.g., moving contact/observer or a changing ray path channel), bias is introduced during the long contact observation interval and the problem is much more difficult to solve. Observation of the contact must be limited to a brief time interval over which the process may be considered locally stationary. In this case, CLMA systems provide what may be considered "short memory" or "snapshot" clues, which may yield imprecise estimates of contact location and motion. However, with a succession of such time observation intervals, the system's temporal processor can remove the biasing non-stationarity in the problem. It does this by superimposing the repeated short-memory estimates to enhance the invariant contact parameters in the problem, ultimately developing a well-defined estimate of the contact's location and motion.

In its totality, then, contact localization and motion estimation constitute a process that is mathematically non-linear and geometrically nonstationary in terms of contact/observer. It is a process not amenable to optimum global system synthesis, as evidenced by most literature in the field which generally deals with optimization of subsystems as realistic conditions are introduced. Systems providing optimal performance have been developed, but these are only for idealized conditions such as stationary contact and observer, Gaussian signal and noise, and extended contact observation times.

AUTOMATED INFORMATION PROCESSING

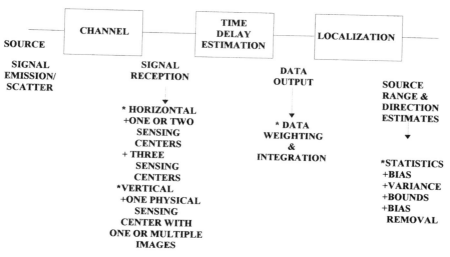

Figure 17.18 Multiple cross-sections of the problem.

A. General Classes of Contact State Estimation Problems

Several types of contact state estimation problems are seen in the literature available on the subject. These may be grouped into general classes on the basis of stationarity, linearity, and observability; that is, they can be grouped according to the degree of relative or apparent contact/observer motion, the complexity of their solution equations (linear or non-linear), and the extent to which a contact is observable (i.e., the extent to which an observer can realize a unique solution from the available data). As mentioned in the introduction, the problem involving a stationary contact and observer has been dealt with in preceding sections. However, for a nonstationary contact and observer, the problems increase in complexity in the following order:

- Linear solution to problems with the contact's state observable over each observation (sampling) interval.
- Linear solution to problems with the contact's state observable only after multiple observation (sampling) intervals.
- Linear solution to problems with the contact's state observable only after multiple observation (sampling) intervals, and only with motion constraints placed on contact and observer.
- Nonlinear solution to problems with the contact's state observable over each observation (sampling) interval.
- Nonlinear solution to problems with the contact's state observable only after multiple observation (sampling) intervals.

- Nonlinear solution to problems with the contact's state observable only after multiple observation (sampling) intervals, and only with motion constraints placed on contact and observer.

These classes of problems are especially difficult to solve when there is a mismatch between physical processes and modeled processes, or when there are large errors in observed parameters. In practice, the latter classes are more commonly encountered than the former. In analyzing the various CLMA problems likely to be encountered, several general statements hold true:

- Linear problems lend themselves readily to optimal estimation with resulting minimum mean square estimation error.
- Nonlinearity increases the complexity and the issues involved in structuring an algorithmic estimator.
- Increased contact observability tends to improve the quality of an estimate and speeds estimator convergence.
- Redundant observation (sampling) is required to reduce the adverse effect of measurement noise.
- Constraints on observer/contact motion encumber the estimation process by delaying estimator convergence, lowering the quality of estimates, and upgrading the ability of the estimator to adapt to mismatches between modeled and physical processes.

Two types of applications exist depending on whether observation of the contact is by active or passive means. In the active case, the contact is ensonified or illuminated by a signal emitted from the observer, and observations are made on the backscattered returns from the contact. In the passive case, the contact itself is an emitter whose signal is received at the observer. For the active case, contact localization and motion estimation fall into classes at the beginning of the preceding list; for the passive case, the problem falls into the classes predominantly at the end of the list.

In either case, the observer is linked to the contact through the intervening propagating medium. When analyzed, the medium is seen to have distinguishable ray paths lying within the usable beam patterns of both contact and observer. Distinguishability here refers to the difference in path lengths measured relative to a reference path or reference time. Each difference in path length is reflected in the time delay incurred by the signal as it propagates through the different paths. In the active case, time delay refers to the difference in arrival time between the reference emission time and reception of a return. Another time delay is discerned at the beginning and end of a period of a sinusoidal signal. These various time delays constitute the basic measurements that a time delay processor extracts from received signals. The desired contact state information is embedded within each time delay, which is characterized by the ray path structure within the channel.

B. Errors and Their Filtering

1. Causes of Errors

A homogeneous and noiseless ray path channel yields a direct functional dependence between the time delay vector and the contact state vector where solution for the contact state is straightforward. In practice, perfect or nearly perfect observations are rarely available. Vector errors are usually introduced due to the time delay measuring system: mismodeling of the environmental factors in the channel, mismodeling of the contact's motion, or inaccurate monitoring of the observer's own motion. Seldom is the spatial gain of a passive system high enough to warrant neglecting these errors.

Regardless of the source of errors, their statistical character influences the selection of a particular contact localization and motion estimation process to filter out the errors. In general, errors are characterized as either biased or unbiased.

2. Characterization of Errors: Biased or Unbiased

In this context, a biased error refers to the tendency of an estimated value to deviate from the true value in one predominant direction. Biased errors may be constant or variable over a number of contact observation intervals. Constant bias may be due to differential dispersion in the channel path (as between a volume and a bottom-reflected path), ray path curvature, or a non-Gaussian distribution of time delay estimates from the time delay processor for a low signal-to-noise ratio or low relative signal-to-noise bandwidth. Constant bias may be transient (as due to a contact maneuver) or persistent (as with a mismodel of the channel's ray path curvature). Once a transient bias is recognized, adaptive control of the process noise may be successfully applied. This may amount to effective re-initialization of the problem with some a priori information on the contact's range. If a higher-order motion model (one that allows estimation of a possible contact maneuver) is used, the estimator is more prone to instability, especially when only large unbiased errors are present. To deal with persistent variable bias requires a model of the process. If available, parameter estimation and process identification may be carried out with diminishing success if the estimation problem belongs to the later classes.

Unbiased errors may have Gaussian or non-Gaussian distributions. A Gaussian fluctuation of time delays may be due, for instance, to such effects as small perturbations in the channel speed profiles or due to the channel boundaries or to the processing of time delays in the presence of limited noise. Even with a Gaussian error distribution on the time delay estimates, their direct mapping into the desired contact states can result in increasingly non-Gaussian distributions as a function of the contact's range and off-broadside direction to the observer's array. From the estimation point of view, it is preferable to maintain an unbiased Gaussian distribution of errors, since this leads to manageable difficulties in the contact state estimation process. Many of the existing contact localization and motion estimator structures are designed on the basis of best unbiased mean square error reduction criteria.

3. Statistical Filtering of Errors

Even when time delays are estimated with unbiased Gaussian errors as would occur with high signal-to-noise spectra and long observation times (or as may occur following stabilization through windowing, gating, and filtering) direct mapping of the time delays into the contact's state can lead to biases in the estimation process. Reduction of this bias and variance in contact state estimates can be accomplished by judicious use of statistical estimation techniques over sequential and finite observations of the contact signal. The contact state estimator is an expanding memory filter that maps imperfect time delay estimates into the invariant contact trajectory parameters (e.g., constant velocity, initial range) over which smoothing is performed. The smoothing reduces, jointly, the variance and the bias in the estimate of contact kinematic parameters. Such a scheme improves substantially on techniques that process inappropriately mapped time delays, or techniques that directly transform the best time delays available into contact motion estimates. The latter approach is optimum only when stationarity of all elements in the problem can be assumed. For this limiting case, the approach using statistical smoothing converges automatically to the optimum estimates. Yet for generalized cases, it remains a viable approach for moving contacts at long ranges, for contact directions off the array's broadside, and for high time delay variances.

Implicit in the discussion is a requirement for correct statistical descriptions of the error processes at hand. The recovery from an incorrect statistical description in digital systems is aided by use of coupling loops for detecting such an event. The ensuing divergence is bypassed and the processes are routed in a degraded mode until the system recovers. When the traditionally separated signal and data processing stages are interactive, further improvement can take place because system deterioration is usually local and not total.

C. Elements in the Formulation and Solution of CLMA Problems

Three elements need definition in the formulation of a contact's state estimation process. These are encountered regardless of the class that the CLMA problem belongs to and regardless of the application at hand. The three elements are:

- A model of the relation between the contact's state and the observables (i.e., time delays).
- A model of the contact's state, e.g., stationary, constant velocity.

AUTOMATED INFORMATION PROCESSING

- A criterion to filter out errors from the observables and models.

Of the various errors that are encountered, some are due to the time delay estimation process, some are due to the modeling of the channel, some are due to the presumed motion of the contact or the observer or both, and some are generated by the form of the data processing structure. Regardless of the error sources, filtering of unbiased errors has been dealt with collectively using varied estimation techniques. These include linear minimum variance, least squares, weighted least squares, maximum likelihood, and Bayes estimators. Performance of the resulting estimation procedures varies depending upon the available statistical descriptors. For Gaussian error distributions, the linear minimum variance estimates results agree with many of the others. In addition, non-linear problems can be fitted through linearization, and minimum variance estimators can accommodate such cases with little or no knowledge of the probability density function of the errors. This latter characteristic explains the widespread use of linearization techniques since, more often than not, a probability distribution is merely conjectured.

As noted, the estimation problem begins by hypothesizing the functional relationship between received time delays and contact's state descriptors. Sensor diversity and channel diversity must be taken into account, since the number of measured time delays depends on the number of spatially separated sensors and on the number of intersecting ray paths in the channel. Two time delays at each observation interval are needed to provide passively positional information on the contact. Synthetic diversity must also be considered; this refers to the orderly assembly of time delays estimated over successive observation intervals to enhance the available estimates and to provide otherwise unavailable estimates. With time delay-only that yields at each instant a single contact's direction, the ranging relationship between moving contact/observer is quite circuitous and requires a series of time delay measurements combined with an observer velocity change.

Notwithstanding the relational complexity, alignment of snapshot estimates of contact localization and motion requires a modeling of the nominal underlying processes. This calls for hypothesizing a dynamic model of the contact. Mismatches between real and modeled phenomena lead to biased errors, and estimates of these errors must be made along with estimates of the contact's motion. Bias estimation remains a difficult problem, and bias due to the contact's nominal motion is presumed to be predominantly constant in velocity interspersed with arbitrary maneuvers. The modeling presumes this type of motion with added unbiased perturbations to account for deviations on that motion. The perturbation input levels are varied to reflect the credibility in the evolution of the motion models. This control process is used in relation to the functional dependence of the contact's states upon observed time delays. Even when the contact's motion model is inadequate, the evolution of the time delays has been modeled locally through nominal, low-order, polynomial expansions that prove helpful over a limited number of time delay estimates.

Given the contact's dynamic model and the functional dependence of its state on the measured time delays, a criterion for "best" estimation of the contact's states is chosen which yields the estimator structure. If a choice is made to minimize the average mean square error between estimated and true contact states, the procedure is a straightforward mathematical one applicable to varied situations. Other means to minimize errors, such as the maximum likelihood technique, can lead to insurmountable analytical difficulties for non-Gaussian statistics. The characteristic of the residual error between estimated and measured time delays is applied to weigh the adjustments on the contact's states estimates until satisfactory minimization of the error is obtained. The residual error contains the cumulative error characteristics (biased and unbiased) which are sifted, either by an operator or automatically, so that the estimation process is conducted only on the data error characteristics that the estimator is designed to handle. Residual error characterization remains an active area of research; one in which detection of the bias has been stressed. Much of the attention has centered on adaptation to biasing caused by contact maneuvers. However, increasing attention is being paid to biasing due to sensor positioning and environmental effects, and also on the effects of certain types of random errors.

The general motion estimation problem of contacts is a difficult but important problem. The approach has been to derive models which are of sufficient general utility but not too cumbersome for practical utilization:

- First, the contact motion is considered to have a constant velocity; more sophisticated dynamic models build on that nominal motion assumption.
- The channel is considered effectively homogeneous; the effects of more complicated channels are mapped, when possible, on the simpler homogeneity assumption.
- The time delay estimator provides a noisy measurements sequence with zero mean Gaussian noise, based on observations of some aspect of the contact motion; extension to this postulated case gives added complications.

The first objective is a characterization in a coordinate system of a) *the constant velocity contact motion model*, and b) *the variety of measurement models*. In choosing a coordinate system, it has been found that it is better to formulate the problem so that the observations where the noise is, are set in a linear relation to the contact's states.

A top level description of the problem and the variety of observations are given in Figure 17.19. Basically, the variety of possible observations, given a sensor suite, relate to the interconnection of:

- The plane(s) in which the environmental ray paths lie.

AUTOMATED INFORMATION PROCESSING

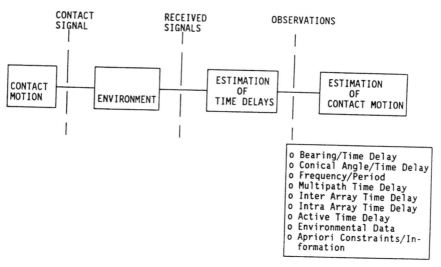

Figure 17.19 Estimation of contact motion from a variety of observations.

- The plane containing the observation sensors' axis or array normal.

For instance, when the three planes correspond, we have a bearings-only motion analysis problem for a two-sensor suite. Varied approaches exist when applying estimation theory to the tracking problem:

- Bayesian estimation requires a complete knowledge, often unavailable, of cost function and probability structures.
- With maximum likelihood, the parameters to be estimated are unknown but not random, and it is not required to know their a priori probability densities.
- With linear minimum variance estimation, no probability densities are assumed, only the first and second order moments are known.
- Finally, deterministic optimization can be applied without a priori stochastic assumptions.

For certain recurring performance indices in the applications, many of the approaches lead to equivalent results. The basic elements of the solution include:

- Selection of candidate data points (objects) based on some attribute, i.e., signal exceeding a certain level in a detection gate (cluster).
- Formulation of dynamic and observation models.
- Determination of the valid data points and their association with parent tracks, i.e., snapshot to snapshot association of multiple frequencies in a clustered/multisource background.

- Filtering algorithm for tight estimate of tracks.

The preceding concepts extend to generalized contact tracking problems of multiple physical contacts in a cluttered environment. Basically, the process is as follows:

- Contacts give rise to data. The measured data convey attributes, e.g., location and non-location types that permit differentiation between the contacts of interests and false alarms. Data with similar characteristics are processed together first, followed by those with dissimilar characteristics.
- The attributes are quantitative and/or qualitative. The quantitative ones are processed via statistical and non-statistical or heuristic approaches. The qualitative ones are used for screening and gating possible situations and for track merging.
- Time tagged data emanating from a particular contact constitute a report. The receipt or development of a report indicates that enough contact-like qualities are detected to justify reporting it.
- A series of reports are used to develop tracks. Tracks are a set of curves based on the association and filtering of other series of reports where each subset is judged to have emanated from a particular contact. Similarly, a false alarm or provisional set is prepared for those reports that are judged to have emanated from false or non-contacts of interest.
- Data association with contact and/or raypaths association is especially important in the complex environment due to the presence of multipath situations.
- Data partitioning assigns reports to track sets or false alarm sets. Unrestricted partitioning is a report-oriented association technique to update the tracks where all reports and all tracks are reconsidered against all possible hypotheses. Restricted partitioning treats separately each new report against an established track set. On the other hand, there is the nearest neighbor approach where reports and tracks are associated based on their statistical closeness.
- Data filtering is based on the best, e.g., least-square error fit of the data elements associated with a given track.
- Overall, an association and tracking algorithm includes track initiations, track terminations, yields track evolutions, contacts number and resolution, and some measures of goodness.

In the complex environment, the tracking filter encounters certain dominant problems. This is principally caused by measurements of uncertain origin where not only data/contact but also data/path associations have not been set. It is due to:

AUTOMATED INFORMATION PROCESSING

- Complicated raypath structure linking contact to sensor.
- Other contacts in the sensors' receiving area.
- Poor sensor quality.
- Lack of identifying contact characteristics.
- Crossing contacts.
- Poor sensor viewing geometry.
- Time late data.

The last item can render the a priori assumption of constant contact velocity questionable, since a contact is just as likely to conduct maneuvers over the long span of time between data arrivals. Then, a higher order dynamic model or a random walk model is appropriate to match this present situation more closely. At times, parallel filters of constant and non-constant contact velocity are exercised in order to identify the situation at hand.

D. Expert System Concepts for Contact Localization and Tracking With Uncertain Conditions

So far, algorithmic techniques have been presented to resolve fluctuations in signal and data processing solutions in the channel environment. Expert system approaches are much needed to deal with algorithmically uncharacterized fluctuations and to provide not only significant performance improvement despite these uncertainties but also to execute expert level tasks heretofore undone, quickly and reliably. An effective system configuration is one that integrates the algorithms of signal/channel analysis and pattern recognition to the tools of artificial intelligence.

In signal and data processing, good progress has been made in deriving CLMA approaches. The outputs of such approaches typically are range, depth, course, and speed of a contact. However, capabilities and performances of current CLMA approaches are severely limited by being unable to utilize various sources of knowledge. For those, they rely almost entirely on human experts to reduce those uncertainties that do not belong to a limiting set of assumptions, i.e., Gaussian errors, constant contact velocity, etc. Even under the above restrictive assumptions, CLMA approaches do not support the extraction of information from minimum/maximum type of data and process solely single valued data. In many domains, interval types of data are available and their inclusion in the decision process can prove very helpful to the estimation process. Interval type data are encountered near the threshold of signal processors, i.e., contact in a given spatial area, are dictated by the physical capabilities of the contact, i.e., contact maximum speed or are yielded by the environmental considerations, i.e., convergence zone size, channel depths, etc. More importantly, the existing hypothesis testing for validating assumptions is rudimentary.

A knowledge-based expert system that performs on those data types is much needed. It should provide not only major improvements in performance, but also an intelligent automated system that performs the expert level tasks quickly and reliably. In fact, the information derived from signal, data, and environmental processing can be inputted to an expert system which should use such information as well as a priori knowledge of contacts and environments, along with knowledge gained (learned) from system operations.

Both artificial intelligence and pattern recognition play very important roles in building expert systems. We elaborate on the elements for constructing a contact localization and tracking system.

1. Expert Systems

An expert system is a computing system which implements organized knowledge on some specific area of human expertise. Basically, the expert system consists of knowledge (data, rules) and control strategy (how and in what order). Data are found at varied levels. Rules use data at one level to infer data at another level. Currently, most expert systems employ IF...THEN rules to represent knowledge. The control strategy refers to a mechanism for manipulating these rules to form inferences, make interpretations or decisions, etc.

Some important features in a typical expert system are:

- The use of rules with certainty factors.
- A backward chaining inference path starting with a goal and using applicable operators until an appropriate earlier state is reached or the system backtracks.
- An explanation facility to support the inference given by the system.
- The inclusion of structured knowledge and mathematical relationships with the production rules.
- The incorporation of time dependence to depict trends in the growth of system states.
- A method in constraint propagation for limiting search.
- A procedure for conflict resolution.
- An ordered approach for blind search that does not rely on a knowledge input.
- A graph searching method that uses heuristic knowledge to focus the search space.
- An approach for satisfying multiple goals, with each goal satisfied independently and then integrating.

2. An Expert System Structural Elements

In building an expert system, a growth layered path may be the safe way to achieve the complete set of features. As such, the entry level expert system may not follow precisely the construction of other mature expert systems. As the expert system grows, it progressively meets the demands for multisensor integration, multi-contact tracking, and other complex situations in difficult channel environments.

The configuration of the system integrates the tools of signal/channel analysis, pattern recognition, and artificial intelligence (Figure 17.20). The signal/channel analysis function encompasses the available algorithmic techniques in detection, estimation, tracking, channel representation or their outputs.

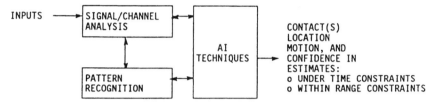

Figure 17.20 Expert system approach.

The pattern recognition function contains template matching, cluster analysis, feature extraction, and delivers the needed classifiers of the uncertainties that the signal/channel analysis block is attempting to reduce. The artificial intelligence block performs the supervision/management function and contains the non-algorithmic problem resolution tools and the mapping of beliefs into the system operation. It performs the task assignment functions as to:

- Independent solution of problems germane to each specified domain with similar characteristics.
- Joint integration of solutions through shared domains.
- Negotiated solution among conflicting domains/experts.

Such a cooperative system improves on the approaches that:

- Are compartmented and have non-interacting domains.
- Allow a limited number of uncertainty characteristics.
- Have little hypothesis testing.
- Have little to no min-max information processing.
- Take advantage of little to no heuristics.

Given the "right" conditions or appropriate uncertainties, algorithmically based systems give the best possible answers, since it is known how to design optimally with those conditions in mind. In practice, two issues are faced with those systems. First, the "right" conditions do not always exist, and second, it is not always known when they do not exist. The aim of a combined algorithmic/heuristic based system is to preserve the "best solution" goal when conditions allow, but to produce, otherwise, satisfactory answers relying on experience, dispositions, and the like.

The ultimate purpose in constructing an expert system then is to enable reasoning under varied and dynamic uncertainties, in order to infer the desired information on the contact, along with confidence limits and under time constraints. The three blocks in Figure 17.20 work, where each is applicable, on reducing the possible uncertainties encountered in CLMA problems using different approaches. The uncertainties are triggered by:

- Errors in signal processing output, i.e., noisy time delays, direction angles, frequencies, etc.
- Errors in sensors' position monitoring as in towed or drifting sensors.
- Errors in data/ray path association as in a multipath channel.
- Errors in data/contact association as in a multicontact situation.
- Errors in environmental data processing as in mapping of time delays into a contact position.
- Errors in the modeling process, i.e., contact motion, hypothesis setting, etc.

An accounting of the uncertainties and their characters must be kept by the system. This function is critical to their successful reduction by the system. The character of an uncertainty determines whether available algorithmic techniques and/or heuristic techniques should be applied to minimize its effect. Much of the algorithmic works rely on the presence of a type of uncertainty and generally presumes a Gaussian unbiased characteristic. Then the need exists for some hypothesis and revision mechanism to avoid catastrophic system failure due to mismatch of assumed models and input data. Also needed is an expert system that deals with uncertainties beyond the Gaussian type. A mature system would accept suppositions, give results, and check for contradictions. Thus, it will reduce the number and variety of residual uncertainties that people have to cope with at present.

The knowledge domain of the system would contain observations, conclusions, and rules about the contacts' and sensors' characteristics, channel performance prediction, signal analysis, etc. Specifically, signal analysis contains signal processing and data processing techniques and algorithms and/or their outputs, such as generalized correlation, cepstrum, complex demodulation, spectral analysis, maximum likelihood estimation, energy detection, weighted least square estimation, etc. Included in the signal processing are the adaptive windows and gates applied in the varied domains, e.g., frequency, time, space. Characteristics of contacts include a priori known data and signal classifications such as transients, narrowband, and broadband signals. The channel performance prediction develops metrics to the spatial cells apportioning the channel that links the sensors to the surrounding environment.

Various levels of production rules are generated. The simple production rules codify and introduce the heuristic information into the decision process. The more powerful rules are designed to perform algorithms' selection, inclusion of mathematical relationships, on-line adaptations, and consistency checks across the global system. They test postulates of invariance whether deterministic or statistical. They check on statistical measures such as bias, trend, and on distributions. Thus, the more complicated rules test whether the data fit the assumptions on the model. The expert system should initially deal with a fixed number of situations and then grow by including new rules and modifying the existing rules, if necessary, to a more generalized expert system. The system should continue to expand the knowledge domain and the list of rules to deal with new situations through learning and human interaction. Present CLMA systems are predominantly based on a presumed fixed set of operating conditions. But in practice, variants to those conditions are the main cause of catastrophic system failures, and there is ample gain in developing a self-adjusting system.

The ultimate strategy is to reconcile among weight and integrate over a collection of specialized knowledge domains and decisions. The need for the approach lies in the desire to accommodate the widest variety of practical situations where a priori information is generally lacking. Too often signal analysis systems are compartmented, and there is little or haphazard communication from one to another. Each sub-system by itself has natural performance limits which can be enhanced through sub-system interactions. The overall strategy is as follows:

- Using each knowledge domain, break the uncertainty in the physical space into regions.
- Search for common regions across knowledge domains.
- Lower sizes of most probable regions by fusing knowledge domains in an algorithm.

The subject of fusion or integration plays a major role in system growth.

3. *Layered Integration*

Layered fusion works in several levels on the identified uncertainties, within and across domains. The first level involves a global search while the succeeding levels involve increasingly more local searches, to refine and eliminate unlikely possibilities within a domain. In multi-domain fusion, we have, for example, fusion of time delays from different sensors and/or fusion of environmental prediction results to point to a certain contact. In one domain, for instance, the correlation output shows a number of peaks, from which the highest peaks above a threshold are picked. In the other domain, environmental considerations yield metrics or distance measures within the channel featured by appropriate sets of time delays. Channel boundaries are defined by contact type, e.g., surface sensor spatial coverage, contact depth or height capability, etc. The combined outputs from the two lead to an enumeration or a smaller number of possible range values. By integrating information from different levels and across domains, unlikely possibilities can be minimized and quite often a final decision or estimate can be obtained.

As an illustration, a reverberant channel is now discussed. The types of raypath structure linking the contact to the receiver are not known initially. The channel space, as bounded by receiver/contact capabilities, is divided into sub-spaces by the environmental processor characterized by a known set of time delays. Each sub-space defines a position in the channel and is either uniquely identified by the set of time delays attached to it or in conjunction with the neighboring sets of time delays attached to the contiguous sub-spaces. The sub-space enclosing the emitter is not known a priori. Pattern recognition techniques are applied to match the time delay patterns from the signal processor with those attached to the sub-spaces, and the recognition of the emitter's approximate location. Refinement of the estimate is then possible once data/path type association is established. If more than one type of ray is available, hypothesis testing would be used on the candidate types to ferret out the uncertainty. In environments with complicated raypath structure, forward-backward chaining may have to be applied. When contact, data, and path associations are established, algorithmic passive tracking techniques are well suited to refine the contact motion estimates within and across sub-spaces. Thus, integration of all primitive data is beneficial. When sensors' data have disparate uncertainty characteristics, then merger of the higher level estimates, such as range, should be done.

This section delineates the system elements for realistic contact tracking. A design concept is presented taking into account the structured and distinguishing as well as unstructured and fuzzy characteristics of sensors, contacts, signal analysis/pattern recognition systems and their environments.

In building an expert system, one is aiming to capitalize on the beneficial characteristics of two processors, the "human processor" who is slow, inaccurate, but innovative, and the computer which is fast, accurate, but inflexible. The ultimate system would hopefully replace the existing inflexibility in the machine with the innovation of the expert(s). Eventually, the system inputs would be sensor reports, task requests, and/or feedback. The system output through data, event, or goal driven inferences would provide a situation description, an assessment, and sensors' management approach.

VI. GENERALIZED PREDICTION, PROCESSING, WAVE PROPAGATION, AND VIBRATION METHOD

Traditionally, separate treatments are given to wave generation and propagation, vibration, signal, and data processing. Books dedicated to each of these subjects abound. In practice, however, prediction and processing of wave

AUTOMATED INFORMATION PROCESSING

and vibration go hand in hand. Prediction of the wave field from solution of the wave equation enables the judicious placement in space of the processing equipment where the received signal is enhanced against the noise. Processing of the received time function is then conducted to combat the noise and estimate the desired signal.

Joint prediction of propagation, vibration, and processing can be performed using a common approach that capitalizes on the underlying similarity in both problems when formulated via an integral approach. Prediction of the wave field and vibration calls for the solution of the Fredholm integral equation when formulated as such. Synthesis of the processor also calls for the solution of an integral equation of the same class as the one encountered in the propagation and vibration problems.

Some distinctive features of the approach make it attractive from analytical and computational points of view. For application to either prediction or processing, the procedure calls for splitting the total problem into two parts:

- One part has an analytical solution which is expressible in terms of known functions.
- The other part has the solution developed by a series of successive approximations which are continued until the contribution of additional iterations is negligible.

Using such an approach, the problem of general variable systems in space and/or time can be treated. These systems are subjected to general forcing functions as embodied by distributed source functions and two-point boundary conditions. Such a description encompasses many outstanding problems in a variety of applications to wave and vibration. Synthesis of an optimum filter for detection or tracking, wave propagation in varied channels, in layered ducts, in spatial filters, in arrays and lenses, vibration of dynamical structures with inhomogeneous members such as beams, rods, chains,..., all reduce to the analysis of the previously defined problem.

A general solution to such a problem is difficult. In treating it analytically, a very limited number of variabilities may be analyzed in terms of known functions. Iterative perturbation methods (Neumann series) are particularly helpful when the problem under consideration closely resembles one solvable in terms of known functions. Otherwise, the process is so cumbersome as to be intractable. Specifically, the Fredholm series is a valid solution regardless of the perturbation size. Unfortunately, the terms in the series involve multiple integrations over determinants, both of order N, where the value of N increases in step with the perturbation size. Furthermore, this form of solution is not amenable to an implementation on the computer. The loss of accuracy and the required computer time and storage present serious problems when very large determinants and multiple integrations are to be evaluated. An alternative method converts, at the outset, the differential equation and its boundary conditions or the equivalent integral equation to a set of algebraic equations. This conversion is equivalent to the direct approximation of the distributed system by a lumped system. There remains the problem of solving a coupled system of algebraic equations by a known method. Although a solution is now practical, it is by no means straightforward for systems with a large number of degrees of freedom. The exact methods of solution, such as the Gauss elimination, suffer from loss in accuracy and excessive computer time. The iterative procedures are inherently free of such inaccuracies, but they are applicable to systems with small perturbations. The Neumann series for distributed systems and the Jacobi iterative procedures for the lumped systems require respectively that the norm of the kernel and of the matrix be less than one.

To accrue the benefits of iterative techniques while mitigating the divergence problems experienced during large perturbations, the Hassab iterative approach can be applied. Basically, the boundary value formulation is transformed into the solution of an initial value problem plus the determination of a constant. It is an unified approach to the solution of both forced and unforced systems for problems in wave propagation, vibration, signal, and data processing. Arbitrary source functions, general variability, and general boundary conditions are permitted. The impulse response is determined. A procedure for the generation of the system's normal modes falls out. The condition of symmetry on the Green function is relaxed. Further, the treatment deals with physical systems that yield a not strictly bilinear Green function.

The general problem is depicted in Figure 17.21. Facets of the general problem can be described by a representative set of difference, differential, and/or integral operators. Discretization of integral operators yields matrix operators where continuous variables lead to discrete matrix elements. Ease of solution varies:

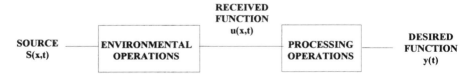

Figure 17.21 Basic operations in sound and vibration problems.

- Linear discrete systems have a finite number of degrees of freedom, have properties independent of time, and lead to linear differential or difference equations with constant coefficients or Volterra integral equations. Their solution is simple for prescribed initial conditions. Network theory leads to such formulation. Digital signal processing deals with linear difference equations. They form the starting point for describing discrete linear systems, i.e., frequency response, Z-transform, geometrical interpretation via the Z-plane.
- In analyzing wave propagation and vibration problems as well as estimation problems with optimum weighting, the crux of the difficulties resides in the handling of systems with variable coefficients

AUTOMATED INFORMATION PROCESSING

and arbitrary source and boundary conditions. Those problems lead invariably to Fredholm integral equations or the Helmholtz differential equation with large perturbations. A transformation followed by iteration leads to a unified approach to treat those problems. In estimation problems where the linear mean square error weighting applies, the unified approach leads to an initial value problem with updates against the latest measurement only due to its product kernel characteristic. This representation is particularly desirable since it is amenable to joint analytical studies and numerical calculations.

This versatile approach systematically reduces to quadrature problems in vibration, wave propagation, signal prediction, and processing. The solution procedure is iterative where the contribution of each additional iteration vs. the total contribution of all the previous iterations gives an assessment of solution quality, if iteration is interrupted at that point. This is an attractive feature in a real-time system, where a solution may begin from a given familiar approximation, and then proceed to correct that approximation and to improve on it so as to satisfy the opposing constraints of time vs. solution validity and accuracy. Here, these iterations are guaranteed to converge to the unique solution. The calculations are recursive in character and are thus compatible with machine operations. Overall, numerical accuracy is increased by a factor of two, the storage requirement is reduced by nearly one half, the number of operations is reduced by a (2/3)n and 2n for forced and unforced systems, respectively, where n is the matrix size. The procedure is applicable to high order problems that yield a kth-order differential equation. Then one solves for $(k - 1)$ unknown constants instead of only one. The approach is also applicable to a vector wave equation, where the vector field is representable by a scalar potential and a vector potential whose components are separable. In acoustics, this allows the treatment of shear waves in solids, i.e., the ocean bottom, in addition to the longitudinal waves. In electromagnetics, this allows the treatment of coupled vector waves as in the ionosphere.

VII. SIGNAL PREDICTION IN THE PRESENCE OF MULTIPLICATIVE NOISE

In preceding sections, we have considered the problem of signal prediction and estimation in the presence of input additive noise. We have encountered cross-terms between signal and noise only at the output of nonlinear and memoryless processing. Now we extend our considerations to the case of multiplicative signal and noise at the input along with linear memory operators. Two approaches are applied:

- An analytical solution that improves on first order approximations, e.g., Rayleigh-Gans Born (R-G-B).

- A numerical solution that capitalizes on the techniques of the preceding section.

The interest in multiplicative signal and noise for waves propagating through random media is due to the importance of scatter propagation in communication or remote sensing, on the one hand, and the limitations imposed upon tracking and guidance systems for vehicles on the other. When a source emits a perfectly steady signal, the signal received, even at moderate ranges, may display wave level fluctuations. This phenomenon, akin to scintillation of stars or fluctuations of radio signals received via a scatter link, is due to time variations of inhomogeneities in the medium. These variations modulate the incident wave leading to amplitude and phase fluctuations of the resultant field; this causes amplification and fading of the wave field. A number of different physical phenomena cause inhomogeneities in the channel. There are, for instance, inherent temperature fluctuations that contaminate a great deal of existing data; there may also be turbulence in the body of the channel and on its surface. All these phenomena make a channel inhomogeneous and produce intensity fluctuations of the wave field.

Two types of inhomogeneities are encountered, deterministic and random. The concept of randomness is introduced because of the complex variation of the medium properties that makes it difficult to analyze its effect in a more direct way. Physically, an inhomogeneity is said to be deterministic if an identical measurement is performed many times, and the results obtained are always alike. However, if all conditions under the control of the experimenter remain the same, and the results continually differ from each other, the inhomogeneity is said to be random.

The early investigations on the scattering phenomena used the ray concept where the energy is visualized as traveling along ray paths that obey Snell's Law and Fermat's Principle. For large distances, however, the energy scattered off the raypath by inhomogeneities in the medium is often of importance; these energies are not accounted for in the ray treatment.

To account for off-ray scattering, the single scatterer or Rayleigh-Gans-Born approximation is used. This method is valid for a small volume and a small perturbation in the refractive index. Thus, a Rayleigh approximation breaks down for long range propagation, because the inhomogeneities display a cumulative scattering effect, despite only small perturbations in the refractive index; for example, if a localized volume of inhomogeneity can scatter energy from a transmitter to a receiver, as supposed by the R-G-B approximation, then an inhomogeneous volume can act as a new transmitter scattering energy via other inhomogeneous volumes to the receiver.

Methods for analytically extending the above techniques to include multiple scattering effects have been developed for a homogeneous background and homogeneous correlation function. Analytical extensions to an inhomogeneous background have been made in the limit of small scale fluctuations as well as when the correlation function is inhomogeneous inside the scattering

AUTOMATED INFORMATION PROCESSING

volume, and its mean characteristics vary smoothly. For more general randomness, application of the unified approach successively each time against one realization of the medium, leads to an ensemble description over which mean and variance of the coherent and incoherent fields can be deduced.

This section introduces analytical and numerical methods to describe the desired information when it is multiplicative with some random effect and is also submitted to a linear operator with memory. This problem is physically relevant for information propagating in random media. For description of the information, the first and second order statistical measures are derived. Those statistical descriptors are needed when designing processors which are receiving information that has propagated through such random media.

VIII. DISCUSSIONS

Processing rests on the search for invariant properties as it attempts to extract information under risky and uncertain conditions. The total gain in a processing system is derived from spatial or array gain (size, number, and placement of sensors, etc.), from accounting for propagation channel features and from spectral and/or temporal processing gain (signal, data). The joint activities in spatial, spectral, and temporal processing aim to perform some or all of the following functions to develop a total picture of the outside world:

- Is anybody/anything of interest there; i.e., detection?
- Who is he/what is it; i.e., identification/classification?
- Is he/it a new detection or an update on an old one; i.e., association/correlation?
- Where is he/it; Where is he/it going; i.e., localization/motion analysis?
- What does he/it intend to do; i.e., response/reaction?

Practical considerations, such as array stabilization, cost, spatial coherence, and platform dimension limit the achievable spatial gain, while the temporal dimension is limited by the requirement to yield an acceptable solution within a given time lapse. When the elements of the problem are stationary, the temporal processing reduces to a continual integration over the observed source clues until desired accuracy in the solution is obtained. When the elements of the problem are nonstationary, i.e., moving source/receiver, changing channel characteristics, etc., the source observation is limited to a basic observation time interval, over which the process may be considered locally stationary. The system then gives a short memory snapshot or single look estimates, i.e., its lowest level of performance, and yields at times a fuzzy picture of the estimates. With a succession of basic time observation intervals, the temporal processor basically removes the nonstationarity in the problem and superimposes the repeated fuzzy pictures to reinforce the invariant parameters in the problem and to develop a well-defined picture of the desired information.

Processing (signal/data) reduces many pieces of data into comprehensible information for humans. It eliminates non-informative data, creates simplification, and develops data groups that sort out the tangle of factors. Processing based on invariance performs:

- Confirmation as to the presence of certain elements of data.
- Summary of that data by finding its lowest or highest number, a representative number, i.e., mean, median, and mode, and indicating its variability, i.e., variance characteristic or feature. When the feature or characteristics persist over new data, tracking, prediction, and confirmation of the feature are enhanced.
- Identification of data that form a pattern thus enhancing classification/categorization.

18 PROGRAM MANAGEMENT

I. INTRODUCTION

Management in any kind of endeavor involves the basic functions of planning, organizing, activating, and controlling with the processes linked together through decision making and communication, as explained in Part I. The system and computer concepts and approaches of Parts II–III provide for an effective infrastructure to execute the management in varied endeavors; Part II defines the patterns for systematic design and development; and Part III focuses on computer system technology pertinent to its design, development, operation, and use.

Endeavors, be they in aerospace, banking, commerce, industry..., are of the regular and repetitive type and/or of the unique type; examples of such endeavors are given in the chapters of Part IV. An endeavor is undertaken to either gain market share, win in overall business, and/or collect profitability from a stand alone product. A program is an endeavor whose activities are unique or will not be repeated in the same form in the future, are complicated with an extensive number of activities, are constrained by pre-set performance, time, space, and cost; a mini-program is known as a project. Uniqueness, complexity, and constraints render program management a very challenging endeavor that requires careful planning, organizing, activating, and controlling, structured in a system's approach. The starting reference is the contract and it may be internally generated or externally set with a customer.

At the center is the program manager as the responsible individual to make it work; his interfaces are many (Figure 18.1). As in any other system representation, the functions of input, processing, and output are encountered (Figure 18.2). In this formulation, the program manager leads the team that synthesizes the processing function to transform the given inputs into the desired outputs within stated constraints. By its nature, the transformation operator (set of actions) is by no means deterministic, homogeneous in space, constant in time and thus easily derivable at the outset. It is space and time dependent, is subject to time and boundary conditions, is random and inhomogeneous and, thus, its generation

380 SYSTEMS MANAGEMENT: People, Computers, Machines, Materials

Figure 18.1 Program interface management.

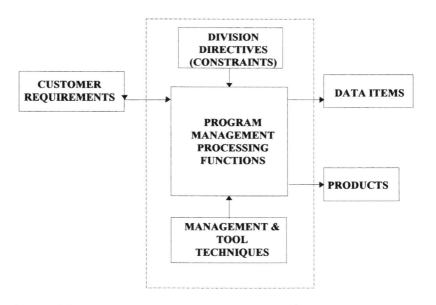

Figure 18.2 Program management input, processing, and output representation.

PROGRAM MANAGEMENT

is at best risky and its estimates dynamic; these require continuous reflection, monitoring, and adaptation to:

- Identify possible future conditions.
- List the possible alternative approaches.
- Set the pay-off associated with each alternative.
- Set the likelihood of possible future conditions.
- Establish a decision criterion to select the best alternative.

However complex sounding it seems, the program manager must think through at the outset the transformation set of actions (plan) which will lead from the contract to the product. In the process he/she decides what needs to be done, when it needs to be done, how much it will cost, and who will do it. While he/she does not decide how it is to be done, he/she sets the controls on the activated actions to assure that the planned actions are successfully implemented (Figure 18.3). Program management reviews provide one vehicle for probing into the actions taken and assessing their utility; if failure is detected or projected, replanning activities are set which may result in revised new authorization, addition or deletion of tasks and resource commitments. Thus, program management plans are dynamic, phased, and incrementally detailed.

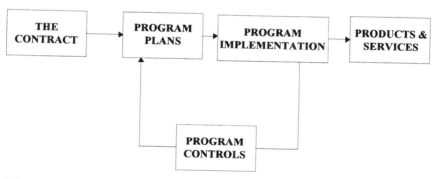

Figure 18.3 Top level program management system.

II. BASIC PROCESSING FLOW IN PROGRAM MANAGEMENT

Program management can be visualized as a virtual corporation; it is scrutinized; it is event-based; it consists of multiple contributing elements from within and outside the real corporation, all networked as a single logical transformation process with definitions of roles, responsibilities, functions, schedules, costs. By its nature, it is a time-dependent set of associations linked in a combination of organizational structures with simple and direct mapping to the ultimate product/service it is required to provide. Whenever projects or events are completed, the virtual corporation is dissolved.

In effecting a program, the plans, their implementations and controls are developed and exercised by the program manager and his/her team (Figure 18.3) to reach stated outputs usually set in terms of a contract. A program management processing flow is depicted in Figure 18.4. It structures and articulates the activities needed to transform the contract requirements into products and services. In this transformation, the contract requirements of performance, cost, schedule are decomposed first to enable implementation then are integrated to provide the expected product/service. The resultant is an integrated process to provide the product/service based on concrete milestones required by the contract. The decomposition or work breakdown structure (WBS) and integration processes answer very basic but critical questions (Figures 18.5 and 18.6):

- Why? It is because of what the contract requires; the contract defines the deliverable products and services and delineates three significant and interdependent parameters of required work performance, schedule, and cost. The requirements are stated in the request for proposal, statement of work, specifications and associated documents; in total, they establish the framework for decomposition–integration to the implementing organizations of the scope of work, technical objectives, schedules, and costs.
- What? The end deliverables are set in the contract statement of work (SOW). Prior to activation, a decomposition or breakdown of the work must be conceived down to the performing level followed by integration of the efforts up to the end deliverables. While the SOW describes what is to be accomplished, the work decomposition or breakdown process provides the structure which depicts graphically the work organization in a flow-down concept, e.g., hierarchical, beginning at the top with the end product/service and descending through the successive levels of details from system, subsystem, component down to the functional activities or tasks which describe the detailed work to be performed, its schedule and cost. The resulting work breakdown structure plans the deliverables of the contract, and frames the performance, cost, and schedule plans into a traceable and integrated total plan.
- How well? A technical performance breakdown or decomposition plan must be framed to depict, in correspondence to the WBS flow down of the integrated products/services to tasks, a flow down of contract specifications through successive levels of details down to the functional activities or tasks which now describe how well the detailed work shall be performed.
- How long/when? A schedule breakdown or decomposition plan must be framed to depict, in correspondence to the WBS flow down

PROGRAM MANAGEMENT

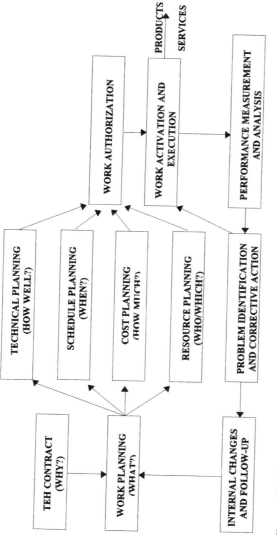

Figure 18.4 Program management process flow.

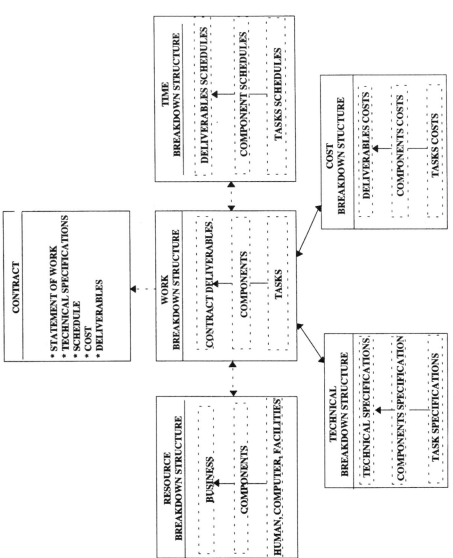

Figure 18.5 Breakdown and integration of contract elements for implementation into deliverables.

PROGRAM MANAGEMENT

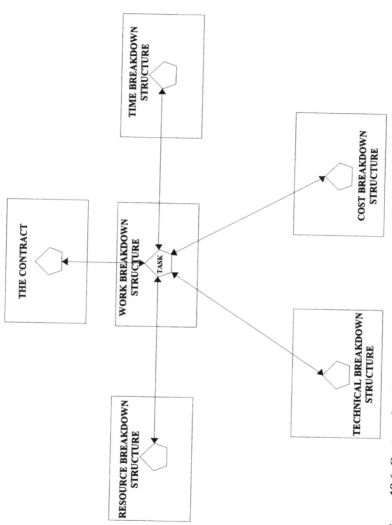

Figure 18.6 Contract breakdown and linking at the lowest levels to depict activation and control.

of product/services to tasks, a flow down of contract schedule of deliverables through successive levels of details down to the functional activities of tasks which now describe when each detailed work shall be performed to support the contract delivery dates.
- How much? Cost breakdown or decomposition is a universal measure of work and a cost plan must be framed to depict, in correspondence to the WBS flow down of products/services to tasks, a flow down of contract price through successive levels of details down to the functional activities of tasks which now describe how much each detailed work shall cost if their summary is to fall below contract price.
- Who? Contract work is executed by humans/computers/machines and a resource organization plan is framed to depict, in correspondence to WBS flow down of products/services to tasks, a flow down of the resources through successive levels of details down to the functional activities of tasks which now describe "who" will perform each detailed work.

In framing the breakdown implementation and integration of the contract elements (work, resource, time, cost, technical), varied structures are used (Figure 18.7).

- Hierarchical structure where each breakdown is usually linked through one-to-many and is only in a downward flow; many-to-many linkages can be handled with a combination of hierarchies.
- Network structure where each breakdown is usually linked through one-to-many, but direction is not only downward and is thus indicated by an arrow.
- Relational structure where the breakdowns are set in a tabular form and linkages are done through repeated key fields.
- Combination of above.

The linkages in the varied structures may be physical or logical through some indexing or mapping mechanism (alphabetical/numerical/attribute). Linkage is essential to the integration process and needs to occur not only within each element (work, resource, time, costs, technical), but also across the decomposed elements where each task plus its allocated resource, its scheduled time, its cost profile, and its expected technical performance represent a mini-contract (Figure 18.8). Like the contract itself, those mini-contracts are negotiated and authorized for activation.

Work execution seldom proceeds according to plans if ever; thus, planning is not a batch process, but rather a recursive one where constant measurements of progress against plans are performed; discrepancies are analyzed, their sources are identified, and corrections are instituted to guide

PROGRAM MANAGEMENT

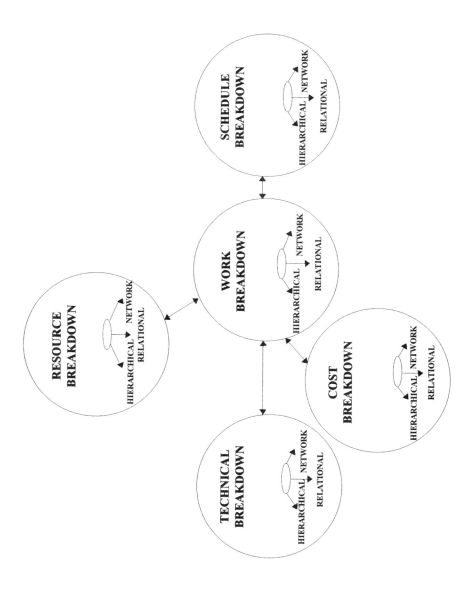

Figure 18.7 Generic breakdown and integration structures.

388 SYSTEMS MANAGEMENT: People, Computers, Machines, Materials

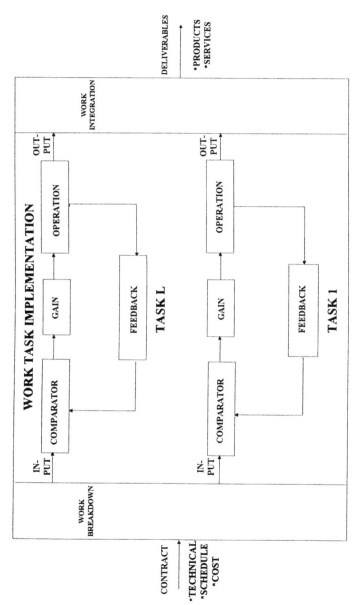

Figure 18.8 Contract execution as execution of integrated large number of tasks (mini contracts), each with its derived technical schedule and cost.

future progress toward satisfaction of the contract. Programs do not lead to systems that can be analytically described, that can be automatically controlled, and whose performances are predictable. The contract breakdown planning process can lead to thousands of tasks, each with its attendant performance, cost, and schedule, all estimated with a wide variety of assumptions, details, and uncertainties, and all in need of further detailing and of continuous monitoring and control. Though many monitoring processes are automated, much of the decision making in control remains within human purview. After all, the size and complexity of the whole operation are such that it is practically impossible to pre-set reliably the most beneficial overall combination of the separate tasks; it is more likely that the most economical alternative for one task will automatically preclude applying the most economical alternative for another task. Throughout it all, program management must conduct sub-optimization while remaining focused on the relevant factors of the problems. One such overall factor is money used to assess work progress where the metrics are:

- (The sum of performed tasks' budgeted dollar value) minus (the sum of performed tasks' actual dollar expenditure); this indicates cost overrun if the metric is negative and under run if positive.
- (The sum of performed tasks' budgeted dollar value) minus (value of those scheduled to be completed by this time); this indicates a lag if this metric is negative or a lead if positive.

III. DETAILED PROGRAM MANAGEMENT IMPLEMENTATION

The work planning process is based on the contract statement of work (SOW) and deliverables. The SOW gives a detailed description of work to be performed, the applicable specifications, standards, and other contractual requirements, as well as the SOW tasks tied to the deliverables. The contract may be generated internally within an enterprise to reach some desired goals or agreed to with an external customer for the delivery of products/services. When planning a program, the broad objectives are initiated by the program manager and detailed by the functional managers under the leadership of the program management team. The targets provided by the program management are expanded upon by the functional organizations into (Figure 18.9):

- Technical performance measures.
- Resource planning to the work package/cost account.
- Organization schedules.
- Work package milestones.

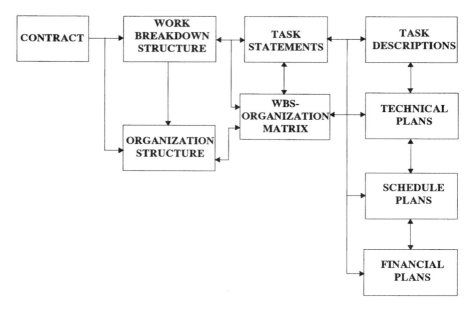

Figure 18.9 Program planning with increased detailing of contract objectives.

The steps in program planning include:

- A work breakdown structure with elements denoting the products to be developed and the work to be executed in order to achieve the desired products.
- Work packaging of specific subproduct/service to be executed by a specific operating unit in the organization.
- WBS-Organization matrix preparation.
- Detailing of task descriptions within work packages.
- Logical phasing network of all WBS elements to show the events' activities and their interrelationships and/or dependencies.
- Time phasing of the logical network to assure a coordinated development of work packages within WBS elements of hardware/software/service.
- Technical performance metrics for setting target parameters for the WBS elements down to their work packages.
- Cost budgets for each work package.
- A cost–schedule control system to estimate, acquire, allocate, and control the resources, e.g., people, material, facilities, money, time.

PROGRAM MANAGEMENT

A. Implementation of Work Breakdown and Organizational Allocation

The work planning process begins by mapping the SOW and deliverables into work packages, called WBS elements. The WBS is a product-oriented family tree-down of hardware, software, and services necessary to achieve the specified product(s).

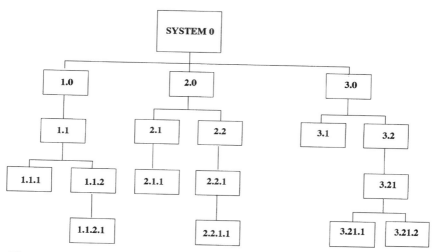

Figure 18.10 Tree-like work breakdown structure.

A WBS can be developed on a tree-like basis (Figure 18.10) or tabular basis (Table 18.1) with a decimal numbering structure to allow the mapping process. In either case, it depicts a progressive passing down of contract objectives from higher to lower levels of management. Its configuration content depends on general principles where:

- Significant products or services with a large dollar value or personnel requirements are represented by an element.
- Element descriptions are meaningful and clear.
- No task is overlooked and related tasks are placed in the proper elements and clearly defined.
- Major responsibility assignments are directly mappable to WBS elements.
- Level of indenture is determined by the dollar size of work packages, personnel requirements, and number of milestones in each task.
- Adopted numbering system is amenable to the integration of the master project schedules with subordinate schedules, assignment of individual responsibilities, accumulation of costs, integration of

Table 18.1 Tabular Work Breakdown Structure

Level 0	Level 1	Level 2	Level 3	Level 4
System	1.0__			
		1.1__		
			1.1.1__	
	2.0__	2.1__		
			2.1.1__	
		2.2__		
			2.2.1__	
				2.2.1.1__
	3.0__	3.1__		
		3.2__		
			3.2.1__	
				3.2.1.1__
				3.2.1.2__

technical performance and reporting, work authorization, performance analysis, and subcontracting.

The benefits of WBS are many and include an integrated management framework for: a) *partitioning and allocation of contract requirements and their integration*; b) *assigning responsibility*; c) *scheduling the work*; d) *estimating costs*; and e) *negotiating the contract*. In the process of work organization planning, the following are set:

- An expanded WBS of the contract proposal with added lower level elements.
- A WBS dictionary to summarize each element's tasks.
- A WBS task–responsibility table that links responsible organizations/individuals to each WBS element.
- A cost account structure that sums up cost in varied ways such as WBS elements, organization units....
- A list of components, hardware, and software to be developed.
- A schedule tree corresponding to the WBS elements and their integration.
- A specification tree corresponding to the WBS elements.

The basic driver is to maintain a clear map between the end items and the work, the specification, scheduling and cost decomposition, and re-integration processes across departmental boundaries.

To describe the work elements in detail, task descriptions are developed to expand the task statements in the contract statement of work which provides the task description outlines. The task descriptions are grouped under the WBS elements numbers, and directed toward a specific organization, are referenced in work authorization documents, are are used as the basis for costing and scheduling. Preparations of task descriptions are authorized based on the WBS-Organization assignment matrix (Table 18.2). Along with task description, expansion of the WBS-Organization matrix serves to activate the team through identification of responsible individuals, their telephone numbers, reviewing schedule and places, etc.

Table 18.2 Mapping of Work Breakdown to Organizational Units (P = Principal, S = Support)

Work Elements (What)	Organization (Who)				
	Program Manager Operating Units	Engineering Operating Units	Manufacturing Operating Units	Finance Operating Units	Sub-contractor Units
	A ... F	A ... K	A ... G	A ... E	A ... D
1.0	P			S	
1.1	P	S	S	S	S
1.1.1	P	S			
1.1.2		P	S		

Cost account sums up the effort to be undertaken by a major organization on a single WBS element. These accounts yield the lowest level in the WBS at which actual costs are collected. They form the intersection point of cost–schedule controls of the work breakdown structure and the organization structure. To exercise management within an organization, work packages are set; these packages are detailed short-span jobs within a cost account identified to accomplish the required work in the WBS element. The work package is a unit of work within one operating organizational unit and is needed to complete a specific job such as a piece of hardware, a software item, a report or well defined service. The size of the work package depends on the needed degree of management visibility to control the work; more importantly, the work within the package should be broken down into work items with clearly measurable milestones; breakdown of milestones down to "inchstones" may be called for depending on the risk involved in failing delivery on a given milestone. In any case, just as undersampling is detrimental, oversampling is not necessarily beneficial especially in nonlinear operations where it may trigger more noise than signal. Like any controlled piece of work, the work package is defined by a scheduled start, completion dates, a scope, and a budget set in terms of man-hours, dollars, or any other measurable units (Figure 18.11).

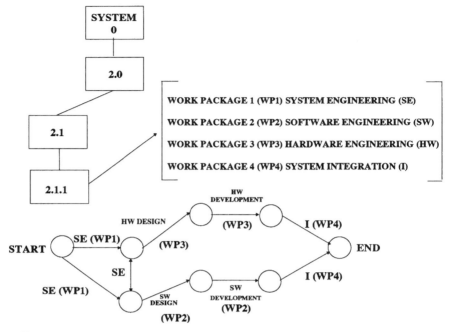

Figure 18.11 Work packages within a WBS element.

B. Technical Management Implementation

To manage the evolution of a product/system from concept to operation, requirements, design, development, integration, and test tasks are defined. A systematic approach to this evolution has been detailed in Part II. The tasks progress along program phases of concept evaluation, demonstration and validation, full-scale development, production, and deployment. In major programs, a series of important technical plans are developed and activated to control the work. Similarity in structure abounds across the plans. The tree-down decomposition and tree-up integration processes apply down to the basic system entities or building blocks. The plans describe in detail the structure in terms of those entities going from the upper levels into lower levels; for each entity, the plan defines the interfaces, flow across organizational boundaries, data characteristics, and processing characteristics in the design. The entity may be software, hardware, or human. In a software/hardware/human structure, for instance, the system functions allocated to each type are set into a number of configuration entities; in turn, each configuration entity is set into lower level or sub-configuration entities; at any level, the interface diagram is described; for each interface, the messages flowing through it are described; and for any message, at any level, the involved entities or sub-entities in processing the message are described. Processing involves unfailingly the

PROGRAM MANAGEMENT

395

entities across the structure in horizontal/diagonal flows in addition to vertical flows. The plans include:

- A system engineering plan which frames the organizational responsibilities and authority, the levels of control for performance as well as design and methods of control, the technical program assurance methods, the schedules for design and technical program reviews, and the control of system documentation. The description of processes to be used as well as the generation of specifications are included. Also included are reliability and maintainability, engineering logistics management, human factors, safety, parts standardization, producibility, quality, and transportability.
- A technical performance plan which establishes system budgets and their partitioning and allocation to lower levels where they are used for progress monitoring and corrective actions. Performance status is gauged through technical control parameters of memory capacity in bytes, response time in seconds, throughput in instructions per second, weight in pounds, size in feet, power in watts, mean time between repairs in minutes, mean time between failures in hours, etc.
- A functional design plan which provides the system description, the partitioning to subsystems and their descriptions, the architecture definition, the schedule with correlation to related plans, simulation and modeling validation, design alternatives, tradeoffs and rationale.
- A software/hardware development plan which covers development methodology, facilities, implementation schedule, organization and resource allocation, standards and conventions, design assurance techniques, documentation, configuration and data management.
- A review plan which provides items to be reviewed, content of the review and who should review. Reviews usually include a system requirement review, system design review, preliminary design review, and critical design review.
- A safety plan which provides for design, product marking, safe testing, handling, transportation, and operation of the system product.
- A test plan which provides the types of tests and test methodology, all traceable to requirements specifications.
- A training plan which provides for curriculum development including training equipment set-up, training facilities, lesson plans, and pilot training sessions.
- A configuration plan which provides baseline development and control and defines procedures and interfaces between the contractor and customer for revisions and releases.
- A quality assurance plan which provides for a quality product through monitoring and audits of software, hardware, and testing.

- A production plan which provides for producibility, test methods, and production flow as well as an associated design to unit production cost plan and life-cycle analysis plan.
- A resource plan which provides for effective utilization of people, facilities, and investments through activation of the right people with the right tools at the right time.
- A risks plan which identifies and classifies a program's technical risks, develops alternate approaches, evaluates cost–schedule risk vs. alternate approaches, establishes a plan and responsibility for retirement of each risk.

Many tools support the technical planning effort. These include the work breakdown structure, specification tree, specification documents, configuration planning documents (equipment, software, firmware), concept design reviews, allocated system budgets, etc.

C. Schedule Management Implementation

Schedules orchestrate the activation process and organize the position in time of events relative to each other; simply, schedules are a time table of the activities expanded on the tasks' development and their integration; schedules summarize the numerous activities undertaken by the organizational units into significant events relatable to the program's major milestones. This summarization process is depicted in Figure 18.12a. The number of schedules are many and varied; a representative list of basic schedules includes: master schedules, (program organization, subcontractors...), major subsystem schedules, milestone schedules, product schedules, detailed WBS element schedules, detailed network schedules, review schedules, detailed work schedules, workarounds special schedules, manufacturing schedules, cost-milestones schedules, data requirements schedules, dependencies schedules... . Major products are probably supported by many organizations and may be derived from multiple WBS elements. Thus the schedules coordinate the diverse activities in the many tasks into composite integrated milestone schedules.

Scheduling is an iterative process that requires many adaptations and revisions, especially in complex product/service development. However, the critical elements in any schedule must always bring out the events critical to the program; they include:

- Contract deliverables.
- Major reviews, dependencies and decision points.
- Completion of major tasks within WBS elements.
- Long lead-time activity procurements.
- Contract billing milestones.
- Critical path milestones.

PROGRAM MANAGEMENT

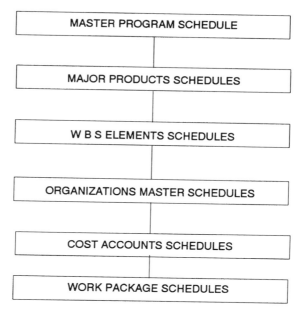

Figure 18.12a Basic types of schedules.

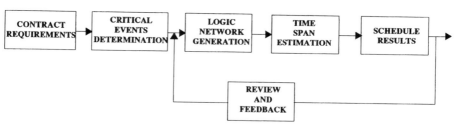

Figure 18.12b The development of scheduling.

The evolution of the scheduling process is depicted in Figure 18.12b. The steps in scheduling proceed by:

- Summarizing key program objectives.
- Defining program activities and products.
- Establishing the program logic network; e.g., the sequences and dependencies of program objectives, activities, and products including requirements, design, development, production, facilitization, subcontracting, etc.
- Estimating gross time spans.
- Relating time span estimates to the events in the logic network completed with functional personnel participation.

- Preparing a preliminary master program schedule and subordinate schedules for WBS elements.
- Verifying the feasibility of the schedules through mapping of time estimates on the events/activities diagram.
- Developing detailed organizational schedules and insuring scheduling support and compatibility among organizations.
- Integrating the schedules with the WBS and the cost–schedule control system.

Unfailingly, schedules are dynamic and iterative rather than static; they gain in detail as they are successively treed-down to the operating level. Most importantly, it is necessary for the composite schedules with their integrated milestones to be coordinated to ensure work completion of the major contract products that may be developed by multiple organizations and described through several WBS elements. Figures 18.12a and 18.12b depict an integrated set of program schedules. There are basically three graphical techniques to frame such schedules:

- In Gantt charts, the project is broken down into tasks with start and completion dates. The result is a graphical depiction that may be annotated to provide visibility into progress to date with accomplishments and expenditures noted against the plan (Figure 18.13a).
- In milestone scheduling, a date is selected when discrete and measurable events in tasks should be completed, thus adding detail to the information given by a basic Gantt chart. Milestones serve as review points and when superimposed on a Gantt chart, they show dependencies within the same task. To show dependencies across tasks, the Gantt chart needs to be modified and, in the process, this leads to network representation.
- In network scheduling, the interrelationships among tasks are brought out to ensure that coordinated work across tasks occur. Now, the immediate predecessor tasks to each task are added to ensure depiction of needed sequencing (Figure 18.13b).

D. Financial Management Implementation

An underlying measure of management is reduced cost; reduced cost returns profit which is the principal objective for the existence of a business. Even when other objectives are sought, creation of capital must still be made by earning a profit in order to satisfy those other goals.

The cost budget framework uses the work breakdown structure where elements of cost are attached to each work element based on task descriptions and schedules. In developing the budget, cost is aggregated by estimates of direct labor costs, overhead, purchased parts, raw materials, subcontracting,

PROGRAM MANAGEMENT

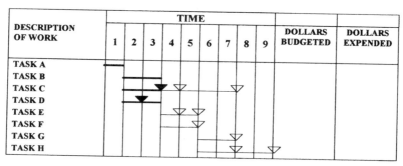

FIGURE 18-13a. Gantt chart. Plan ____ ; Accomplishment ▼ Milestone ▽

Figure 18.13a

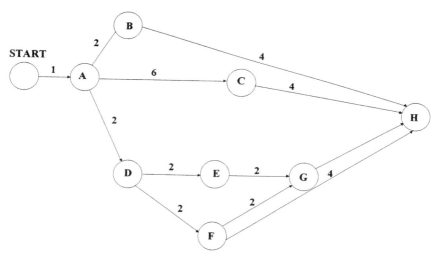

Figure 18.13b Network diagram with dependencies among tasks A to H noted in addition to time to accomplish a task.

materials burden, general and administrative costs, and interest; the price is figured by adding profit to the aggregate cost. The process is top down, bottom up and, most likely, both ways. The top-down process begins from an established competitive price to win, with profit and overhead subtracted to yield target cost; then target cost is allocated down the WBS elements by an experienced team of senior program and functional people supported by costing models. The bottom-up process begins with the responsible individual for each task estimating manpower needs and material requirements; cost estimates are effected by applying hourly material and burden rates. In either process, final estimates are obtained through mutual evaluation between the task responsible individual, his/her management, and the program manager. In competitive markets, cost management must begin with target costs set on designers, developers, and suppliers to guide their tradeoffs; otherwise the appealing cost to the buyers is not factored in upfront, and we end up designing the product, calculating its cost, and then figuring whether it will sell. However the final estimates are arrived at, cost accumulation by the WBS element and organizational responsibility are summarized and reported. Cost by WBS elements satisfies the contract baseline and cost by organization the performance baseline.

With the budget plan in place, the planning cycle is ready for activation. By now we have identified what we plan to do, who will do it, how we plan to do it, how long it will take, and how much it will cost. Integration of those plans provides the basis for the control system that measures actual progress against the plans. A financial system provides:

- Cost, performance, and schedule reports where, per WBS element, the budgeted dollar value of work that has been accomplished is compared to:
  - A budget dollar value of work that should have been accomplished at this juncture of time.
  - An actual dollar value expanded on the accomplished work.
- Variances between:
  - What has been accomplished and what should have been accomplished by now (schedule variance).
  - The budgeted dollar value of what has been accomplished and the actual expenditures (cost variance).
  - The budgeted dollar value at completion time and what is projected now to be the indicated final cost.

The financial system does not manage for you, but it helps you manage. It organizes information, highlights departures from plans, and helps to determine alternative approaches.

Successful control requires enough significant milestones that will help to assess progress. Beside the associated time and cost budgets, these milestones must be defined so that their completion can be recognized by both technical and management personnel. Otherwise, technical progress is not easy to measure; this is usually encountered in programs exploring new horizons where in depth definition, at the onset, of all the milestones cannot be accomplished. The presence of subjective milestones can abound on large innovative programs, and the assessment of progress against projected plans depends in the final analysis on the in-depth experience of the program manager in the subject at hand. Complications are due to uncertainties not only in progress measurement, but also in the reference plan; filtering of such uncertainties is not amenable to algorithmic processing, and estimation/decision-making are prone to risk where a simple one-to-one relationship is nonexistent between the statement of work, the work breakdown structure, the schedule, the budget, and the control system. Thus, human

PROGRAM MANAGEMENT

judgement based on subject knowledge must overlay the algorithmic control system to ferret out realism from misconceptions and to guide the controlled activities toward desired goals.

IV. DISCUSSIONS

As in any kind of endeavor, there are the same basic processes of management, i.e., planning, organizing, activating, controlling, and communicating. With program management, the endeavors differ enough from one to another for things to start anew with every new program. In fact, the commitment by top management to support the program manager needs to be renewed with every program. Despite the stated commitment, it is not unusual that the proper resources he/she requires are not assigned as needed. The first step or obstacle to overcome is in the assembly of competent personnel; the best way to shape and transform the people into a team is to set them up in a single location and ensure unencumbered quick and clear communication; proximity breaks communication barriers and departmental boundaries and develops a sense of identity with the program; breakdown of communication barriers is increasingly realized through computer networking which is creating regional and global proximity. The second step is to give specific direction through work statements, budgets, and schedules. The third step is to measure performance and require compliance. The fourth step is to use corporate memory in design reviews. In every step, the program manager must ensure that all activities add value enabled through simplified and integrated processes where outputs are directly linkable to contract deliverables.

The needs of progressive program phases require tailoring of the team organization to the activities at hand. Ordinarily, activities start primarily with paper work, progress into design, go into software and hardware development, and eventually end in the field operation. Throughout, the program manager responsibilities include:

- Planning the program.
- Assigning and authorizing tasks.
- Activating the program.
- Maintaining continuous technical, schedule, cost, and performance visibility.
- Serving as the prime interface with the customer.
- Ensuring customer acceptance.
- Making a profit.

These responsibilities are executed through others as the program manager interfaces with the varied organizational units (Figure 18.14) and as those units are variably applied during the evolution of the job from contract award to product/service delivery, (Figure 18.15).

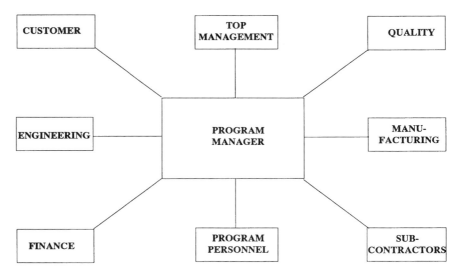

Figure 18.14 Program manager interface diagram.

In discharging his/her responsibilities, the program manager needs in-depth knowledge of the undertaking; he/she should know upfront:

- What the contract says, what is to be done, what resources he/she has; what are the potential pitfalls with the customer, top management, the staff, and the matrix support.
- What leeway is there in cost–schedule–performance functions when dealing with future pitfalls; it is not possible to fix all three functions and one finds that greater knowledge of one will create a greater uncertainty in the other.
- What tools are there to use so as to provide warning, not just history; which ones weigh most heavily on measurable completions; which ones result in a minimum difference with contractual requirements; which ones make only essential demands and are easily comprehended by the whole team.
- What the user's and customer's needs are and their fulfillment through available technologies, alternate system concepts, resulting cost/schedule/performance tradeoffs, and preferred solution for the system, subsystem, equipment, and software (Figure 18.16). The front-end conceptual effort has a small cost since the rate of expenditure per week grows with the later phases of the program; getting bugs out at the earliest possible stage will avoid very costly delays in the production phase where the weekly rate of expenditure is at its maximum.

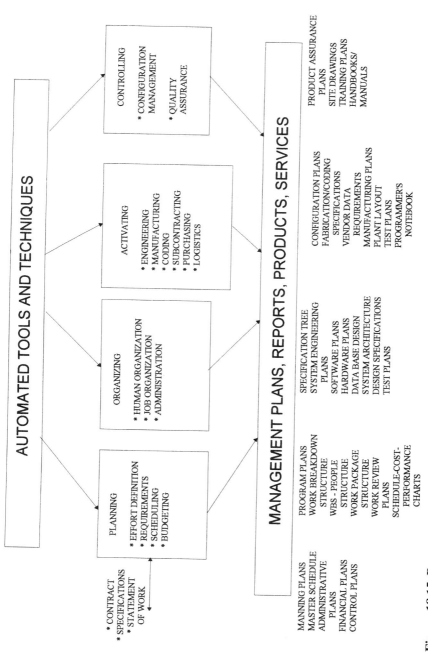

Figure 18.15 Program management and work evolution.

404   SYSTEMS MANAGEMENT: People, Computers, Machines, Materials

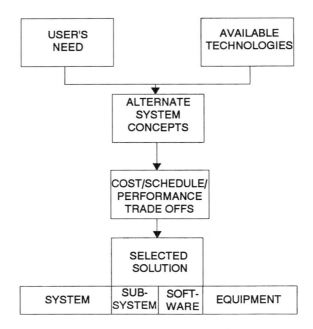

Figure 18.16 Conceptual paths to selected solution.

19 MANAGEMENT OF PRODUCTION AND SERVICE OPERATIONS

I. INTRODUCTION

Management of operations, as explained in Part I, is applied here for the creation of goods and/or provision of services (Figure 19.1). The processing issues and approaches in Parts II–III arise again with the symbolic entities replaced now by their corresponding physical entities. The systems approach in Part II is applied to synthesize the operations. Automation, as explained in Parts II, III, and IV, provides for effective production and service operations of infinitely varied resources. Operations include all direct activities related to the goods and services produced by the system toward that purpose. The main function of management is to participate in and/or direct the system operation through decision making in the areas of:

- Prediction or planning of how much output is needed, of what type, when, where, how... .
- Designing what was predicted or planned, its needed capacity, its geographic location, the layout of the organization be it human, equipment, and/or facilities, and the work flow within/through them.
- Development, installation, and/or operation of the design, the timing, and priority of its activities, the timing and quantity of its inventory in materials, parts, and finished products, and its quality assurance.

The fundamental decisions in an operation pertain to the types of products/goods/services it will deliver. Based on specific choices of products, capacity, location, and facilities, decisions follow. Their principal end goal is productivity. Productivity is a function of technology, decomposition–integration work methods, and their setting in a real-time management system with overall reduced

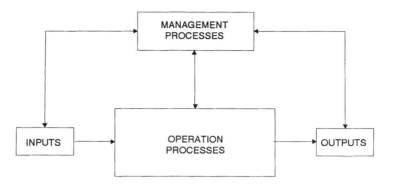

Figure 19.1 Management of operations.

time from input to output. Reduced time translates into innovative processes with low people/equipment idle time, low storage, and low inventory from product order to its delivery. Technology is the smart way to productivity; it improves not only the time but also the quality. Familiar examples illustrate simply the point as we went from the mechanical typewriter to the electric typewriter and then to the word processor, from the carbon copy to various duplicating machines, from handwashing to dishwashers and clothes washers, from the mechanical grass mower to the self-propelled riding mower, from the paint brush to paint rollers and spray guns. There are two phases to production and service operations.

- Planning and design to include capacity planning, location planning, facilities design, product and service design.
- Activation and control to include overall product management, management by product, material management, and quality management.

II. PLANNING AND DESIGN OF PRODUCTION AND SERVICE SYSTEMS

The first two processes of management, planning and organizational design are discussed now. This involves consideration of capacity, location, facilities, and products.

A. Capacity Planning and Design

Design capacity specifies the rate of possible output. Capacity is a major determinant of initial cost. Thus, capacity is balanced against demand requirements. When matched, the operating costs are at the minimum; with over-capacity and under-capacity, decisions must be made to deal with the higher costs (Figure 19.2).

MANAGEMENT OF PRODUCTION AND SERVICE OPERATIONS

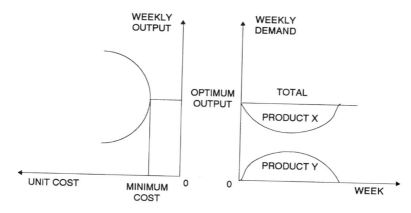

Figure 19.2 Balancing demand with optimum operation.

It is not uncommon that possible output exceeds actual output. Many factors constrain the design capacity to a lower actual level:

- Products/services with design characteristics not amenable to standardization, possessing highly different parts, or with sensitive processes yielding erratic quality.
- Operations on various different tasks requirements resulting in scheduling delays, lack of inventory, and/or irregular quality along with equipment breakdown.
- Human failures triggered by lack of motivation, absenteeism, training or experience, and/or appropriate job content.

Forecasting of demand and transforming it into capacity requirements facilitate the balancing process; this is no different than the balancing process discussed in Chapter 16. Management of probable variations is at the center of a successful operation. The basic demand patterns over time are either constant growth, decline, or cyclical. The basic descriptive indicators of those patterns are set over a specified time frame; those indicators are set in terms of a mean variation, standard deviation, and correlation of the irregularities over the mean. Strong irregularities render the balancing process of demand and capacity harder to make. Possible strategies are available to deal with fluctuations in demands and resulting over-capacity and under-capacity. Alternatives include:

- Using built up inventory, working additional shifts, or subcontracting to deal with temporary over-capacity.
- Planning for complementary demand patterns that tend to balance out unevenness in demand and avoid dips and consequently under-utilization. Throughout it all, demands must be met with an operation capacity to satisfy their requirements. Optimal operation aims to yield

minimum cost per unit for a certain rate of output. Thus, the need to keep the output rate at the minimum unit cost while meeting the demand is the challenge of any successful management operation.

B. Location Planning

There are several factors that influence location planning. Among them are availability of:

- Target markets with their proximity, the costs of distributions, trade barriers, and taxes either local, federal, or import–export.
- Needed raw materials and supplies, their costs, and availability.
- Manpower and its attitude toward work, and its productivity vs. wage scales.
- Support services such as schools, churches, fire, police, housing, and entertainment.
- Transportation accesses to roads, rails, shipping, and air and the costs of movement of either raw materials or finished products.
- Communication networks with connectivity to centers of learning, industry, commerce, and finance.
- Zoned land with low cost and room for expansion.

In assessing the location alternatives, one arrives at both quantitative and qualitative determinants. A top-down process would narrow the alternatives first by country, region, town, then site.

C. Facilities Design

In laying out a facility, three basic elements are addressed: required operations, sequence of operations, and routing. For doing this, there are three basic facilities' layouts:

- Product layouts are favored in highly standardized operations with very similar items or tasks; high standardization leads to continuous processing where work flow proceeds through the successive work stations starting from raw materials or customers to completed product. The advantages are a high rate of output, low unit cost, high utilization of resources, pre-designed routing and scheduling. The disadvantages are susceptibility to failure, inflexibility to change, and need for preventive maintenance.
- Process layouts are favored in highly variable operations where very similar kinds of activities are grouped together in departments. The advantages of this grouping are redundancy in case of failure and flexibility in responding to different processing requirements. The

disadvantages are complex routing and scheduling, inventory control and purchasing, and difficulty in continual utilization of resources, e.g., equipment and high unit cost due to low volume.
- Fixed site layouts are favored when materials, e.g., oil drills or final product, e.g., dam, are too bulky to move from product development site to operation site; they have workers, equipment, and materials brought to the operation site for assembly. These layouts require close scheduling of resources' arrival, monitoring of storage to avoid bottlenecks, and closer supervision of the variety of ongoing activities.

For such layouts, the type of processing found is usually periodic (continuous) in product layouts, and aperiodic (intermittent) in process and fixed site layouts. For a given situation, a combination of the basic layouts is found where the efficiency of product layout and flexibility of the other layouts are captured. This leads to a search for grouping of products or sub-products with similar design characteristics or similar development characteristics into a layout (Figure 19.3). Productivity is enhanced through automation which replaces human functions with equipment functions. Computers control the operation (start and stop), operations sequence, and routing of inputs and outputs among machines. One machine type, the robot, imitates human activity; it consists of three parts, a mechanical arm, a power supply, and a controller. The arm can be equipped with an appropriate device for the task at hand and can be powered in three ways: pneumatically (air driven), hydraulically (fluids under pressure), or electrically which enables it to move to predetermined points or continuously. It is controlled through a set of processing instructions that tell it the details of the operations to be performed. Robots are simple in design when they follow a fixed set of instructions and become more complex as programmable instructions, recognition of objects, and decision making are included in a computer controlled set-up.

In any of the layouts and processes, maximum utilization of the resources is sought to yield a smooth flow of activities without buildup of work at certain points and idleness at others. This is accomplished through task groupings that require equal time from the work stations each grouping is assigned to. This balancing is desirable for the efficient utilization of resources. In balancing a line, work centers are assigned to locations and tasks are assigned to work centers. In setting task flows, each task time is set along with the immediate predecessor tasks. In designing a layout:

- A list of locations is made.
- A list of work centers is made.
- Future work flows between the work centers are projected.
- Assignment of work centers to locations is carried out.
- Transportation costs among the work centers are assessed.
- On-line delays are established.

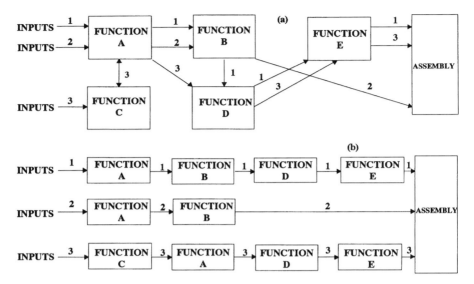

Figure 19.3 Product development through: a) functional layout; b) flexible layout.

Thus, the flow process is depicted to assess the overall sequence of operation, transportation, delay, inspection, and storage steps. This flow process is analogous to that in Part III where material flows replace electrical/symbolic flows.

D. Product Design

There are several factors that influence the design of a product. Among them are the two principal determinants of price:

- Standardization; it starts at the part level and progresses to the module level (grouping of parts) or to the higher levels (grouping of modules). Standardization permits:
  - Module/object re-use or replacement.
  - An increase in productivity through automation of manufacture, assembly, and integration.
  - More ease in diagnosis and remedy.
  - Simplification in training with reduced time and cost.
  - More routine purchasing and control of inventory with fewer types.
  - More improved quality with perfection of fewer types. On the other hand, the cost of change is higher with standardization unless changes are downward compatible.

MANAGEMENT OF PRODUCTION AND SERVICE OPERATIONS

- Reliability; it measures the possible delivery by a product of its intended function. Reliability is specified at a point in time or over a given period of time. Reliability is not deterministic in nature, but rather probabilistic. At a point in time, knowing the probability that the parts will operate and the probability that the product will operate can be computed. When there is no redundancy or backup to any part, the product reliability is the multiplication of all the parts' reliability. With backup to a given part, the effective reliability is the sum of that part's reliability plus the probability of failure for that part multiplied by the reliability of its backup. Over a given period of time, reliability is concerned with product life or length of service. Historical data are used to determine the mean time between failures and the distribution of those failures around the mean; using this data, the reliability of the product at a specified time in the future can be computed.

III. ACTIVATION AND CONTROL OF PRODUCTION AND SERVICE SYSTEMS

The next two processes of management, activation and control, are examined now. This involves special consideration of overall operation, product operation, scheduling, material handling, and quality.

A. Overall Operation Management

Overall operation is based on an integrated demand for products within a few categories without consideration, at this time, of variations in features within a category. A category pulls together goods and services that require a similar capacity (type of employment, equipment, processes, inventory). This initial simplification has for an objective the balancing of the overall demand forecast with supply to reduce associated costs when demand for a specific product is not known yet. The process involves:

- Forecasting overall demand levels for given periods of time.
- Calculating available capacities for single shift, overtime, part-time, subcontracting, inventory (excess), and back order (shortage).
- Calculating unit costs for each option.
- Selecting the best option to satisfy the demand level within the company goals and policies, e.g., profit, stable work force, investment... .

Demand patterns can be influenced by:

- Pricing where reduced rates are set during low demand periods and regular rates during high demand periods for leveling.

- Back orders where delivery is delayed during peak periods and satisfied during slack periods.
- Advertising where demand is created through sampling, communication of product/service features, benefits... .
- Government policy which favors slow down or forbiddance of a product/service through taxation or regulation.
- Complementary demand where excess capacity from one product's slow period is absorbed by delivery of other products/services, thus achieving a level occupation of labor, equipment, and facilities.

Each option entails a cost to the operation of the enterprise and influences its decision-making process. Just as elasticity exists in demand, so does it exist in available capacity. Capacity patterns are influenced by:

- Use of overtime/part-time work to deal with temporary demand excess.
- Use of subcontracting to satisfy a demand for a specific expertise or to alleviate unacceptable slack in back orders.
- Creation of inventories where products or sub-products are completed during slack time for use during peak demand time; this approach is principally used in manufacturing since services are less amenable to storage. In balancing overall demand and capacity patterns, cost is an underlying driver for the choice of strategy.

B. Product Operation Management

In any operation, storage of data, goods, or services implies idle time; idle time is costly since it is non-productive time. A fundamental concern in system operations involves over-stocking and under-stocking which entail decisions in the timing (when) of orders and size (how much) of orders to keep inventory down while keeping operation up. The basic drivers to inventory are:

- Enabling operation; processing of a completed product or service is not instantaneous which unfailingly yields work-in-process; routing and delivery are not instantaneous which unfailingly yields temporary inventory; real-time processes reduce in-process delays and storage delays.
- De-coupling operational steps; otherwise, a misstep in the processes can bring the system down; buffering is used to smooth those hazards out.
- Redundancy; otherwise, any delays in delivery or increases in demand cannot be dealt with reliably; a safety margin protects against unforeseen events.

MANAGEMENT OF PRODUCTION AND SERVICE OPERATIONS

- Response time; otherwise customer demand cannot be met promptly from products in stock, and customer service is degraded.
- Cost; larger lots than needed have a tributary value of usually reducing the cost of service or manufacture.

Types of inventories include raw material, parts, tools, supplies, work in process, and finished goods. Effective tracking of inventories requires:

- A forecast of demands and descriptions of their variability.
- A statement of lead times and descriptions of their variability.
- A statement of inventory overstock (carrying) costs, under-stock (shortage) costs, and ordering costs. Thus, orders are timed to minimize inventory costs (Figure 19.4).

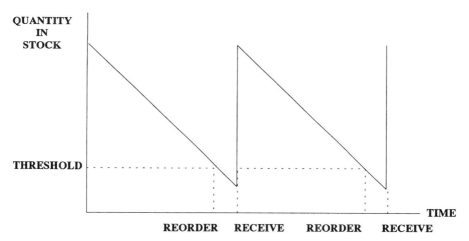

Figure 19.4 Re-ordering is set to allow reception of stock prior to shortage.

C. Scheduling Management

Scheduling creates the map for the activation process; it involves the allocation of work loads to specific work centers and setting the sequence in which the work will be accomplished. Scheduling fixes the timing of specific operations along with the utilization of resources (human, equipment, materials, facilities). It begins with the master schedule which is desegregated into decision points and then specific work assignments. The transformation process yields specific plans for the individual goods and services. In the process, optimization is performed among the conflicting requirements in personnel assignment, use of equipment and facilities, minimization of service response time, inventories, and process times.

Scheduling assigns work loads to specific work centers and determines the sequence of operations. Approaches to scheduling depend on the volume of system output.

- In high volume systems, the loading and sequencing of operations are pre-set during system design and all items proceed practically through the same operations sequencing. Line balancing is the primary goal in ensuring equal work time along the flow line. Disruptions to smooth line operations can be caused by equipment failures, human failures, mishandling, and transitions to different sizes and models. Very seldom a line is devoted to a single product or service, and each change requires slight changes in inputs/processing/outputs, and those variations entail more scheduling efforts.
- In intermediate volume systems, periodic rescheduling of the line takes place where, contrary to job shops, the run size is large, but not large enough to qualify as a high volume system. In such a system, scheduling is mainly concerned with run size and jobs sequencing. The run size must minimize set-up time and inventory costs; for multiple products and dependent set-ups, the scheduling can become complicated. One simple approach computes the run-out time, which is inventory plus possible production divided by demand; the lower the run-out time for an item, the more urgent it is to produce that item.
- In job shops, work-to-order is transacted where orders' arrival cannot be predicted and needed materials, processing time, sequencing or set-up cannot be predicted a priori. In assigning jobs to work or processing centers, minimization of set-up costs, processing time, idle time, and job completion time is sought. Gantt charts depict use of resources over time where time is represented horizontally and resources are represented vertically; the horizontal axis gives the sequence of operations at a resource and vertical axis gives the order of jobs in progress. A typical assignment method of jobs to work centers attempts to maximize profit through performance and efficiency while minimizing cost. Following assignment, sequencing of operations addresses first job priority; priority may be based on:
  - First come, first served.
  - Due date.
  - Preferred customer.
  - Shortest processing time.
  - Time until due date minus remaining time to process.

Following priority, the effectiveness of any adopted sequence is based on a certain average which may be the number of jobs at a work center or completion time and job lateness.

- In service systems, unlike manufacturing systems, customer requests are random and an inventory of services is not possible. Scheduling focuses on providing customer services through efficient employment of a fixed capacity. To control customer waiting time, appointment systems are used to reserve client time. To insure customer resource availability, a reservation system is used. In both, no shows, late arrivals, over-booking, and service failure disrupt the balance between demand and service capabilities. This is aggravated when the scheduling increases in complexity as it involves multiple resources that need to be coordinated.

When scheduling a system, waiting time is tied closely to the number of servers or channels available. Processes are composed of four basic variations (Figure 19.5): a) *single channel* and *single stage*; b) *single channel* and *multiple stages*; c) *multiple channels* and *single stage*; and d) *multiple channels* and *multiple stages*.

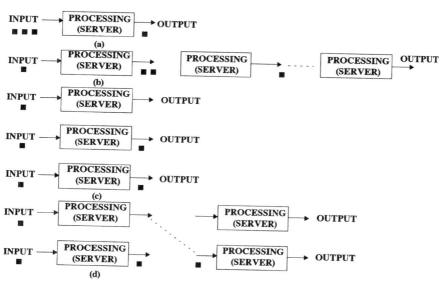

Figure 19.5 Basic variations on serial/parallel processing of products/services, ■.

D. Material Management

Material management describes what is required, when it is required, and how much it costs. Material is a major contributor to cost and its management is essential to cost reduction. Within it, all participants from program managers to engineering, to purchasing through final assembly and test can witness their respective impacts on work flow and productivity. Its inputs are derived from:

- A product breakdown structure which lists the assemblies, sub-assemblies, parts, and raw materials that are required to develop the products; the listing is usually hierarchical.
- An inventory file which profiles in time the varied available or on-order stocks in quantity, type, lead time to re-stock, suppliers, etc.
- A material schedule which is derived from the master schedule for the end products, and the projected time periods when assemblies, sub-assemblies, parts, and raw materials need to be on site.

The material planning outputs define:

- The schedule for planned orders which includes amount and timing.
- The released orders from the plan which define the placed orders and their expected fulfillment.
- The changes to plan which include updates to time and quantity.

The material management system provides:

- Complete management control over acquisition, distribution, and accountability.
- Comprehensive material requirements planning.
- Material tracking.

Its functions begin when a need for material is identified through part number, shop order number, assembly number, quantity and need dates, and end when delivery is made to the ordering activity and costs have been allocated. The functions include (Figure 19.6):

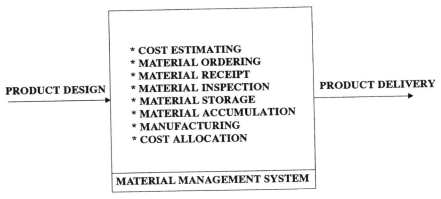

Figure 19.6 Functioning of a material management system.

MANAGEMENT OF PRODUCTION AND SERVICE OPERATIONS 417

- Procurement activities which are prompted by material requirements obtained from bills of material, advance orders, changes in engineering design and in manufacturing schedules, and repayments of material loans. The requirement quantities and dates of assembly determine whether they can be filled from material in stock or are in the procurement cycle, adjust the shop order records to allocate the available material to the present requirement, and issue the financial transaction for input into the accounting system; such material results from over-buys, changes to previous requirements, quantity buys to minimize costs. When requirements cannot be satisfied, the system accumulates like requirements with advance and replacement orders, calculates lead time, optimum number of buys and economic order quantities, and passes the information to purchasing for execution. During purchasing, it monitors the status based on need dates, flags delays in purchase orders, and updates part requirements.
- Inventory activities which track material motion through receiving and inspection to the storeroom; daily count of receipts, along with variations from requirements (quantity, price) are reported. For common procurement, the system allocates and reports the status of orders by participant. During manufacturing, the system disburses the material on the scheduled date using the build schedules. For changes in requirements, the system reallocates the material, transfers the material between shop orders, monitors shortages and excess quantities and determines residual material at a contract's end to meet new requirements. Thus, the system keeps real-time or daily records on inventory balances before, during, and after manufacturing and flags discrepancies.
- Cost accounting activities which allocate the distribution of material costs among shop orders and accounts; it records the dollar values for requisitions, adjusts the records for new rates, identifies the shop orders participating in receipts, informs accounting of due receipts and keeps track of unpaid receipts. The system keeps track of the value of past balances in stock and allocates credit to participating shop orders. At the close of a contract, the system computes the stocks of material used and the value of surplus material as well as scrapped material.

Material management interacts with other management systems (Figure 19.7); it drives or is driven by:

- A configuration system which maintains the assembly drawing for each product and part information provided by the design and directly inputs to the material management system specific data on the parts.

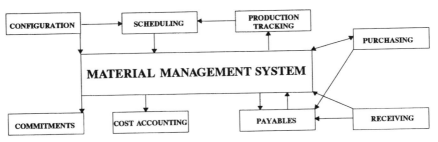

Figure 19.7 Material management system and associated interfaces.

- A scheduling system which provides assembly dates and quantities to the material management system based on the assembly drawing.
- A production tracking system which monitors the progress of assembly accumulations, notes the component part with the worst delivery time for each assembly from the material management system and feeds the status back to the scheduling system.
- A purchasing system which executes the requisitions received from the material management system and feeds back purchase orders placed, change orders issued, requisition status, and allocated shop order costs.
- A commitment system which tracks the dollars committed by the material management system per contract number or shop order number.
- A receiving system which notifies both the material management system and the payable system of received material against a purchase order; it also tracks the material receipts by control number through their movements between stations and informs the material management system.
- A payable system which provides the material management system with purchase order payments and receives from it shop order cost allocations.
- A cost accounting system which tracks the total dollar value of all material transfers between shop orders executed by the material management system.

Material management leads to reduced cost of inventories, accurate requirements for materials, reduced cycle time, accurate accounting, and increased productivity. When inventory data are captured in real time, there is then additional valuable information on the time it takes a specific part to flow through the system, where the delays and bottlenecks are, the actual labor and equipment utilization in a given operation, and end-item scheduling. Automation of large management systems is a necessity, otherwise the sheer magnitude of the job can be overwhelming; for instance, in building or in tearing down and rebuilding a large system during maintenance, e.g., an airplane, one has to deal with more than half a million parts, each at a varied stage of its process cycle; a single component within that system, e.g., a jet engine, may have in excess of three thousand parts and take more than two months to build or tear down and rebuild during maintenance.

In the purchase–supply chain from raw materials to end-users, there is a high cost attached to low response time. In single/multiple purchases, low response is reflected in unacceptably long lead times. Unfailingly, long lead times create higher cost; the payments purchasers make to suppliers are eventually linked to how rapidly the suppliers process their products. It is not unlikely that an order may take minutes to produce while the stated lead time may be weeks, giving a puzzling high ratio. Additionally, purchases are based on forecasts and forecasts may be full of uncertainty thus compounding the need for projected longer lead times. Unfailingly, the cost of inventory can have a wide average range of 0.5% of the price per week to maybe 2% per week due to space occupancy, handling, scrap and obsolescence, damage, inspection, and return of defective material.

E. Quality Management

Quality falls into two categories:

- Quality of design relates to the addition of certain features to a product, be it goods or services, that go beyond the basic product function. For instance, the general purpose of a home is to supply shelter, but there are a variety of choices made as to features, such as elegance, fuel economy, roominess, etc.
- Quality of conformance ensures that the stated quality of design is actually achieved. A total approach to quality starts from customer ideas on features to design, procurement, production, delivery, and maintenance. The selective discriminants by the customer on conforming features are based on a) *appearance*, b) *operation*, c) *reliability*, and d) *service* after delivery. Conformance is practically assured through acceptance sampling of inputs and outputs as well as process control (Figure 19.8). Acceptance sampling tests for conformance before/after the product is processed/completed. Process control tests the process itself to verify not only present quality, but to predict the future quality of the processed product. The cost of sampling, coverage, and frequency should not surpass the cost of defective or non-conforming features; the cost of inspection can become prohibitive if one attempts to catch every defect in every lot. There are varied acceptance/rejection plans; they involve a single sampling of a lot, or double or multiple samplings. Acceptance or rejection usually depends on control limits with upper and lower values on the mean and/or standard deviation.

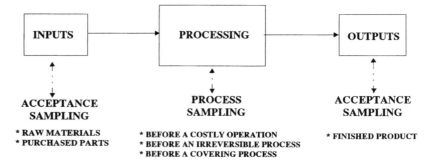

Figure 19.8 Conformance to quality through sampling.

IV. DISCUSSIONS

In its most basic form, operations management seeks to ferret out activities that lengthen time, e.g., idle time, processing time, inventory time; time is money just as sunken money, before its time, costs money.

Operations management falls into two categories:

- That pertaining to the design or planning of a system which produces a certain commodity or service.
- That pertaining to the activation and control of a system to produce a certain commodity or service.

Decisions relating to the design of the system fall into:

- Long-range forecasting and planning for system capacities and locations.
- Selection of processes and equipment to fulfill a given need for a stated investment cost and projected processing cost.
- Location of the system dependent on nearness to market, material cost, tariffs, and taxes.
- Physical facility layout to support basic modes of operation while minimizing overall material handling cost.
- Job design involving the organization of work and integration of human–computer–machine to produce optimally designed jobs.

Decisions relating to the activation and control of the system fall into:

- Forecasting of day-to-day and short-term operational needs.
- Aggregate planning to set basic production rates and employment levels.

MANAGEMENT OF PRODUCTION AND SERVICE OPERATIONS

- Inventory control to maintain the levels needed to provide the needed service at a given cost.
- Scheduling and control of available resources to make best use of them, ensured through information feedback and appropriate control.
- Maintenance of equipment to avoid equipment breakdowns during operation and downtime.
- Quality control to balance the cost of passing defective parts and services against the cost control.
- Cost containment to balance labor, material, and overhead costs.

20 FINANCIAL MANAGEMENT

I. INTRODUCTION

The resource under consideration now, money and its management, is the key process in the operation of any firm. Objective appraisal of a firm's management is unfailingly tied to the firm's financial position, its profitability. Profitability is unfailingly enhanced through introduction of appropriate technology, automation processes, and simplified decomposition–integration processes. Organizations, in addition to developing high quality products and maintaining high standards, must eventually make a profit or at least break even to remain in existence; this applies to business, governmental, or religious organizations.

Financial management involves the planning, organizing, activating, controlling, and communication processes explained in Part I but applied now to the firm's monetary operations. Basically, it secures the needed funds at minimum cost, invests surplus funds for best returns, and provides the accounting services needed to plan and control the operation. Its activities include:

- Financial planning which forecasts capital expenditure funds, sets working capital requirements, projects money-market conditions, predicts results of alternative actions on financial conditions, and conceives the best methods for funds acquisition and investment.
- Tax planning which assesses the tax impact on projected activities and seeks to minimize its load on revenues.
- Funds management which establishes credit policies, collects due funds, provides for financial protection against hazards, and negotiates financing agreements with creditors and investors.
- Cost accounting which tracks costs incurred in relation to work performed, maintains accounts receivables and issues bills and accounts payable, issues payments, and audits financial records against standards and procedures.

- Budgeting which projects profit results against performance, develops planned costs for operations, and products/services development, measures actual results against plans and assesses causes of variances.

Financial services, where fast response and split second reliability can be worth millions of dollars, are adopting the latest in real-time system processes and technologies. For effective financial operation, automated implementation of major parts of the activities is required as developed in Parts II, III, and IV. Automation provides a comprehensive business application where it captures the data once at the source and shares it across applications and thus moves the financial numbers out of the finance department and into the hands of the managers who have a direct influence on the outcome and a stake in the results. This calls for an understanding of the relationship between functions and processes. This includes all details of business transactions where one must specify exactly what data are to be transferred, where, when, and to whom across the organization. This access to data may be enabled through use of three-tier client-server architecture that uncouples the clients or user interfaces from applications servers containing the logic and database servers containing the data. The clients and the two servers sit each on its network and communicate with each other through some remote function calls or messages using some protocol.

II. FINANCIAL SYSTEM

A system's concept of a firm, as explained in Part II, is that of an input–output process in which certain inputs (labor, capital, land, material) are transformed into outputs to be sold on the market. The inputs and outputs are valued at market prices, and the difference between revenues and costs determines the profits; thus profit may be increased by growing the revenues and/or cutting costs of labor, material, capital, etc. The transformation process is subjected to limitations by the firm's internal nature and external environment. A firm's internal organization chart presents the set of authority relationships among its managers. One way of grouping activities is by functions. The financial function deals with the acquisition of funds; finance decisions involve liquidity and solvency. The relative relations of finance function to other functions are given by the firm's system of operation. The system is the collection of activities, operations, or components that interact with one another toward achieving some objectives. The basic elements of a system can be framed as:

- The base elements which represent the place where physical transformation takes place in an organization such as the production operation.

FINANCIAL MANAGEMENT

- The control elements which represent the mechanism that yields the rules, performance standards, and performance measures under which the transformation in the base system takes place; the mechanism consists of variables under the control of the internal system and of external influences over which the internal system has no control (Figure 20.1).
- The financial elements which represent the monetary value of the organization involving financial decision processes such as capital budgeting, cost of capital, and working capital management.

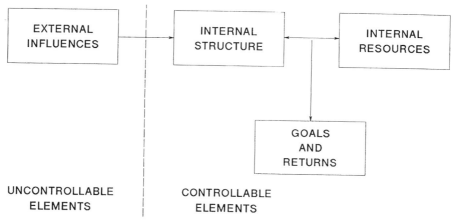

Figure 20.1 Optimizing controllable elements to match returns to goals through finding the right internal structure and the assignment of available resources in the face of uncontrollable external influences.

The flow within the system is of three types (Figure 20.2):

- Physical transforms to include operations undergone by the goods and services.
- Information transforms to include flows in quality, volume, and response time about the goods and services.
- Financial transforms to include sales of goods and services, profit margin on sales, and return on investment.

Modern financial theory draws upon four interactive areas where:

- Management theory defines the organizational structure and the place and purpose of each function within the structure.
- Economic theory defines the conditions under which shareholders' wealth will be maximized; positive determinants of wealth are

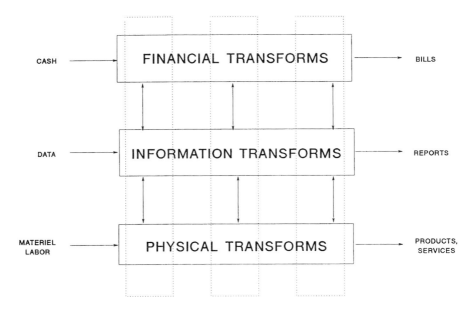

Figure 20.2 Internal transforms within a business system.

profitability and income flows, while a negative determinant is risk; risk is the hazard of not fulfilling the goals and possible loss of resources.
- System analysis defines the control process whereby goals are set, actions are taken, feedback loops are used to compare actual performances with set standards, and adjustments are made to the system where deviations exceed prescribed limits.
- Behavioral science stresses that major decision making is done by humans not automatons and that human behavior must be properly motivated.

These four areas provide descriptive models of the firm's environment based on which prescriptive models are applied to maximize shareholder wealth. Primary financial decisions begin by choosing the line of business in which to operate, the product-market mix. When this choice is taken, profitability and risk are affected by decisions on the size of the business, the extent of debt incurred, the liquidity, and equipment and assets invested. When it comes to profitability, it must be recognized that the government, along with stockholders, is the largest receiver of business profits. To this effect, the government applies monetary and fiscal policies to influence business activities:

- Monetary policy deals with actions to influence the availability and cost of credit.

FINANCIAL MANAGEMENT

- Fiscal policy deals with actions to influence the level of economic activity through the amount and composition of government receipts through taxes, tariffs, and expenditures. The tax environment is set through personal income taxing and corporate taxing.

III. FINANCIAL PLANNING AND CONTROL

The financial manager plans the business's future financial requirements in accordance with a financial analysis of existing strengths and weaknesses followed by forecasting, generation of funds and their budgeting.

A. Financial Analysis

Financial analysis addresses different types of relationships in a firm and compares them to a representative industry average and/or trend for assessment of position:

- Liquidity ratios measure short-term solvency, the ability to pay maturing short-term obligations; one such ratio is the current ratio of cash, accounts receivables, inventories, and marketable securities to payable accounts, maturing notes, and wages; industry average is at 2.5.
- Leverage ratios measure equity or owner supplied funds, the extent to which the firm has been financed by debt; ratios include total debts to total assets, with industry average of 1/3; profit before taxes plus interest charges to interest charges, with industry average of 8; income available for meeting fixed charges to fixed charges, with industry average of 4.
- Activity ratios measure the resources utilization effectiveness; ratios include sales to inventory, with industry average of 10; accounts receivables to sales per day, with industry average of 20; sales to fixed assets, with industry average of 5.
- Profitability ratios measure the resultant effectiveness of objectives, policies, and decisions; ratios include net profit after taxes to sales, with industry average of .04; net profit after taxes to total assets, with industry average of 0.1; and net profit after taxes to net worth, with industry average of 0.15.

Some of the preceding ratios are related to one another and their respective industry average is very approximate. Additional analysis tools include:

- Break even analysis relates fixed costs, variable costs, and total revenues to indicate the level of sales needed for the firm to operate at a profit.

- Funds statement indicates sources of cash and how it was used.

When making short-range forecasts, the principal tool is the cash budget which sets when funds will be needed and the amount. For long-range forecasts, the regression method is often used. Four common approaches are used in forecasting:

- Extrapolations which include moving averages and time series analysis of fluctuational patterns such as trends, seasonal variations, cyclical variations, and irregular forces.
- Barometric techniques which include leading indicators or predictors of future activities, lagging indicators, and coincident or in-phase indicators.
- Opinion polling which includes economic forecasting and sales forecasting.

Incorporated in any forecast are the concepts of risk and uncertainty.

B. Capital Budgeting

Capital budgeting deals with the process of planning expenditures whose returns extend for a long term beyond one year, e.g., land, buildings, equipment, and research. Capital budgeting requires decisions under risk/uncertainty and deals, on the one hand, with supply of capital or its cost to the firm and, on the other hand, the investment opportunities open to the firm. Unfailingly, management policies strive to reduce the cost of capital. Typical classes of decision problems in capital budgeting include expansion decisions, replacement decisions, and buy or lease decisions. Several procedures are used for assessing the relative benefits of opportunities:

- Payback sets the number of years required to return the original investment.
- Net present value defines the present value of future returns discounted at the cost of capital minus the cost of investment.
- Internal rate of return defines the interest rate that equates the present value of future returns to the investment outlay.
- Profitability ratio defines the present value of future benefits divided by the present value of the investment outlay.

When the procedures give conflicting results, the net present value is relied on if capital rationing is not applied; then profitability ratio is used. Throughout it all, risk assessment must overlay any quantification of benefits. Risk is due to the variability of returns on a project. Variability may be objectively or subjectively described in terms of probability distribution, mean, bias, variance, or correlation. To mitigate risk, projects diversification is applied where correlation in returns is purposefully low. Once the approximate financial requirements are set, the next task involves the generation of funds.

C. Financial Leverage

Financial leverage uses debt to increase the rates of return on net worth over the returns available on assets. This is how business financial requirements are met through raising capital through a hybrid of long- and short-term debt, preferred stock, common stock, retained earnings, liquidation, etc. Leveraging introduces risk when the returns on assets are less than the cost of debt, thus reducing the returns on equity. In risky leveraging, the cost of capital or discount rate is integrally higher in the absence of other expected rates of return, e.g., dividends, growth in stock price. The means of raising capital is thus decided upon through a comparative cost of capital. Thus a business debt policy, cost of capital and dividend policy are linked; dividends are remaining dollars from earnings after investment requirements are satisfied.

D. Working Capital

Planning of current assets and current liabilities leads to working capital management. Working capital relates to investment in short-term assets such as cash, short-term securities, accounts receivables, and inventories. Short-term assets have a direct relationship to sales; though necessary, they form non-earning assets. A balanced control needs to take place between collection policies and credit policies, terms of sales; otherwise, the carrying cost of tied up funds can become forbidding. Like account receivables, inventories, e.g., raw materials, work in process and finished goods, form non-earning assets; usually, inventory-to-sales ratios fall in the 10 to 20 percent range, and inventory-to-assets ratios fall in the 15 to 30 percent range. Determinants of inventory are level-of-sales, time length of the production process, and the durability or style factor of the end product. There is always an economic ordering quantity for a product which is recognized through two sets of costs:

- Carrying cost which rises as inventory holdings increase.
- Ordering cost and stock-out cost fall as inventory holdings rise.

E. Short-Term Financing

Short-term credit is debt scheduled for repayment within one year; three major sources of credit exist:

- Trade credit happens during business transactions where cash payment is not usually made on delivery but within thirty days from time of purchase.
- Bank credit on a line-of-credit established prior to the time needed.
- Commercial paper sold in a broad market based on the firm's reputation to meet its obligations.

In addition to the above unsecured loans, short-term secured loans are obtained using short-term credit, such as inventories and accounts receivables.

F. Intermediate Term Financing

Intermediate term credit is debt scheduled for repayment in more than one year but less than fifteen years; three major sources of credit exist:

- Term loan is debt scheduled for payment in up to fifteen years, is retired through systematic repayments of principal and interest over the life of the loan, may have a variable or fixed interest rate, may contain protective covenants of the lender with stipulations for the borrower to limit his debt ratio, is secured through a given asset.
- Conditional sales contracts where title to the asset is held by the lender while the borrower is completing payment in installments of principal plus interest.
- Lease financing is debt incurred against equipment rental with the advantage of a smaller downpayment by the borrower.

G. Long-Term Financing

Long-term financing is obtained through stocks and bonds, planned, underwritten and sold on the open market to ultimate investors. Stocks are a favored form of financing when sales and profit patterns are volatile, when stock prices and interest rates appear to be high, and when the debt ratio is considered high for the line of business. Other forms of financing include a bond, a long-term promissory note, and options such as warrants and convertibles.

H. Mergers

Mergers have played an important role in a firm's growth. Growth has been obtained through external acquisitions or internal development. Mergers provide economics of scale in:

- Research where certain types require large commitments.

- Operating economics where average cost per unit of output is reduced with larger plants and distribution systems, as well as top management consolidation.
- Speed of expansion where ready growth is secured.
- Financing where it is easier with a merger rather than internal expansion.
- Reduced risk where a going business is acquired rather than created.
- Market control where the combination of strengths and the balance of weaknesses are effected.

The price of a merger is influenced by present and expected future earnings after the merger, the market price for such a business, the book value, revenue and profit, and the net current assets or liquidity. In addition to quantitative factors, other qualitative or synergistic considerations enter into influencing the price.

IV. DISCUSSIONS

Firms like any other systems have a life cycle; they are born, they are nourished, they grow and flourish, they decline and eventually die. The underlying cause of failure is bad management which triggers capital dry-up. Avenues to capital vary with the life cycle. Raising capital is tailored to the firm's stage of life:

- At inception, capital may be raised through owners, friends, relatives, customers, suppliers, and/or venture investors.
- In the survival stage, capital may be raised through banks, government and small business programs, and/or leasing.
- During growth, capital may be raised through institutions, new partners, and/or profits.
- During expansion, capital may be raised through stocks, profits, joint ventures, and/or licensing.
- At maturity, capital may be raised through divesture of components, cash flow, and/or international financing.

To delay the onset of saturation, contraction, decline, and/or liquidation, old products/services must be replaced and new products or services must be introduced. There are two types of expansion, vertical and horizontal. Vertical expansion involves the addition to the company's work of more stages in the production process, from raw materials to sales; horizontal expansion involves entry into new product areas closely related to the ongoing product line. Here, companies like individuals, do not have unlimited options; they do a fixed number of things extremely well since expertise in one area can be transferred successfully to a limited number of areas. To increase the product's likelihood of success, a business program must be developed to include a strategic plan, a financial plan, and an operating plan that couples the technological programs of research, advanced development and prototyping, and product/service design to the commercial programs of product/service suitability to the company, likelihood of profit, and methods of introduction. Reviews, appraisals, corrections or adjustments are made to the projected success requirements of the product at the various developmental stages.

BIBLIOGRAPHY

Part I: Management and Systems

Abercrombie, M. L. Johnson, *The Anatomy of Judgement.* London: Hutchinson, 1967.

Andrews, K. R., *The Concept of Corporate Strategy,* revised edition, Homewood, IL: Richard D. Irwin, 1980.

Ansoff, H. I., *Corporate Strategy.* New York: McGraw-Hill, 1965.

Barnard, C. L., *The Function of the Executive.* Cambridge, MA: Harvard University Press, 1966.

Christensen, C. R., Berg, N. A., Salter, M. S., and Stevenson, H. H., *Policy Formulation and Administration,* 8th edition. Homewood, IL: Richard D. Irwin, 1985.

Drucker, P. F., *Management.* New York: Harper and Row, 1972.

Fayol, H., *General and Industrial Management,* translator, C. Storrs. New York: Pitman, 1972.

Gibson, J. L., Ivancevich, J. M., and Donnelly, J. H., *Organizations.* Plano, TX: Business Publications, 1988.

Haimann, T. and Scoll, W. G., *Management in the Modern Organization.* Boston: Houghton Mifflin, 1973.

Hampton, D. R., Summer, C. E., and Webber, R. H., *Organizational Behavior and the Practice of Management.* Glenview, IL: Scott, Foreman and Company, 1982.

Hatten, K. J. and Hatten, M. L., *Strategic Development, Analysis and Action.* Englewood Cliffs, NJ: Prentice Hall, 1987.

Janis, I. L. and Mann, L., *Decision Making.* New York: The Free Press, 1977.

McCaskey, M. B., *The Executive Challenge: Managing Change and Ambiguity.* Boston: Pitman, 1982.

Mintzberg, H., *The Nature of Managerial Work.* New York: Harper and Row, 1973.

Ohmae, K., *The Mind of the Strategist: The Art of Japanese Business.* New York: McGraw Hill, 1982.

Pfeffer, J. and Salancik, G., *The External Control of Organizations, A Resource Dependence Perspective.* New York: Harper and Row, 1978.

Quinn, J. B., *Strategies for Change, Logical Incrementalism.* Homewood, IL: Richard D. Irwin, 1980.

Ross, R. S., *Persuasion: Communication and Interpersonal Relations.* Englewood Cliffs, NJ: Prentice Hall, 1974.

Selznick, P., *Leadership in Administration: A Sociological Interpretation.* New York: Harper and Row, 1957.

Simon, H. A., *Administrative Behavior: A Study of Decision Making Processes in Administrative Organizations,* 3rd edition. New York: The Free Press, 1976.
Sloan, A. P., *My Years with General Motors.* New York: Doubleday, 1963.
Stoner, J. A., *Management.* Englewood Cliffs, NJ: Prentice Hall, 1974.
Sun, T., *The Art of War,* S. B. Griffith, translator. New York, Oxford University Press, 1963.
Thompson, J. D., *Organizations in Action.* New York: McGraw-Hill, 1967.
Uyterhoeven, H. E., Jr., Ackerman, R. E., and Rosenblum, J. W., *Strategy and Organization.* Homewood, IL: Richard D. Irwin, 1977.
Vancil, R. F., *Decentralization: Managerial Ambiguity by Design.* Homewood, IL: Dow Jones-Irwin, 1979.
Vickers, Sir G., *The Art of Judgment, A Study of Policy Making.* New York: Basic Books, 1965.

Part II: Systems Synthesis and Automation

Alter, S. L., *Decision Support Systems: Current Practice and Continuing Challenges.* Reading, MA: Addison Wesley, 1980.
Anderson, L. G. and Settle, R. F., *Benefit-Cost Analysis: A Practical Guide.* Lexington, MA: Lexington Books, 1977.
Anthony, R. N., *Planning and Control Systems: A Framework for Analysis.* Cambridge, MA: Harvard University Press, 1965.
Beam, W. R., *Command, Control and Communications Systems Engineering.* New York: McGraw-Hill, 1989.
Blokdijk, *Planning and Design of Information Systems.* London: Academic Press, 1987.
Blumenthal, S., *Management Information Systems: A Framework for Planning and Development.* Englewood Cliffs, NJ: Prentice Hall, 1969.
Calvez, J. P., *Embedded Real-Time Systems.* New York: John Wiley & Sons, 1993.
Checkland, P., *Systems Thinking, Systems Practice.* New York: John Wiley & Sons, 1981.
Churchman, C. W., *The Design of Inquiring Systems.* New York: Basic Books, 1971.
Churchman, C. W., *The Systems Approach.* New York; Dell, 1968.
Chen, P. P., Editor, *Entity-Relationship Approach to Systems Analysis and Design.* Amsterdam: North Holland Publishing, 1980.
Couger, J., Daniel, Colter, M. A., and Knopp, R. W., *Advanced System Development Feasibility Techniques.* New York: John Wiley & Sons, 1982.
Davis, G. B. and Olson, M. H., *Management Information Systems.* New York: McGraw-Hill, 1985.
Delaney, W. and Vaccari, E., *Dynamic Models and Discrete Event Simulation.* New York: Marcel Dekker, 1989.
DeMarco, T., *Structured Analysis and System Specifications.* New York: Yourdon, 1978.
Dorf, R. C., *Modern Control Systems.* Reading, MA: Wesley Publishing, 1974.
Emery, J. C., *Organizational Planning and Control Systems.* New York: Macmillan, 1969.
Emery, F. E., Editor, *Systems Thinking.* Baltimore: Penguin, 1969.
Farina, A. and Studer, F. A., *Radar Data Processing.* New York: John Wiley & Sons, 1985.

BIBLIOGRAPHY

Flavin, M., *Fundamental Concepts of Information Modeling.* New York: Yourdon, 1981.

Gane, C. and Sarson, T., *Structured Systems Analysis: Tools and Techniques.* Englewood Cliffs, NJ: Prentice Hall, 1979.

Galbraith, J. R., *Organizational Design: An Information Processing View.* Reading, MA: Addison Wesley, 1973.

Ginzberg, M. J., Reitman, W., and Stohy, E. A., Editors, *Decision Support Systems.* New York: Elsevier, 1982.

Glorioso, R. M. and Osorio, F. C., *Engineering Intelligent Systems.* Bedford, MA: Digital Press, 1980.

Hare, V. C., Jr., *Systems Analysis: A Diagnostic Approach.* New York: Harcourt, Brace and World, 1967.

Holloway, C. A., *Decision Making Under Uncertainty.* Englewood Cliffs, NJ: Prentice Hall, 1979.

Hassab, J. C., *Underwater Signal and Data Processing.* Boca Raton, FL: CRC Press, 1989.

Hatley, D. J. and Pirbhai, I. A., *Strategies for Real-Time System Specification.* New York: Dorset House, 1987.

Hai, G. F., Turner, W. S., and Cashwell, L. F., *System Development Methodology.* Amsterdam: North Holland Publishing, 1978.

Hofstede, G. H., *The Game of Budgetary Control.* Assen, The Netherlands: Royal VanGorcum, 1967.

Keen, P. G. W. and Scott-Morton, M., *Decision Support Systems: An Organizational Perspective.* Reading, MA: Addison Wesley, 1978.

Lawson, H. W., Jr., *Understanding Computer Systems.* Rockville, MD: Computer Science Press, 1982.

Leslie, R. E., *Systems Analysis and Design.* Englewood Cliffs, NJ: Prentice Hall, 1986.

London, K., *The People Side of Systems.* New York: McGraw Hill, 1976.

Lundeberg, Mats, Goldkuhl, Goran, Nilsson, and Anders, *Information Systems Development: A Systematic Approach.* Englewood Cliffs, NJ: Prentice Hall, 1980.

Markus, M. L., *Systems in Organizations: Bugs and Features.* Boston: Pitman, 1984.

Miller, J. G., *Living Systems.* New York: John Wiley & Sons, 1978.

Mintzberg, H., *The Structuring of Organizations.* Englewood Cliffs, NJ: Prentice Hall, 1979.

Myers, G. J., *Composite/Structured Design.* New York: Van Nostrand Reinhold, 1978.

Norman, D. A. and Draper, S. W., *Software User Centered System Design.* Hillsdale, NJ: LEA Publishing, 1986.

Polya, G., *How to Solve It.* Princeton, NJ: Princeton University Press, 1973.

Raisbeck, G., *Information Theory.* Cambridge, MA: MIT Press, 1964.

Rothery, B., *The Art of Systems Analysis.* Englewood Cliffs, NJ: Prentice Hall, 1975.

Sayles, L. R. and Chandler, M. K., *Managing Large Systems.* New York: Harper and Row, 1971.

Shannon, C. E. and Weaver, W., *The Mathematical Theory of Communication.* Urbana, IL: University of Illinois Press, 1949.

Stalk, G. and Hout, T. M., *Competing Against Time.* New York: Macmillan, 1990.

Sol, H. G., Editor, *Processes/Tools for Decision Support Systems.* New York: Elsevier, 1983.

Taha, H., *Simulation Modeling and Simnet.* Englewood Cliffs, NJ: Prentice Hall, 1988.
Taylor, D. A., *Object-Oriented Technology: A Manager's Guide.* Reading, MA: Addison Wesley, 1990.
Thierauf, R. J., *Systems Analysis and Design of Real-Time Management Information Systems.* Englewood Cliffs, NJ: Prentice Hall, 1975.
Van Gundy, A. B., *Techniques of Structured Problem Solving.* New York: Van Nostrand Reinhold, 1981.
Von Bertalanffy, L., *General Systems Theory: Foundations, Development, Applications.* New York: George Braziller, 1968.
Wallace, R., Stockenberg, J., and Charette, R. A., *Unified Methodology for Developing Systems.* New York: McGraw-Hill, 1987.
Weinberg, G., *An Introduction to General Systems Theory.* New York: John Wiley & Sons, 1975.
Weiner, N., *Cybernetics, or Control and Communication in the Animal and the Machine.* New York: John Wiley & Sons, 1948.
Yourden, E., *Modern Structured Analysis.* Englewood Cliffs, NJ: Yourdon Press/Prentice Hall, 1989.
Yourden, E. and Constantine, L. L., *Structured Design: Fundamentals of a Discipline of Computer Program and System Design.* Englewood Cliffs, NJ: Prentice Hall, 1979.

Part III: Computer Systems and Automation

Abbott, R. J., *An Integrated Approach to Software Development.* New York: John Wiley & Sons, 1986.
Aho, A. V., Sethi, R., and Ullman, T. D., *Compilers: Principles, Techniques and Tools.* Reading, MA: Addison Wesley, 1985.
Barney, G. C., *Intelligent Instrumentation—Microprocessor Applications in Measurement and Control.* Englewood Cliffs, NJ: Prentice Hall, 1985.
Baron, N., *Computer Languages.* Garden City, NY: Doubleday, 1986.
Ben-Arir, M., *Principles of Concurrent and Distributed Programming.* Englewood Cliffs, NJ: Prentice Hall, 1990.
Black, U. D., *Computer Networks: Protocols, Standards and Interfaces.* Englewood Cliffs, NJ: Prentice Hall, 1993.
Brathwaite, K. S., *Information Engineering.* Boca Raton, FL: CRC Press, 1991.
Chapman, P. L., *Databases.* New York: Methuen, 1985.
Charette, R. N., *Software Engineering Risk Analysis and Management.* New York: McGraw Hill, 1989.
Chorafas, D. N., *Intelligent Networks.* Boca Raton, FL: CRC Press, 1991.
Chorafas, D. N., *Fourth and Fifth Generation Programming Language.* New York: McGraw Hill, 1986.
Chou, W., Editor, *Computer Communications.* Englewood Cliffs, NJ: Prentice Hall, 1983.
Claybrooke, B., *File Management Techniques.* New York: John Wiley & Sons, 1983.
Couger, J. D. and Sawacki, R. A., *Motivating and Managing Computer Personnel,* New York: John Wiley & Sons, 1980.
Cypser, R. J., *Communications Architecture for Distributed Systems.* Reading, Mass: Addison Wesley, 1978.

BIBLIOGRAPHY 437

Date, C. J., *An Introduction to Data Base Systems.* Reading MA: Addison Wesley, 1981.
Dijkstra, E. K., *A Discipline of Programming.* Englewood Cliffs, NJ: Prentice Hall, 1976.
Draffan, I. W. and Poole, I., Editors, *Distributed Data Bases.* Cambridge: Cambridge University Press, 1980.
Dromey, R. G., *How to Solve It by Computer.* Englewood Cliffs, NJ: Prentice Hall, 1982.
Everest, G. C., *Database Management: Objectives, System Functions and Administration.* New York: McGraw Hill, 1985.
Feigenbaum, E. and McCorduck, P., *The Fifth Generation.* Reading, MA: Addison Wesley, 1983.
Flores, I., *Database Architecture.* New York, Van Nostrand Reinhold, 1981.
Forester, T., Editor, *The Microelectronics Revolution: The Complete Guide to the New Technology and Its Impact on Society.* Cambridge, MA: MIT Press, 1981.
Foster, C. and Iberall, T., *Computer Architecture,* 3rd edition. New York: Van Nostrand Reinhold, 1989.
Freeman, D. E. and Perry, O. R., *Input/Output Design.* Rochelle Park, NY: Hayden Book, 1977.
Fulcher, J., *An Introduction to Microcomputer Systems.* Reading, MA: Addison Wesley, 1989.
Gillenson, M. L., *Database, Step by Step.* New York: John Wiley & Sons, 1985.
Hearn, D. and Baker, M. P., *Computer Graphics.* Englewood Cliffs, NJ: Prentice Hall, 1986.
Hiltz, S. R. and Turoff, M., *The Network Nation.* Reading, MA: Addison Wesley, 1978.
Hoare, C. A. and Perrott, R. H., Editors, *Operating Systems Techniques.* New York: Academic Press, 1972.
Hofeditz, C., *Computers and Data Processing Made Simple.* Garden City, NY: Doubleday, 1979.
Hwang, K. and Briggs, F., *Computer Architecture and Parallel Processing.* New York: McGraw Hill, 1984.
Johansen, R., *Teleconferencing and Beyond: Communications in the Office of the Future.* New York: McGraw Hill, 1984.
Kain, K. Y., *Computer Architecture: Software and Hardware.* Englewood Cliffs, NJ: Prentice Hall, 1989.
Kavanagh, P., Editor, *Distributed Network Computing.* Boca Raton, FL: CRC Press, 1992.
Kent, W., *Data and Reality.* Amsterdam: North Holland Publishing, 1978.
Kim, K., Editor, *Distributed Computer Systems.* Boca Raton, FL: CRC Press, 1992.
Klinger, A., Fu, K. S., and Kunii, T. L., Editors, *Data Structures, Computer Graphics and Pattern Recognition.* New York: Academic Press, 1977.
Langefors, B. and Sundgren, B., *Information Systems Architecture.* New York: Petrocelli/Charter, 1975.
Laurie, E. J., *Computers, Automation, and Society.* Homewood IL: Richard D. Irwin, 1979.
Liebowitz, B. N. and Carson, J. N., *Multiple Processor Systems for Real-Time Applications.* Englewood Cliffs, NJ: Prentice Hall, 1985.
Lin, W. C., *Handbook of Digital System Design,* 2nd edition, Boca Raton, FL: CRC Press, 1991.

Lister, A. M., *Fundamentals of Operating Systems.* London: Macmillan Press, 1975.
Longley, D. and Shain, M., *Data and Computer Security.* Boca Raton, FL: CRC Press, 1990.
Loomis, M., *Data Management and File Processing.* Englewood Cliffs, NJ: Prentice Hall, 1983.
Lorin, H. and Deitel, H. M., *Operating Systems.* Reading, MA: Addison Wesley, 1981.
Lucus, H. C., Jr., *Implementation: The Key to Successful Information Systems.* New York: Columbia University Press, 1981.
Lynch, M., *Computer Numerical Control: Advanced Techniques.* New York: McGraw Hill, 1993.
McFarlan, F. W. and McKenney, J. L., *Corporate Information Systems Management: The Issues Facing Senior Executives.* Homewood, IL: Richard D. Irwin, 1983.
Mano, M. M., *Computer System Architecture,* 2nd edition. Englewood Cliffs, NJ: Prentice Hall, 1982.
Martin, D., *Database Design and Implementation.* New York: Van Nostrand Reinhold, 1980.
Martin, J., *Computer Networks and Distributed Processing.* Englewood Cliffs, NJ: Prentice Hall, 1981.
Martin, J., *Application Development without Programmers.* Englewood Cliffs, NJ: Prentice Hall, 1982.
Miller, M. J. and Ahamed, S. V., *Digital Transmission Systems and Networks.* Rockville, MD: Computer Science Press, 1988.
Milutinovie, V. M., Editor, *High Level Language Computer Architecture.* MD: Computer Science Press, 1989.
Moreau, R., *The Computer Comes of Age: The People, the Hardware and the Software.* Cambridge, MA: MIT Press, 1984.
Mumford, E. and Sackman, H., Editors, *Human Choice and Computers.* Amsterdam: North Holland Publishing, 1975.
Naisbitt, J., *Megatrends.* New York: Warner, 1982.
Nora, S. and Minc, A., *The Computerization of Society.* Cambridge, MA: MIT Press, 1980.
Pooch, U. W., Editor, *Computer Engineering.* Boca Raton, FL: CRC Press, 1991.
Pooch, U. W., Machuel, D., and McCarin, J., *Telecommunications and Networking.* Boca Raton, FL: CRC Press, 1991.
Pratt, T. W., *Programming Languages: Design and Implementation.* Englewood Cliffs, NJ: Prentice Hall, 1984.
Pressman, R. S., *Software Engineering: A Practitioner's Approach.* New York: McGraw Hill, 1982.
Puppe, F., *Systematic Introduction to Expert Systems: Knowledge Representations and Problem Solving Methods.* New York: Springer Verlag, 1993.
Rafiquzzaman, M., *Microprocessors and Microcomputer-Based System Design.* Boca Raton, FL: CRC Press, 1990.
Raiffa, H., *Decision Analysis: Choice Under Uncertainty.* Reading, MA: Addison Wesley, 1974.
Sanders, D. H. and Birkin, S. J., *Computers and Management in a Changing Society.* New York: McGraw Hill, 1980.
Schwartz, M., *Telecommunication Networks: Protocol, Modeling and Analysis.* Reading, MA: Addison Wesley, 1987.

Shneiderman, B., *Software Psychology: Human Factors in Computer and Information Systems.* Cambridge, MA: Winthrop, 1980.
Siegel, S. G. and Bryan, W. L., *Software Product Assurance.* New York: Elsevier, 1988.
Sprague, R. J., Jr. and Carlson, E. D., *Building Effective Decision Support Systems.* Englewood Cliffs, NJ: Prentice Hall, 1982.
Stallings, W., *Local Area Networks: An Introduction.* New York: Macmillan, 1983.
Sutcliffe, A., *Human-Computer Interface Design.* New York: Springer-Verlag, 1989.
Tausworthe, K. C., *Standardized Development of Computer Software.* Englewood Cliffs, NJ: Prentice Hall, 1977.
Taylor, P. M., *Understanding Robotics.* Boca Raton, FL: CRC Press, 1990.
Tooley, M., *Data Communications Pocket Book.* Boca Raton, FL: CRC Press, 1990.
Thurber, K. J. and Masson, G. M., *Distributed Processor Communication Architecture.* Lexington, MA: D.C. Heath, 1979.
Toy, W. and Zee, B., *Computer Hardware/Software Architecture.* Englewood Cliffs, NJ: Prentice Hall, 1986.
Tsichritzis, D. C. and Lochovsky, F. H., *Data Base Management Systems.* New York: Academic Press, 1978.
Uhlig, R. P., Farber, D. J., and Bair, J. H., *The Office of the Future.* New York: North Holland Publishing, 1979.
Vassiliou, Y., Editor, *Human Factors in Interactive Computer Systems.* Norwood, NJ: Ablex Publishing, 1984.
Vick, C. R. and Ramamoothy, C. V., *Handbook of Software Engineering.* New York: Van Nostrand Reinhold, 1984.
Waite, W. and Goos, G., *Compiler Construction.* New York: Springer Verlag, 1984.
Whisler, T. L., *The Impact of Computers on Organizations.* New York: Praeger, 1970.
Whitaker, J., *Maintaining Electronic Systems.* Boca Raton, FL: CRC Press, 1991.
Woodson, W. E., *Human Factors Design Handbook.* New York: McGraw Hill, 1981.

Part IV: Application Systems Management

Abell, D. F. and Hammond, J. S., *Strategic Market Planning: Problems and Analyical Approaches.* Englewood Cliffs, NJ: Prentice Hall, 1979.
Ackerman, R. W., *The Social Challenge to Business.* Cambridge, MA: Harvard University Press, 1975.
Anthony, R. N. and Reece, J. S., *Management Accounting.* Homewood, IL: Richard D. Irwin, 1975.
Blanchard, B. S., *Design and Manage to Life Cycle Cost.* Portland, OR: M/A Press, 1978.
Bright, J. R. and Schoeman, M. E., Editor, *A Guide to Practical Technological Forecasting.* Englewood Cliffs, NJ: Prentice Hall, 1973.
Buffa, E., *Operations Management,* 3rd edition. New York: John Wiley & Sons, 1972.
Burch, J., *Information Systems: Theory and Practice.* New York: John Wiley & Sons, 1983.
Burns, T. and Stalker, D. M., *The Management of Innovation.* London: Tavistock, 1959.
Chandler, A. D., Jr., *Strategy and Structure: Chapters in the History of the American Industrial Enterprise.* Cambridge, MA: MIT Press, 1962.

Chase, R. and Aquilano, N., *Production and Operations Management.* Homewood, IL: Richard D. Irwin, 1981.

Chen, C. H., Editor, *Digital Waveform Processing and Recognition,* Boca Raton, FL: CRC Press, 1982.

Cleland, D. I. and King, W. R., Editors, *Project Management Handbook.* New York: Van Nostrand Reinhold Co., 1983.

Cyert, R. and March, J., *A Behavioral Theory of a Firm.* Englewood Cliffs, NJ: Prentice Hall, 1963.

Drucker, P. F., *Managing for the Future.* New York: Tralley Books/Penguin Books, 1992.

Duncan, A. J., *Quality Control and Industrial Statistics,* 4th edition. Homewood, IL: Richard D. Irwin, 1974.

Ein-Dor, P., *Information Systems Management.* New York: Elsevier, 1989.

Elliott, D. F., Editor, *Handbook of Digital Signal Processing Engineering Applications.* London: Academic Press, 1987.

Fayol, H., *General and Industrial Management,* translator, C. Storrs. New York: Pitman 1972. First French edition, 1916.

Feher, K., Editor, *Advanced Digital Communications System and Signal Processing Techniques.* Englewood Cliffs, NJ: Prentice Hall, 1987.

Fu, K. S., Ganzalez, R. C., and Lee, C. S., *Robotics: Control, Sensing, Vision and Intelligence.* New York, McGraw Hill, 1987.

Grant, E. L. and Ireson, W. G., *Principles of Engineering Economy, 5th edition.* New York: Ronald Press, 1970.

Gravens, D. W., Hills, G. S., and Woodruff, R. B., *Marketing Decision Making: Concepts and Strategy.* Homewood, IL: Richard D. Irwin, 1976.

Gray, E. R., Editor, *Business Policy and Strategy: Selected Readings.* Austin, TX: Austin Press, 1979.

Gumpert, D. E., Editor, *The Marketing Renaissance.* New York: John Wiley & Sons, 1985.

Hassab, J. C., *Underwater Signal and Data Processing.* Boca Raton, FL: CRC Press, 1989.

Hax, A. and Majleef, N. S., *Strategic Management: An Integrated Perspective.* Englewood Cliffs, NJ: Prentice Hall, 1984.

Higgins, J. M., *Cases in Contemporary Business.* Chicago: Dryden Press, 1982.

Jones, J. J., *Logistic Support Analysis Handbook.* Blue Ridge Summitt, PA: TAB Books, 1989.

Kallman, E. A., *Information Systems for Planning and Decisionmaking.* New York: Van Nostrand Reinhold, 1981.

Kenney, M., *Beyond Mass Production: The Japanese System and Its Transfer to the U.S.* New York: Oxford University Press, 1993.

Kerzner, H., *Project Management.* New York: Van Nostrand Reinhold, 1989.

Lewis, B. L., Kretschmer, F. F., and Shelton, W., *Aspects of Radar Signal Processing.* MA: Artech House, 1986.

Mamdani, E. H. and Gaines, B. R., Editor, *Fuzzy Reasoning and Its Applications.* Academic Press, 1982.

Mauser, F. F. and Schawrtz, D. J., *American Business: An Introduction.* New York: Harcourt Brace Jovanovich, 1974.

Meigs, W. B. and Meigs, R. F., *Financial Accounting.* New York: McGraw Hill, 1979.

BIBLIOGRAPHY

Metzger, P. W., *Managing a Programming Project.* Englewood Cliffs, NJ: Prentice Hall, 1973.

Moder, J., Davis, E. W., and Phillips, C., *Project Management with CPM and PERT.* New York: Van Nostrand Reinhold, 1983.

Nathanson, F. E., Reilley, J. P., and Cohen, M. N., *Radar Design Principles.* New York: McGraw Hill, 1990.

Nickerson, C. B., *Accounting Handbook for Non Accountants.* New York: Van Nostrand Reinhold, 1986.

Pascale, R. T. and Athos, A. G., *The Art of Japanese Management: Applications for American Executives.* New York: Simon and Schuster, 1981.

Porter, M. E., *Competitive Strategy: Techniques for Analyzing Industries and Competitors.* New York: The Free Press, 1980.

Post, J. E., *Corporate Behavior and Social Change.* Reston, VA: Reston Publishing, 1978.

Roelofs, H. M. and Houseman, G. L., *The American Political System.* New York: Macmillan, 1983.

Salter, M. S., *Diversification Through Acquisition: Strategies for Creating Economic Value.* New York: The Free Press, 1979.

Schmidt, T. W. and Taylor, R. E., *Simulation and Analysis of Industrial Systems.* Homewood, IL: Richard D. Irwin, 1970.

Shahinpoor, M. A., *Robot Engineering Textbook.* New York: Harper and Row, 1987.

Skolnik, M. I., Editor, *Radar Applications.* New York: IEEE Press, 1987.

Stevenson, W. J., *Production/Operations Management.* Homewood, IL: Richard D. Irwin, 1986.

Sumanth, D. J., *Productivity Engineering and Management.* New York: McGraw Hill, 1984.

Vollman, T. E., Berry, W. L., and Whybark, D. C., *Manufacturing Planning and Control Systems.* Homewood, IL: Richard D. Irwin, 1984.

Voss, C., *Just-in-Time: A Global Status Report.* New York: IFS, 1989.

Walters, J. R. and Nidsen, N. R., *Crafting Knowledge Based Systems.* New York: John Wiley & Sons, 1988.

Wegman, E. J. and Smith, J. G., Editors, *Statistical Signal Processing.* New York: Marcel Dekker, 1984.

Wunnicke, D., *Corporate Financial Risk Management: Practical Techniques of Financial Engineering.* New York: John Wiley & Sons, 1992.

INDEX

A

Access, random, 269
Access environment, 186
Acoustic noise, 329–330
Action tables, 126
Activating process, 10
Activation
   of conflicting and changing activities, 38–42
   defined, 37–38
   of human and automated systems, 42–45
Activation processes, 37–47
   authoritative, 37–38
   consultative, 38
   participative, 38
Activators' characteristics, 43–44
Activity ratios, 427
Address busses, 162
Address markers, 221
Algol, 174, see also Language(s)
Algorithms, query decomposition, 248–249
Allocation, 21
   of devices, 202
   of functions, 51, 258–259
   register, 177
   storage, 177, 180
Alteration of files, 234
ALU (Arithmetic Logic Unit), 152
Ambiguities of work, 41
Analysis phase, 85

Analyzers
   front-end, 177
   lexical, 177
   semantic, 177
   syntactic, 177
AND output, 151, 152
Animation, 211, see also Graphics
Applications defined, 6, 258–259
Applications management
   automated management systems, 285–325
      architecture of, 304–317
      automated management functions in, 302–304
      basic elements in, 288–300
      discussion, 317–325
      real-time control in, 286–288
      system configuration in, 300–302
   financial management, 421–432
      discussion, 431–432
      financial planning and control, 427–431
      financial system and, 424–427
   information processing in risk and uncertainty, 327–378
   production and service
      activation and control, 411–419
      discussions, 420–421
      planning and design, 406–411
   program management, 379–407
      basic processing flow, 381–389
      implementation, 389–401
      financial management, 398–401

schedule management, 396–401
technical management, 394–396
work breakdown and
organizational allocation,
391–394
Application specific integrated circuits
(APICs), 160, 162
Application systems, 141
Application utilization, 263
Architecture
of automated management systems,
304–317
implementation approaches,
309–313
requirements, 304–309
resources management, 313–317
of computer systems, 139–143
database systems, 233–236
hardware, 139
levels of abstraction in, 149
of multiprocessing systems, 258–264
software, 139
system functional, 169
system hardware, 169
system software, 169
Arithmetic instructions, 137
Arithmetic Logic Unit (ALU), 152, 156,
157–158
Array data, 175
Array of records, 175
Arrays, 178–179, 335, 338
Artificial intelligence, 369, see also
Expert systems
ASICs (application specific integrated
circuits), 160, 162
Assemblers, 174
Assembly language programmer,
265
Assembly level portability, 189
Asynchronous control, 134
Asynchronous operating systems,
265
Atmospheric noise, 330–332
Attributes, 242
Auditability, 242
Authoritative activation processes, 37–38
Automated management systems,
285–325
architecture of, 304–317

automated management functions in,
302–304
basic elements in, 288–300
discussion, 317–325
real-time control in, 286–288
system configuration in, 300–302
Automation
computer system architecture, 139–145,
see also Computer systems
database management systems,
217–253
architecture, 233–238
benefits of, 225
characteristics, 225–228
design, 238–243
discussions, 250–254
distributed processing, 245–250
for multiprocessing in real-time,
273–275
organization, 228–233
principles and background,
217–219
storage: file organization and access
methods, 219–227
system operation with, 243–245
defined, 8
discussions, 145
implementation of, 95
multiprocessing for real-time
applications, 255–281
background and principles,
255–258
communication systems, 267–270
database management systems,
273–275
discussions, 277–281
operating systems, 270–273
reliability and recovery systems,
275–277
software and hardware views,
264–267
systems architecture, 258–264
processes in, 132–139
control, 132
instruction handling, 137–139
logic, 132–137
uniprocessing, see also Uniprocessing;
Uniprocessors
applications, 170–171

INDEX

445

hardware systems, 149–170
    discussions, 169–170
    interfacing, 160–168
    organization and activation, 155–160
    uniprocessor basic elements, 150–154
    input/output management in, 197–215
    management processes in, 171–196
Availability, 88, 257–258
    of input/output devices, 214

B

Back end compilers, 177–178
Back end processors, 297
Backup and recovery, 225, 242
Balance sheet budget, 58
Balancing, 313
Barometric forecasting, 428
Bases of power, 46–47
Batch processing, 198
Bayesian estimation, 365
Behavioral science, 426
Behaviors of activators, 43
Benchmarking, 263
Bias, square of the, 355–356
Biased errors, 361
Bits defined, 204, 217
Block diagrams, 80, 83
Blocks, 176, 204, 266
    defined, 204, 218
    size of, 221
Boolean data, 175
Boundaries, 86
Boundary conditions, 102–104
Bounding procedures, 98–100
Brainstorming, 22
Break-even analysis, 427
B-tree oriented indexes, 222–224
Budgeting, 20, 55–62, 398–400
    capital, 428–429
Budgets
    balance sheet, 58
    discussions, 62–63
    effectiveness of, 61
    financial, 55–58
    information, 61–62

quality, 53, 61
quantity, 61
time, 53, 58–60
Buffering, 204
Bus connection, 269
Bus integration, 143
Bus linking, 302
Busses
    address, 162
    control, 162
    data, 162
    parallel, 162
    serial, 162–163
    shared, 269
Bussing, 141, 162–163
Byte defined, 204

C

Caches, 153
CAD/CAM (computer aided design and manufacturing), 210–211, see also Graphics
Calculations, 297
    defined, 245
    nonrecursive, 297
    recursive, 297
Call statements, 229
Capacity, 88, 123
Capacity planning, 406–408
Capital, working, 429
Capital budgeting, 428–429
Cassettes, 165
Catalog processing, 209
Catalogs, 208–209
Cathode ray tubes (CRTs), 165, 211
Centralized control, 249–250
Central organization, 261
Central processing unit (CPU), see CPU
Chained files, 265
Chaining of data, 207–208
Change, 42, 51
Channels of communication, 69
Characters defined, 217
Checkpoint dumps, 275
Checkpoints, 119
Checks (tests), 119–123
Check sums, 166

Chips, 311
Chosen plans, 23
Circuit switching, 268
CLMA systems
    contact tracking, 366
    definition and function, 357–358
    elements in problem formulation and solution, 362–367
    error and error filtering, 361–362
    expert systems for uncertainty conditions, 367–372
    general classes of problems, 359–360
    stationarity in, 358
Cobol, 174, see also Language(s)
Code generation, 177–178
Coercion as base of power, 46–47
Communication, 64–74
    among processes, 136
    barriers to effective, 73
    channels of, 69
    CPU in, 153
    in decision-making, 70–71
    definition and function of, 65–66
    highlights in history of, 68
    horizontal, 31
    improvement of, 70
    inter-application, 263–264
    interprocessor, 273
    means of, 69–70
    in multiprocessing systems, 267–270
    multitask, 272
    operating system, 187–188
    within operating system, 190
    patterns of, 69
    process components of, 65–68
    in real-time systems, 111
    in system sharing, 117
    tenets of effective, 73–74
Communication overhead, 255
Comparison testing, 168
Compilation, 180–182
Compiler implementations, 178–180
Compilers, 171, 174, 176–182, 304
    back end, 177–178
    front end, 177
    implementation models, 178–180
    instruction execution, 181–182
    instruction set design, 180–182

Complex instruction sets, 181
Complexity, 51, 84
Computer aided design and manufacturing (CAD/CAM), 210–211, see also Graphics
Computer animation, 211
Computer basic units, 152–154
Computer data summarization, 210
Computer graphics, 210–213, see also Graphics
Computer image processing, 211
Computers
    as compared with humans, 44–45
    CPU, 137
    defined, 6, 8
    Direct Memory Access (DMA), 137
    displacement of humans by, 47
    functions within system, 88
    go-between function of, 322
    processes performed by, 131
    working registers, 137–138
Computer systems, architecture of, 139–143, 149
Conceptual simulation, 101
Concurrency, 225, 242
Concurrent processes, 187
Conditional control instructions, 139
Conditions, boundary, 102–104
Configuration, 300–302
Configuration plans, 395
Conflict
    minimization of, 42
    negotiation and solution procedures, 42
    sources of, 38–42
    in system goals, 98
Connectivity, 92–94, 270
Consistency, 249–250
Constant velocity constant motion model, 364
Constraints, 23
Consultative activation processes, 38
Contact localization and motion analysis systems, see CLMA systems
Contamination of communication, 73
Context switching, 190
Contracts, see Statement of work

INDEX

Control, 49–63, see also System control
   asynchronous, 134
   centralized, 249–250
   criteria for well-formulated system, 287
   decentralized, 250
   flexibility in, 49
   as means not end, 49
   post-processing, 52
   for process cooperation, 136
   real-time
      in automated systems, 286–288
      predictive, 51
   resource, 273
   screening, 51
   steps in implementing, 51
   synchronous, 134
   system, 262
Control busses, 162
Control failure, 167
Control flows, 94
Control instructions, 139
   conditional, 139
Controlling process, 10
Control operations, 152–153
Control structures, 185
Control systems
   real-time, 125–126
   software, 304
Control unit, 156, 158–159
Coordination, 30
Corporate levels of system control, 54
Cost
   initial versus ongoing, 123
   optimum design versus cost of manufacture, 258
   as principal driver, 96
Cost account, 393
Cost accounting, 417, 418
Cost budget, 398–400
CPU, 137
   components of, 151–154
   defined, 142
   function of, 150
   input/output management, 197–198
   memory access in, 141
CPU overhead, 255
Cramer-Rao inequality, 355
Cramer-Rao lower bound, 355
Crashes, 123
Critical path, 58–60
Cross parity, 167
CRTs (cathode ray tubes), 165, 211
Customer decomposition, 29
Cycles, 22
Cyclic redundancy check (CRC) characters, 166–167

D

Damping, 332
Data
   as compared with information, 61–62
   types of, 175
   units of, 204–205
Data abstractions, 185
Data administration, 243
Database management systems, 217–253
   architecture, 233–238
   benefits of, 225
   characteristics, 225–228
   design, 238–243
   discussions, 250–254
   distributed processing, 245–250
   for multiprocessing in real-time, 273–275
   organization, 228–233
      hierarchical, 229
      network, 230
      pseudo-relational, 232–233
      relational, 230–232
   organizational approaches to, 225
   points of comparison in, 231–232
   principles and background, 217–219
   storage: file organization and access methods, 219–227
   system operation with, 243–245
Database processing, 259
Database requirements, 263
Databases
   defined, 218
   hierarchical, 229
   network, 230
   pseudo-relational, 232–233
   relational, 230–232

Database systems, 140
Data busses, 162
Data collection, 23
Data decomposition, 127–128
Data dictionary, 227, 242–243
Data dumps, 275
Data files, 208–209, see also File components; Files
Data flow diagrams, 237
Data independence, 215, 238
Data joins, 231–233
Data-level parallelism, 143
Data managers, see also Database management
  advantages of, 225
  lower level, 219
  middle level, 218
  top level, 218
Data normalization, 239
Data objects, 175
  implementations, 178
  types of, 175
Data security, 238, 241
Data staging, 203
Data storage, 177, 180, 203–204, see also Memory
  files, 208
    catalog, 208–209
    chain tables, 208
    labels, 208
    pages, 208
    records, 208
    volume table of contents, 209
Data strings, 175
Data structures, 185
Data transport, 141
De-buggers, 174
Decentralization, 31–32
Decentralized control, 250
Decimal data, 175
Decision-making
  communication and, 70–71
  as management process, 10–11
  in production/service management, 405–406
  searches and, 101
  system planning and, 15–24
Decision matrixes, 83

Decisions
  non-programmed, 20
  programmed, 20
Decision structures, 127
Decomposition, see also Work breakdown structure
  by customer, 29
  data, 127–128
  by equipment, 29
  by function, 29
  functional, 126–127, 313
  geographic, 29
  in implementation, 95–96
  information extraction by, 338–345
  of multiprocesses, 258–262
  by process, 29
  query decomposition algorithm, 248–249
  in real-time systems, 126–127
  in signal location and velocity estimation, 356–372, see also CLMA systems
  spatial, 336–343
  spectral, 343
  stages of, 262–264
  temporal, 344–345
  by time, 29
Decoupling, 127
Dedicated function organization, 272
Deductive logic, 79
Demand forecasting, 407, 411–412
Descriptors for simulation, 102–104
Design, see Systems also design
  product, 410–411
  quality of, 419
Design effort/modification, 231–232
Design methodologies, 123–124
Design phase, 85–88
Design processes, 124–127, see also Real-time systems
Design requirements, 262–263
Deterministic optimization, 365
Development
  human, 53
  organization, 53
  product, 53
Development phases of systems, 85–96
Device allocation, 202

INDEX

Device independence, 214–215
Device inventory, 202–203
Diagnosis, 276
Digital analyzers, 167–168
Digital logic, 151–152
Direct Memory Access (DMA), 137, 153, 163, 165
Direct organization, 205–207
Directories, 250
Direct tests, 105–106
Discounted messages, 73
Discretization, 374
Disks, 165, 221
  file organization on, 221–222
  format of, 221
Display devices, 211–213
Displays
  predictive techniques for, 299
  transformation techniques for, 299–300
Distinguishability, 360
Distributed file systems, 273
Distributed organization, 261
Distributed processing, 245–250
Diversity conditions, 346–347
Doppler shifted signals, 344
DRAM (dynamic RAM), 159–160
Drawings, 80
Dumps
  checkpoint, 275
  data, 275
Dynamic RAM (DRAM), 159–160

E

Economic theory, 425–426
Editors, 174
Effectiveness, 88
Efficiency of operating systems, 188–191
Electromagnetic noise, 330–332
Elements (transactions), representation of, 83–84
Elimination, simplification by, 99–100
Embedded statements, 227
Emission, failed, 73
Emulation, 168
  in-circuit, 168
Encoder—decoder mismatch, 73

Entering access environments, 186
Entities, 242
Entity relationship, 241
Environment implementations, 179–180
Equational representations, 79
Equipment, decomposition by, 29
Error detection, 276
  uniprocessors, 166–168
Errors
  biased, 361
  in signal processing, 361–362
  unbiased, 361–362
  uncertainties triggered by, 370
Events (transactions), 92
Event tables, 126
Exception handling, 121
Execution
  pseudoparallel, 116
  serial, 116
Exiting access environments, 186
Expense components of financial budgets, 58
Experience/expectations
  of activated, 43–44
  of activators, 43
Expertise as base of power, 46
Expert systems
  defined, 368
  for tracking in uncertainty conditions, 367–372
External tests, 105
Extrapolation forecasting, 428

F

Facilities design, 408–410
Failed emission, 73
Failure, 123, 167–168
  control, 167
  hard, 276
  major sources of, 275
  power, 167
  sensor, 167
  transient, 276
  uniprocessor, 167
Fault detection, 119–121, 193–194
Fault recovery, 195

Faults
  hardware, 191
  procedural, 191
  software, 191
Faults and failures, 51
Fault tolerance, 257–258
Fault-tolerant processing systems, 191–194
Feedback, 70, 125
Feedback processing loop, 285
Fermat's principle, 376
Fields
  defined, 217
  key, 218
File access, 166
File components, 208
  catalog, 208–209
  chain tables, 208
  labels, 208
  pages, 208
  records, 208
  volume table of contents, 209
File management, 166
File organization, 221–222
Files
  chained, 265
  defined, 204–205, 218
  in distributed processing, 248
  divided by function, 261
  divided by related functions, 261
  indexed, 222–223, 266
  linear, 222
  logical, 232–233
  opening/closing, 201–202
  organization and access methods, 219–227
  physical, 232–233
  shared, 188
  temporary, 210
File systems, distributed, 273
File utilization, 263
Filtering
  notch, 73
  of signal errors, 362
Financial analysis, 427–428
Financial budgets, 55–58
  expense components of, 58
  profit components of, 58
  revenue components of, 58

Financial leverage, 429
Financial management, 421–432
  discussions, 431–432
  financial planning and control, 427–431
    capital budgeting, 428–429
    financial analysis, 427–428
    financial leverage, 429
    financing
      intermediate term, 430
      long-term, 430
      short-term, 429–430
      mergers, 430–431
      working capital, 429
    financial system and, 424–427
    program management for, 398–401
Financial statement, 53, 58
Financing
  intermediate term, 430
  long-term, 430
  short-term, 429–430
Finite state systems, 126
First-order systems, 84
Fiscal policy, 426
Flat file (pseudo-relational) databases, 226, 232–233
Flip-flop memory, 153
Flow, for program management, 381–389
Flow charts, 80
Flow-chart symbols, 298
Flow control
  implementations, 179
  of operator objects, 175–176
Flow graphs, 80
Forecasting, 21–22
  financial, 428
Formal communication, 69
Formats, 221
Fortran, 173, see also Language(s)
Fredholm series, 373, 375
Frequency of use, 234
Front-end analyzers, 177
Front end compilers, 177
Front-end processors, 297
Functional abstractions, 185–188
Functional decomposition, 29, 126–127
Functional design plan, 395
Functionality, 91

INDEX

Functional requirements, 262–263
Functional specification, 289–290
Functional work breakdown, 290
Functions, partitioning and allocation of, 51, 258–259
Funds statement, 428

G

Gantt charts, 398
Gating, 151–152, 345–346, 352
   range, 356
Gaussian distributions, 361–362
General motion estimation problems, 364
Geographic decomposition, 29
Goals
   conflicts among, 42, 98
   defining of, 98
   framing of, 98
Graphics
   display devices, 211–213
   examples of interactive, 210–211
   general-purpose routines, 213
   peripherals interfacing with, 213
Graphics routines, 213
Graphic tablets, 213
Green function, 374
Group actuators, 45
Group builders, 45
Group detractors, 45
Grouping, 99–100, 127
Group productivity, 45
Groups, composition of, 45

H

Hamming code, 166–167
Hard failure, 276
Hardware, 172, see also Computers
   architecture, 139
   for multiprocessing systems, 264, 266–267
Hardware faults, 191
Hardware redundancy, 192
Hardware register set, 153
Hardware resources management, 185
Hardware stack, 153
Hashing, 225, 230

Hassab iterative, 374
Helmholtz equation, 375
Hierarchical communication, 69
Hierarchical databases, 229
   pointer-based, 229
   sequential hierarchy, 229
   structure, 225
Hierarchical data structure, 239
Hierarchical management, 218
Hierarchical organization, 26, 28
Hierarchical system configuration, 300
High order languages (HOLs), 149, 172–176, 174–175, see also Language(s)
Homogeneous propagation, 347
Horizontal communication, 31
Horizontal organization, 272
Horizontal passive observation problems, 353–354
Human–computer communication, 176–182, see also Compilers
Human–computer input/output, 210–214, see also Graphics
Human development measures, 53
Human operator, 215–216
Human processors, as compared with computers, 44–45
Human requirements, 263
Humans
   displacement by computers, 47
   functions within system, 88
   processes performed by, 131
Human simulation, 102

I

Identification, resource, 274–275
IF ... THEN rules, 368
Implementation, 88–96
   of automation, 95
   constituents encountered in, 92–94
   control flows in, 94–96
   decomposition in, 95–96
   elements tracked during, 91
Implementation phase, 88–94
Implementations
   compiler, 178–180
   data objects, 178
   environment, 179–180

exceptions, 180
flow control, 179
operators, 179
storage allocation, 180
In-circuit emulation, 168
Independence
  of data, 225, 238
  of elements, 258
Indexed files, 222–223, 266
Indexed organization, 205
Indexes
  B-tree oriented, 222–224
  multilevel hardware oriented, 222
  simple, 222
Indexing, 244–245
Indirect tests, 106
Inductive logic, 79
Ineffective systems, 106–107, see also System simplification
Influencing, 43
Informal communication, 69
Information
  budgeting and, 61–62
  as compared with data, 61–62
  criteria for conveying, 73–74
  system complexity and, 84
Information handling, 107
Information processing, 327–378
  actual and lower bound localization accuracies, 353–356
  basic signal processing, 345–352
    source, channel, and processor, 347–351
    time intervals and, 345–347
    weighting, 352
      by gating, 352
      by windowing, 352
  discussions, 366–378
  generalized prediction, processing, wave propagation, and vibration method, 372–376
  information extraction, 338–345
    by spatial decomposition, 338–343
    by spectral decomposition, 343
    by temporal decomposition, 344–345
  location and velocity of contacts, 356–372
    contact sire estimation, 359–360
    errors and filtering, 360–362
    expert systems in, 367–372
    formulation and solution of CLMA problems, 362–367
  as signal and data processing, 327
  signal noise and channel outputs, 329–332
  signal prediction with multiplicative noise, 376–377
  structures for, 332–338
Information structures, 185
Inhomogeneity, 376
Input, 96
  testing of, 105
Input devices, 299
Input/output channels, 200
Input/output control, 189
Input/output controller, 154
Input/output device requirement, 200–201
Input/output instructions, 137
Input/output management, in multiprocessing, 266–267
Input/output matrixes, 83
Input/output processing
  data organization, 204–210
    chaining, 207–208
    direct, 205–207
    indexed, 205
    linked list, 207
    sequential, 205
    storage, 208–210, see also Data storage
    units of data, 204–205
  operating characteristics, 214–215
  processes involved in, 198–204
    buffering, 204
    data staging, 203
    data storage, 203–204
    device allocation, 202
    device inventory, 202–203
    I/O channels, 200
    I/O device requirement, 200–201
    I/O supervisor, 199
    I/O system controls, 200
    opening/closing files, 201–202
    scheduling, 202
Input/output scheduling, 163–164
  uniprocessor systems, 163

INDEX 453

Input/output supervisor, 199
Input/output system controls, 200
Input/output unit, 157, 160
Instruction execution, 181–182
Instruction executions, 181–182
Instruction handling, 137–139
Instruction-level parallelism, 143
Instruction mix, 182
Instructions, 95
Instruction set design, 180–182
Instruction sets, 181
  complex, 181
  reduced, 181
Integer data, 175
Integration, 30
  horizontal, 31
  layered, 371–372
  round-robin, 31
  of work units, 143
Integrity, 215
  of data, 225
  of information, 73
Interactive graphics, 210–214, see also Graphics
Inter-application communication, 263–264
Interconnect technology, 313
Interdependence, as source of conflict, 41
Interface requirements, 262
Interfaces, 86, 91, see also Graphics
  operating system, 189
  smart, 302
Interfacing, 144
  uniprocessor systems, 160–168
    busing techniques, 162–163
    error detection, correction, fail-soft, maintenance, 166–168
    input/output scheduling, 163–164
    peripherals, 164–166
Intermodule synchronization, 186–187
Internal rate of return, 428
Internal tests, 105
Internetworking, 270
Interprocess communication, 273
Interprocessor communication, 273
Interrupts, 163, 190
  priority, 304
  simple, 303
  vectoral, 304

Inventory, 417
Iteration, 127
Iterative flow control, 175

J

Job-level parallelism, 143
Joining, 231, 232–233
Joysticks, 213

K

Keyboards, 164, 213
Key fields
  defined, 218
  organization of, 218

L

Labels, 208
Language(s), 70
  assembly, 174
  commercial computer, 173–174
  computer, 196
  defined, 174–175
  descriptive, 173
  elements composing, 195–196
  elements of high-order, 173–176
  high order (HOL), 174–175
  machine, 174
  object-oriented, 173
  procedural, 173
  structured query, 231
Layered integration, 371–372
Layered protocols, 267–268
Layouts
  process, 408–409
  product, 408
Leadership, 43
Learning systems, 84
LEDs, 164, 211
Legitimacy as base of power, 46
Leverage ratios, 427
Lexical analyzers, 177
Light emitting diodes (LEDs), 164, 211
Linear discrete systems, 374
Linear files, 222
Linear minimum variance estimation, 365

Linkage
  in multiprocessing, 259–261
  in program management, 386–388
Linked list organization, 207
Linkers, 174
Links, 188
Liquidity ratios, 427
Load sharing, 335
Localizability, 79
Localization accuracies, 353–356
Location planning, 408
Locked records, 274
Logic
  deductive, 79
  digital, 151–152
  inductive, 79
Logical files, 232–233
Logical operator objects, 175
Logic analysis, 118–119
Logic (digital) analyzers, 167–168
Logic planning, 134–137

M

Macros, 174
Main memory, 153
Maintenance, 22, 167–168
Maintenance files, 234
Managed system defined, 6–7
Management
  decision-making and, 10–11
  defined, 5
  function of, 9–10
  hierarchical, 218
  memory, 188
  processor, 188
  span of, 32–33
  systems approach to, 3–13
    decision-making, 10–11
    discussions, 11–12
    mechanization and automation, 4–9
    process types and management functions, 9–10
    real-time systems, 3–4
Management processes, 131–147
  in automated systems, see also Automated systems
  controlling, 10
  defined and classified, 5–6
  as elements of system, 77–78
  organizing, 9–10
  planning, 9–10
  system communication, 65–74
    discussion, 73–74
    process, 66–73
  system control, 49–63, see also Budgeting; System control
    budgeting, 55–62
    discussions, 62–63
    process of, 50–54
  systems activation, 10, 37–47
    of conflicting and changing activities, 38–42
    defined, 37–38
    discussions, 45–47
    of human and automated systems, 42–45
  systems organization, 25–36
    coordination and integration of work, 29–31
    decomposition (breakdown) of work, 29, 95–96
    design of organizations, 32–33
    discussions, 33–35
    organization process and structure, 26–29
    span of management and decentralization, 31
  systems planning, 15–24
    classes of problems in, 21–22
    decision-making and, 19–21
    discussion, 22–24
    external premises, 16
    process of, 16–19
Management theory, 425
Managers
  data, see Data managers
  of global issues, 272
  of local issues, 272
  for multiprocessing, 270–272
  operating system, 304
  system services, 303–304
Manual operations, 6
Mapping
  of functional requirements, 264
  memory, 255
  of time delay estimates, 356
Market share, 53

INDEX

Mass storage peripherals, 154
Master-slave organization, 272
Material management, 415–419
Matrixes, 83
  decision, 83
  input/output, 83
Maximum likelihood estimation, 365
Mean time between failures (MTBF), 257–258
Mean time to repair (MTTR), 257–258
Mechanization defined, 6–8, 317
Media of communication, 73
Memory, see also Data storage
  cache, 153
  defined, 142–143
  hardware register set, 153
  hardware stack, 153
  main, 153, 265
  primary, 220
  random access (RAM), 141, 159–160
  read-only (ROM), 141, 149, 159
  register set, 265
  secondary, 220–221
  semiconductor, 153–154
  shared, 187, 272
Memory access, 141
Memory management, 188, 266
Memory management hardware, 190
Memory mapping, 255
Memory sharing, 116–117, 187, 272
Memory space assignment, 265
Memory unit, 156, 159–160
Memory utilization, 263
Menus, 245
Mergers, 430–431
Messages, 187
  discounted, 73
Message switching, 268
Microprograms, 139–140
Milestone scheduling, 398
Military applications, 324–378, see also Information processing
Mismatch, encoder—decoder, 73
Modeling, 80, 104–105
Modular construction, 107
Modules, 176
  package, 176
  task, 176

Monetary policy, 426
Monitoring
  in system simplification, 104
  time, 21
Motion analysis, see CLMA systems
Motives, 43
Mouse, 164
MTBF (mean time between failures), 257–258
MTTR (mean time to repair), 257–258
Multimeters, 167
Multipath passive observation problems, 354–355
Multipath propagation, 347
Multiple scattering, 376–377
Multiplicative noise, 375–377
Multipoint linking, 302
Multiprocessing, 198
  for real-time applications, 255–281
    background and principles, 255–258
    communication systems, 267–270
    database management systems, 273–275
    discussions, 277–281
    operating systems, 270–273
    reliability and recovery systems, 275–277
    software and hardware views, 264–267
    systems architecture, 258–264
Multiprocessors, 184
Multiprogramming, 187, 198, 303
Multiprogramming software, 265
Multitasking, 265, 272, 273, 303

N

Negotiation, 42
Net present value, 428
Network databases, 230
Network data structure, 225, 239
Networking, 141, see also Multiprocessing
  comparative assessment of, 270
Network integration, 143
Network operating systems, 184

Network organization, 26
Network scheduling, 398
Network system configuration, 300
Neumann series, 374
Noise
    acoustic, 329–330
    electromagnetic, 330–332
    multiplicative, 375–377
Nonrecursive calculation, 297
Non-stationarity, 343, 344–345, 359–360
Notch filtering, 73
NOT output, 151, 152
Numeric operator objects, 175

O

Objectives, 17
Object level portability, 189
Objects
    data, 175
    operator, 175–176
Off-ray scattering, 376
Opening/closing files, 201–202
Open loop processing, 285
Operating system manager, 304
Operating system processing, 259
Operating systems, 140, 174
    asynchronous, 265
    dedicated function organization, 272
    functions of, 182–184
        input/output, 197–198
    horizontal organization, 272
    master-slave organization, 272
    multiprocessing, 270–273
    multiprocessor, 184
    network, 184
    operating modes, 198, see also Input/output management
    organization and activation, 184–188
    portability and efficiency, 188–191
    real-time, 183
    structures, 272–273
    synchronous, 265
    transaction, 184
Operation management
    overall, 411–412
    product, 412–413

Operator objects, 175–176
Operators implementations, 179
Opinion polling, 428
Organization
    database, 261
        central, 261
        distributed, 261
    of files, 219–227
    of operating systems, 272–273
    process of, 26
    structure of, 26–29
Organizational allocation, 391–394
Organizational depiction, 27
Organizational structure, 30
    hierarchical, 26, 28
    network, 26
    relational, 27
Organization charts, 80
Organization development measures, 53
Organizations
    characteristics and variable growth, 34
    design of, 32–33
Organizing in conflict minimization, 42
Organizing process, 9–10
OR output, 151, 152
Output, 96, see also Input/output
    testing of, 105
Output devices, 299
Overall operation management, 411–412
Overhead
    communication, 255
    CPU, 255
Overhead utilization, 264
Overlay storage, 299

P

Package modules, 176
Packaging, 311
Packet switching, 268
Pages, 208
Parallel busses, 162
Parallelism, 258, 259, 313
    levels of, 143
Parallel system configuration, 300, 302
Parity, 166
    cross, 167

INDEX

Participative activation processes, 38
Partitioning, 21, 275
  of functions, 51, 258–259
Pascal, 174, see also Language(s)
Passive observation problems
  horizontal, 353–354
  multipath, 354–355
Pattern recognition, 369
Payback, 428
Performance, 91, 238
Performance evaluation, 51
Performance indices in signal
    processing, 365–366
Performance measurement, 51
Performance measures, 267
Performance requirements, 262
Peripherals, 137, 164–166
  cassettes, 165
  cathode ray tubes (CRTs), 165
  disks, 165
  file access, 166
  file management, 166
  for graphics interfaces, 213
  keyboards, 164
  LEDs, 164
  mass storage, 154
  mouse, 164
  in multiprocessing, 267
  stepper motors, 164
  teletypes, 164
  track 00, 166
  uniprocessor systems, 164
Physical files, 232–233
Picture displays, 212, see also Graphics
Pipes, 187
Pixels, 299
Planning, 406–411
  basic steps of, 22–23
  capacity, 406–408
  in conflict minimization, 42
  facilities, 408–410
  financial, 427–431, see also Financial
      management
  product, 410–411
Planning process, 9–10, 15–19, see also
    System planning
Plans
  chosen, 23
  criteria for, 19

  single-use, 17–18
  for technical management
      implementation, 395–396
Point-to-point integration, 143
Point-to-point linking, 302
Point-to-point networking, 269
Policy, financial, 426–427
Polling, 163
Portability
  assembly level, 189
  object level, 189
  of operating system, 188–189
  of operating systems, 188–191
  source level, 189
Ports, 188
Post-processing control, 52
Power, bases of, 46–47
Power failure, 167
Power signal to noise ratio, 332
Predictive real-time control, 51
Premises, 23
Primary memory, 220
Priority condition, 102
Priority interrupts, 304
Privileges, 186
Probing, 105, 354–355
Problem definition, 22
Procedural faults, 191
Procedures, 86
Process, 92
  decomposition by, 29
  defined, 132
Processes
  concurrent, 187
  defined, 8
  management, See Management
      processes
  in real-time systems, 111
Processing
  database, 259
  functions involved in, 297–298
  operating system, 259
  resource management, 259
  as step in planning, 23
  traffic, 259
Process layouts, 408–409
Processor management, 188
Processors, see also Multiprocessors
  back-end, 297

front-end, 297
interconnection, 259–261
loosely and tightly coupled, 264
selection, 259
Processor sharing, 115–116
Process state, 134, 266
Process synchronization, 273
Procurement, 417
Product design, 410–411
Product development measures, 53
Production and service management
  activation and control, 411–419
    material management, 415–419
    overall operation management, 411–412
    product operation management, 412–413
    quality management, 419–420
    scheduling management, 413–415
  discussions, 420–421
  planning and design, 406–411
    capacity, 406–408
    facilities, 408–410
    product, 410–411
Production plans, 396
Productivity, 11, 23
  group, 45
  measurement of, 53
  physical and mental factors of, 43
Product layouts, 408
Product operation management, 412–413
Profitability, 53
Profitability ratio, 427, 428
Profit components of financial budgets, 58
Program defined, 8
Program environment, 171
Program management, 379–407
  basic processing flow, 381–389
  implementation, 389–401
    financial management, 398–401
    schedule management, 396–401
    technical management, 394–396
    work breakdown and organizational allocation, 391–394
  interface management, 379, 380
  transformation operator in, 379–381
Program manager, 379–381
  transformation contract of, 381

Programming
  ease of, 238
  languages for, 173–176
  programs for, 174
Programs, 171–172, see also Language(s); Software
  compiler, 176–182
Project managers, 31
Propagation, signal, 347–352
  homogeneous, 347
  multipath, 347
Propagation delay, 267
Protection processes, 265
Protocols, 141
  layered, 267–268
Prototyping, 22, 100
Pseudoparallel execution, 116
Pseudo-relational databases, 226, 232–233
Pseudo-relational data structure, 239

Q

Qualification, 91
Quality
  of conformance, 419
  of design, 419
Quality assurance plans, 395
Quality budgets, 53, 61
Quality management, 419–420
Quantity budgets, 61
Query decomposition algorithm, 248–249
Query language, structured, 231
Query processing, 248–249
Query statements, 227
Queue administration, 189–190
Queues, 179

R

Radar systems, 329, 333, see also Information processing
RAM (random access memory), 141, 159–160
Random access, 269
Random access memory (RAM), 141, 159–160
Randomness, 22

INDEX

Random organization of fields, 218
Random scattering, 377
Range gating, 356
Ratios
  financial, 427
  of system to resources, 98
Rayleigh-Gans-Born approximation, 375–377
Read-only memory (ROM), 141, 149, 159
Read/write operations, 162
Real data, 175
Real-time applications, multiprocessing for, 255–281
  background and principles, 255–258
  communication systems, 267–270
  database management systems, 273–275
  discussions, 277–281
  operating systems, 270–273
  reliability and recovery systems, 275–277
  software and hardware views, 264–267
  systems architecture, 258–264
Real-time clock, 190
Real-time control, 286–288
  predictive, 51
Real-time operating systems, 183
Real-time systems, 109–129
  basic elements, 110–113
  design methodology, 123–124
  design processes, 124–128
    control system process, 125–126
    finite state system process, 126
    functional decomposition, 126–127
  discussions, 128–129
  implementation, 113–117
  importance of time, 109–110
  real-time defined, 4–5
  reliability, 118–123
  scheduling, 117–118
Recognition of communications/information, 74
Record data, 175

Records, 266
  chaining of, 207–208
  defined, 204, 218
  format for, 221
  locked, 274
Recoverability, 214–215
Recovery
  in multiprocessing systems, 275–277
  software, 276–277
  system, 276
Recovery processes, 119–123
Recursive calculation, 297
Reduced instruction sets, 181
Redundancy, 70, 238, 258, 275, 276
  cyclic, 166–167
  hardware, 192
  software, 192–193
  time, 193
Referral as base of power, 46
Register allocation, 177
Register insertion, 270
Relational databases, 225, 230–232
Relational operator objects, 175
Relational organization, 27
Relations, types of, 248
Relationships, 242–243
Reliability, 257–258, 262, 275
  of input/output devices, 214
  in multiprocessing systems, 275–277
  of products, 411
  of real-time systems, 118–123
Reliability requirements, 263
Replacement, 22
Representation
  as base of power, 46
  of system, 79–85, see also System representation
Requirements
  computer, 263–264
  functional, 262–263
Resolution, 344–345
Resource control, 273
Resource identification, 274–275
Resource management, 313–317
Resource management processing, 259
Resource plans, 396

Resource pooling, 3–4
Resources
  boundaries of, 77–78
  defined, 5
  optimal employment of, 23–24
  as source of conflict, 38
  system as grouping of, 11
Resources management
  hardware, 185
  software, 185
Response, 255–256
Retrieval, 234, 244, 248
Revenue components of financial budgets, 58
Review plans, 395
Reward as base of power, 46
R-G-B (Rayleigh-Gans-Born) approximation, 375–377
Rights, 186
Ring linking, 302
Ring networking, 269
Risk and uncertainty, 327–378, see also Information processing
Risks plans, 396
Rolling back, 121–123
ROM (read-only memory), 141, 149, 159
Round-robin integration, 31
Routes
  system, 102
  work, 102
Routing, 58
R relations, 246, 249

S

Safety plans, 395
Scalability, 258
Scalar models, 178
Scattering, 376–377
  multiple, 376–377
  off-ray, 376
  random, 377
Schedules
  basic types of, 397
  graphical techniques for, 398
Scheduling, 21–22, 58–60, 102
  development of, 397–398
  input/output, 163–164, 202
  uniprocessor systems, 163
  in multiprocessing, 266
  program management for, 396–401
  real-time, 117–118
Scheduling management, 413–414, 413–415
Screening control, 51
Search and estimation, 22
Search processes, 100–101
Secondary memory, 220–221
Second-order systems, 84
Security, 225
  data, 238, 241
Self diagnostics, 168
Semantic analyzers, 177
Semiconductor memory, 153–154
SEND commands, 273
Sensor failure, 167
Sequencing, 21–22, 127
Sequential flow control, 175
Sequential organization, 205
  of fields, 218
Serial busses, 162–163
Serial execution, 116
Serial system configuration, 302
Serviceability, 214
Service operation management, 405–419, see also Production and service management
Services, 91
Service systems, 415
Shared busses, 269
Shared files, 188
Shared memory, 187, 226–227, 272
Sharing, 115–116
  memory, 116–117, 226–227, 272
  processor, 115–116
Signal processing, 327–378, see also Information processing
  actual and lower bound localization accuracies, 353–356
  basic
    source, channel, and processor, 347–351
    weighting, 352
      by gating, 352
      by windowing, 352
  discussions, 366–378

INDEX
461

generalized prediction, processing, wave propagation, and vibration method, 372–376
location and velocity of contacts, 356–372
 contact site estimation, 359–360
 errors and filtering, 360–362
 expert systems in, 367–372
 formulation and solution of CLMA problems, 362–367
 signal prediction with multiplicative noise, 376–377
 time intervals and, 345–347
Signal propagation, 347–352
Signals
 defined, 329
 Doppler shifted, 344
Signal to noise ratio, 332
Signature analysis systems, 168
Simple interrupts, 303
Simplification, 97–108, see also System simplification
Simulation, 22, 101–105
 advantages of, 101
 conceptual, 101
 descriptors for, 102–104
 in error testing, 168
 human, 102
 modeling in, 104–105
 principal elements in, 101–102
 symbolic, 102
Size, 51
Smart interfaces, 302
Snell's law, 376
Sockets, 273
Software, 172–173
 abstractions in
  data, 185
  functional, 185–188
 architecture, 139
 control system, 304
 for multiprocessing systems, 264, 265–266
 portability of, 188–189
 recovery of, 276–277
 utility system, 304
Software faults, 191
Software/hardware development plan, 395

Software redundancy, 192–193
Software resources management, 185
Sonar systems, 329, 333, see also Information processing
Sorting, 244–245
Source level portability, 189
Source signature, 347
Space inventory, 210
Span of management, 31–32
Sparing, 123
Spatial condition, 102
Spatial decomposition, 336–343
Spatial resolution, 345
Spectral decomposition, 343
Square of the bias, 355–356
SRAM (static RAM), 160
S relations, 246, 249
Standardization of products, 410
Standard processor utilization, 263–264
Standard setting, 51
Star linking, 302
Statement of work
 contents of, 382–386
 implementation of, 388–390
Statements
 call, 229
 embedded, 227
 query, 227
State tables, 126
Static RAM (SRAM), 160
Stationarity/non-stationarity, 343, 344–345
Statistical bounding, 100
Statistical error filtering, 362
Stepper motors, 164
Storage
 overlay, 299
 virtual, 299
Storage allocation, 177, 180
Storage allocation implementations, 180
Stored responses, 168
Strapped linking, 302
Stress testing, 118
String data, 175
String operator objects, 175
Structured models, 178
Structured query language, 231
Structuring, 290–291

Subroutines, 174
Substitution, bounding by, 100
Supplementation, 115
Switching, 268–269
  circuit, 268
  message, 268
  packet, 268
Symbolic simulation, 102
Symbol manipulation, 137
Symbols, 84–85
  flow-chart, 298
Symmetry, bounding by, 100
Synchronization
  of computer architecture, 143
  intermodule, 186–187
  in multiprocessing, 266
  of multiprocessing databases, 274
  of multiprocessors, 184
  process, 273
  in real-time systems, 111
  in system sharing, 117
Synchronous control, 134
Synchronous operating systems, 265
Syntactic analyzers, 177
Synthesis, See Systems synthesis
System(s)
  as composed of objects, 78–79
  composition of, 9
  defined, 5
  development phases of, 85–96
    analysis, 85
    definition, 85
    design, 85–88
    implementation, 88–94
  elements of, 78–79
  as grouping of resources, 11
  managed, 6–7
  mature and complex, 8–9
  natural and human-made, 35
System activation, 37–47, see also Activation processes
  of conflicting and changing activities, 38–42
  defined, 37–38
  discussions, 45–47
  of human and automated systems, 42–45
  learning procedures during, 38
System analysis, 426

System block diagram, 80
System communication, 65–74, see also Communication
  discussion, 73–74
  process, 66–73
System control, 4–63, 262
  budgeting, 55–62
    discussions, 62–63
    effectiveness of budgets, 61
    financial budgets, 55–58
    information and, 61–62
    quantity and quality budgets, 61
    time budgets, 58–60
  corporate levels of, 54
  levels of, 52–54
  need for, 51
  process of, 50–54
  types of, 51–52
System description, 77–96
  discussions, 96
  elements of system, 77–78
  system analysis and design, 85–88
  system implementation, 88–96
  system representation, 78–85
System engineering plan, 395
System flow graphs, 80
System organization, 25–36
  coordination and integration of work, 29–31
  decomposition (breakdown) of work, 29, 95–96, see also Decomposition
  design of organizations, 32–33
  discussions, 33–35
  organization process and structure, 26–29
  span of management and decentralization, 31
System planning, 15–24
  classes of problems in, 21–22
  decision-making and, 19–21
  discussions, 22–24
  external premises, 16
  process of, 16–19
System procedures, 86
System representation, 79–85
  complexity and, 84
  development phases and, 85
  elements and events in, 83–84

symbols for, 84–85
tools for, 80–83
types of, 79–80
Systems analysis, 85
Systems analysts, 236–237
Systems approach, 3, 4
Systems definition, 85
Systems design, 85–88, see also Design
database systems, 236–243
System services managers, 303–304
Systems implementation, 88–94
System simplification
approaches to, 97–98
discussion, 108
ineffective systems
characteristics, 106–107
improvement, 107
processes of, 98–106
search processes, 100–101
simulation processes, 101–105
test processes, 105–106
top-down, eliminating and grouping, 98–100
Systems managers, 11
Systems synthesis
in automated systems, 131–147, see also Automated systems
discussions, 96
elements of system, 77–78
real-time systems, 109–129
basic elements, 110–113
design methodology, 123–124
design processes, 124–128
discussions, 128–129
implementation, 113–117
importance of time, 109–110
reliability, 118–123
scheduling, 117–118
system analysis and design, 85–88
system description, 77–96
system implementation, 88–96
system representation, 78–85
system simplification, 97–108
discussions, 108
improvement of ineffective systems, 106–107

processes of, 98–106
System standards, 51–52
System timing, 86–88

T

Tables, 83
action, 126
event, 126
state, 126
Task grouping, 127
Tasking, 127
Task modules, 176
Task relevance, 43
Task states, 266
Technical management implementation, 394–396
Technical performance plan, 395
Teletypes, 164
Temporal decomposition, 29, 127, 344–345
Temporal grouping, 127
Testing
comparison, 168
stress, 118
Test plans, 395
Tests
direct, 105–106
external, 105
indirect, 106
internal, 105
properties enabled by, 106
real-time checks, 119–123
Thermal noise, 330
Thread-level parallelism, 143
Thresholding, 100
Throughput, 267
Tiles (pixels), 299
Time
as basis of desired characteristics, 255–256
decomposition by, 29
significance of, 109–111
Time budgets, 53, 58–60
Time condition, 102
Time delay estimation, 346–347
Time intervals, 345–346
Time monitoring, 21
Time redundancy, 193

Time series analyses, 22
Timing, system, 86–88
Token passing, 269
Tool kits, 174
Top-down simplification methods, 98–100
Touch panels, 213
Touch pens, 213
Track 00, 166
Track descriptor, 221
Track formats, 221
Traffic processing, 259
Training plans, 395
Transaction operating systems, 184
Transactions (events), 92
 representation of, 83–84
Transfer flow control, 175–176
Transformation, 96, 375
 bounding by, 100
Transient failure, 276
Transportation
 as compared with communication, 66
 highlights in history of, 68
Trends, 22
Type checking, 186

U

Uncertainty and risk, 327–378, see also Information processing
Uniprocessing
 input/output, 197–215, see also Input/output management
  data organization, 204–210
  discussions, 215–216
  between human and computer, 210–214
  operating characteristics, 214–215
  processes of, 198–204
 management processes in, 171–196
  discussions, 194–196
  elements of high-order languages, 172–176
  fault-tolerant processing system, 191–194
  general principles, 171–172

 human–computer communication:
  compiler organization, 176–182
  management processes in, see also Compilers
  operating system
   functions, 182–184
   organization and activation, 184–188
   portability and efficiency, 188–191
Uniprocessors
 basic elements of, 150–154
  computer basic units, 152–154
  digital logic, 151–152
 essential organization and activation, 155–160
 failure of, 167
 interfacing of systems, 160–168
  busing techniques, 162–163
  error detection, correction, fail-soft, maintenance, 166–168
  input/output scheduling, 163–164
  peripherals, 164–166
Units of data, 204–205
Updates, 231–232, 248, 249
Use, frequency of, 234
Utility system software, 304
Utilization, processor, 263–264

V

Variables, bounding of, 100
Variety of measurement models, 364
Vectoral interrupts, 304
Vibration, 374–375
Virtual storage, 299
Visualization, 80
Volume table of contents, 209

W

Wave fields, see Signal processing
Wave propagation, 374–375
Wave velocities, 332
Weighting of signals, 352
Windowing, 352
Words defined, 217

Work
    ambiguities of, 41–42
    breakdown of (decomposition), 29, 95–96, see also Decomposition
    coordination and integration of, 30–31
Work breakdown structure
    formulation of, 382–386
    implementation of, 391–394
Work centers, 86, 102
Work division, see Decomposition
Working capital, 429
Working registers, 137–138
Work loads, 102
Work routes, 102
Work unit integration, 143
Written descriptions, 80

Z

Zeroth-order systems, 84